Modeling and Optimization of Optical Communication Networks

Scrivener Publishing
100 Cummings Center, Suite 541J
Beverly, MA 01915-6106

Publishers at Scrivener
Martin Scrivener (martin@scrivenerpublishing.com)
Phillip Carmical (pcarmical@scrivenerpublishing.com)

Modeling and Optimization of Optical Communication Networks

Edited by
Chandra Singh
Rathishchandra R. Gatti
K.V.S.S.S.S. Sairam
and
Ashish Singh

Scrivener Publishing

WILEY

This edition first published 2023 by John Wiley & Sons, Inc., 111 River Street, Hoboken, NJ 07030, USA and Scrivener Publishing LLC, 100 Cummings Center, Suite 541J, Beverly, MA 01915, USA

For more information about Scrivener publications please visit www.scrivenerpublishing.com.

Wiley Global Headquarters
111 River Street, Hoboken, NJ 07030, USA

For details of our global editorial offices, customer services, and more information about Wiley products visit us at www.wiley.com.

Library of Congress Cataloging-in-Publication Data

ISBN 978-1-119-83920-0

Cover image: Circuit Board, Kiosk88 | Dreamstime.com
Cover design by Kris Hackerott

Set in size of 11pt and Minion Pro by Manila Typesetting Company, Makati, Philippines

Printed in the USA

10 9 8 7 6 5 4 3 2 1

Contents

Preface

About the Book

The focus of this book is on the key technologies associated with modelling and optimization of optical communication networks. This book provides a basis for discussing open principles, methods and research problems in the modelling of optical communication networks. It also provides a systematic overview of the state-of-the-art research efforts and potential research directions to deal with optical communication networks. It also simultaneously focuses on extending the limits of currently used systems encompassing optical and wireless domains and explores novel research on wireless and optical techniques and systems, describing practical implementation activities, results and issues.

Key Features of Book

This Book serves like a handbook on applications for both academia and industry. It includes the detailed discussions on real world case studies on Trends & Technologies associated with Modelling of Optical Communication Networks. This Book also describes several numerical models and algorithms for simulation and optimization of optical communication networks. Modelling & Optimization presents several opportunities for automating operations and introducing intelligent decision making in network planning and in dynamic control and management of network resources, including issues like connection establishment, self- configuration and self-optimization, through prediction and estimation by utilizing present network state and historical data.

This book provides a basis for discussing open principles, methods and research problems in Modelling of Optical Communication Networks - It focuses on extending the limits of currently used systems encompassing optical and wireless domains, and explores the latest developments in applications like photonics, high speed communication systems and networks, visible light communication, nano-photonics, wireless, and MIMO systems.

Organization of the Book

Chapters are organized as Concept of Optical communication & Networking.

Chapter 1 Investigation on Optical Sensors for Heart Rate Monitoring

This chapter focuses on how the Optical Heart Rate Monitoring device operates, the components involved, and issues and challenges faced while monitoring the heart rate. Most wearable optical heart rate monitoring devices use photoplethysmography (PPG) to compute the heartbeat.

Chapter 2 Adopting a Fusion Approach for Optical Amplification

The chapter focuses on the current state-of-the-art hybrid optical amplifier design, theoretical background, and various inline configurations. In HOAs, main concerns, including other achieved channel capacity, cross-talk, gain uniformity, and transitory consequences, have been discussed.

Chapter 3 Optical Sensors

This chapter presents an extensive review of various optical fiber sensors, their principles, and an up-to-date overview of their applications.

Chapter 4 Defective and Failure Sensors Detection and Removal in Wireless Sensor Network

The Scanning Algorithm for Cut Tracking (SCT) and the Elimination of Faulty Sensor (EFS) algorithms are used in this proposed chapter. The SCT Algorithm will be used to track the state of all sensors in this chapter. This approach is very scalable because the workload does not rise as the number of sensors grows.

Chapter 5 Optical Fiber and Prime Optical Devices for Optical Communication

This chapter focuses on the Usage of fiber optic cable for optical communication has paved way in establishing links over long distances with higher data rates, light weight, higher security and with low transmission loss.

Chapter 6 Evaluation of Lower Layer Parameters in Body Area Networks

This chapter includes various interference parameters for throughput and the reasons for loss of packets. Also comparison of different radio models in terms of power is analysed. While in hardware section, using Bluetooth Low Energy (BLE) device the sensor data is received in android application (app). Also demonstration of various protocols in the field of Internet of Things (IoT) is presented and finally, Messsage Queue Telemetry Transport (MQTT) protocol and Hyper Text Transfer Protocol Secure (HTTPS) protocol are implemented to read sensor data in the cloud. To make the utilization more reliable, an app is designed for this specific application of WBAN.

Chapter 7 Analysing a Microstrip Antenna Sensor Design for Achieving Biocompatibility
This chapter discusses about the different design structure like Split Ring Resonator (SRR), Circular Ring Resonator (CRR) and Triangular Ring Resonator. The designed antenna perform accurately without harming any any muscle tissue.

Chapter 8 Photonic Crystal Based Routers for all Optical Communication Networks
In this chapter, we focus our topic exploring the optical network component - router using photonic crystals, which will be a perfect candidate to be integrated in Photonic integrated circuits (PIC) for optical communication and networking systems. Different configurations of photonic crystal based routers have been detailed and reviewed based on the performance.

Chapter 9 Fiber Optic Communication: Evolution, Technology, Recent Developments, and Future Trends
This chapter discusses fiber-optic communication systems and their fundamental technologies. It also discusses current developments as well as technological trends for the foreseeable future.

Chapter 10 Difficulties of Fiber Optic Setup and Maintenance in a Developing Nation
This chapter focuses on the difficulties associated with fibre cable deployment in India, with a particular emphasis on the economic, regulatory, and managerial difficulties. It is possible that external causes, such as dig-ups during road building, are a result of the problems associated with frequent fibre cutting. A lack of fibre deployment and management regulatory guidelines and policies poses a significant challenge to fiber management in the region.

Chapter 11 Machine Learning-Enabled Flexible Optical Transport Networks
This chapter overviews the various existing solutions for optimizing the Space Division Multiplexed-Elastic Optical Networks (SDM-EONs). Firstly, in view of enabling the realization of SDM-EONs enabled by the development of appropriate fiber solution to ensure long haul signal transmission.

Chapter 12 Role of Wavelength Division Multiplexing in Optical Communication

This chapter focuses on Normal WDM, Coarse WDM, and Dense WDM are the three wavelength patterns used in WDM systems. The data transmission speed increases as a result of the WDM concept.

Chapter 13 Optical Ultra-Sensitive Nanoscale Biosensor Design for Water Analysis

The present research is on the variation of material index of refraction between the normal water and water infected with B. cereus spores. This technique relies on the unique index of refraction as a spectral signature for the B. cereus spores detection. The designed biosensor is Circular and Triangular resonator structures using Photonic Crystal.

Chapter 14 A Study on Connected Cars-V2V Communication

This chapter focuses on the uses a VANET (Vehicular ad hoc networks) for communication purpose. It works 360 degrees at any direction. Exchange of information takes place by adapting a suitable protocol is explained in this chapter.

Chapter 15 Broadband Wireless Network Era in Wireless Communication – Routing Theory and Practices

This chapter aims to provide BWNs with a straight forward roadmap of theoretical context so that they can manage various efficient routing mechanisms.

Chapter 16 Recent Trends in Optical Communication, Challenges and Opportunities

The goal of this chapter is to look at the nonlinearities that arise in optical fibers and the possible solution offered by machine learning techniques for increasing optical fibre communication capacity.

Chapter 17 Photonic Communication Systems and Networks

This chapter focused on Photon is the particle of light, which is extensively used in modern digital communication systems as a signal carrier. In the previous era, electromagnetic waves were used for communication systems development.

Chapter 18 RSA-Based Encryption Approach for Preserving Confidentiality Against Factorization Attacks

This chapter analyses different attacks possible on the proposed system and reports the efficiencyof the proposed system.

Chapter 19 Sailfish Optimizer Algorithm (SFO) for Optimized Clustering in Internet of Things (IoT) Related to Healthcare Industry

The proposed chapter focuses on SFO based energy efficient algorithm is utilized for discovering the CHs ideal situation. The simulations under energy consumed, network lifetime and throughput are carried out.

Chapter 20 Li-Fi Technology and Its Applications

In this era of communication technology, This chapter focuses on LiFi is another and proficient method of remote communication. LiFi utilizes LED to communicate information. The data transmission of information is done remotely.

Chapter 21 Smart Emergency Assistance Using Optics

As the number of accidents are increasing, there has to be some device that can help in providing medical assistance immediately. This gives rise to the need for a system which can assist people in such unprecedented emergencies. We therefore intend to provide a solution by building an emergency assistance system that assists people in getting ambulance services in need.

Editing this book was an incredible chance for which parcel of help was expected from many individuals. We had fine support from our family, friends and fellow members especially we thank the Chairman, Principal, faculty, and fraternity of Sahyadri College of Engineering & Management, Mangaluru & NMAM Institute of Technology, Nitte.

Chandra Singh
Rathishchandra R Gatti
K.V.S.S.S.S. Sairam
Ashish Singh

1

Investigation on Optical Sensors for Heart Rate Monitoring

V. Vijeya Kaveri[1]*, V. Meenakshi[2], N. Kousika[1] and A. Pushpalatha[1]

[1]CSE, Sri Krishna College of Engineering and Technology, Coimbatore, India
[2]EEE, Sathyabama Institute of Science and Technology, Chennai, India

Abstract

This chapter focuses on how the optical heart rate monitoring (OHRM) device operates, the components involved, and issues and challenges faced while monitoring the heart rate. Most wearable optical heart rate monitoring devices use photoplethysmography (PPG) to compute the heartbeat. PPG is shorthand for reflecting out onto the surface and trying to measure the quantity of sunlight scattered by blood circulation. PPG sensors emphasize that the beam joining the body disperses in a familiar sequence when blood circulation patterns change. The PPG sensor's core elements, such as the optoelectronic transmitter, computerized pulse controller, magnetometer, and machine learning, are critical in estimating the heartbeat. PPG assessment in a state of rest (falling asleep, seated, and standing always) is incredibly easy, but measuring PPG throughout the exercise program (workouts, jumping, riding bikes) is challenging. In reality, you will face five significant challenges when using OHRM to create wearable devices such as optical noise, skin tone, cross-over the problem, location of the sensor, and low perfusion. These challenges can be overcome by choosing good opt mechanics and signal extraction algorithms. PPG sensor is used to measure breathing rate, heart rate variability, blood pressure, and cardiac efficiency [3]. OHRM may be used for lifestyle, in-session, and personal health metrics in a real-time scenario. We have also focused on parameters like wearability, accuracy, battery life, and time usage of the device.

Keywords: OHMR, PPG, optical sensor, pulse wave analysis, heart rate, vascular disease, Raynaud's phenomenon, vein

**Corresponding author*: vijerama.kaveri@gmail.com

Chandra Singh, Rathishchandra R Gatti, K.V.S.S.S.S. Sairam and Ashish Singh (eds.) Modeling and Optimization of Optical Communication Networks, (1–10) © 2023 Scrivener Publishing LLC

1.1 Introduction

Photoplethysmography (PPG) is an optoelectronic method for detecting pressure changes in the vasculature of cells [8]. Pulse oximeters, vasculature diagnostic testing, and electronic beat-to-beat cardiac monitoring gadgets are a few commonly produced healthcare devices that use it. The most common element of PPG technology makes only illumination for enlightening the cells and a light detector to evaluate pretty slight seasonal variations in illuminance connected with transformations in blood circulation in the catchment flow rate. PPG is a quasi tissue process that utilizes a red or relatively close beam. Even though its ease of use, the origins of the specific parts of the PPG signal are undisclosed.

1.2 Overview of PPG

1.2.1 PPG Waveform

The 'AC' component of the PPG amplitude is commonly referred to as the peristaltic aspect, and intensity repetition of approximately 1 Hz depends entirely on the heartbeat (Figure 1.1). This AC portion is imposed on top of substantial quasi-DC terms of distribution to cells and the traditional blood volume. This DC element constantly changes in response to breath, vascular behavior, and vasoconstrictive vibrations. As per Allen and Murray, body posture impacts these properties [1]. Both AC and DC can be recovered for forthcoming pulsatile analysis using appropriate electrical filtration and update.

1.2.2 Photoplethysmography Waveforms Based on the Origin of Optical Concern

Reflection, propagation, multiple scattering, absorption, and viewable radioactivity are all operations associated with the communication of sunlight with living tissues (Anderson and others). There have been several studies in electro-optic methods regarding PPG dimensions between 1948 and 1993 by Hertzman and Randall [15].

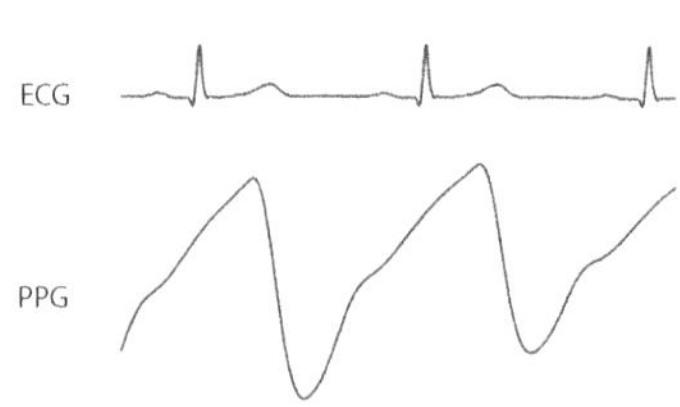

Figure 1.1 PPG signal and corresponding electrocardiogram (ECG).

Researchers identified three significant elements that influence the sensor's illumination: blood density, microvascular movement, and red blood cell alignment (RBC). The orientation implications were validated by recording the respiratory muscle's output voltage waveform from dentine and in a discharge tube. Flow rate adjustments should no longer be an option, and more recent times through using Naslund *et al.* [25], who discovered peristaltic wave patterns in joints. Perfusion is proportional to captured pulses, and the more blood is there, the minimum amount of the emission is ameliorated. The authors in their article [7, 8, 15, 20] has discussed as that attempts to evaluate heartbeats amplitude have often been failure.

The frequency band of the emitted energy is crucial within communication for three reasons [9, 21]: (1) the electro-optic liquid door: cells, mostly water, that also refract sunlight very powerfully in the ultra-violet and more extended electromagnetic frequencies. Melanin consumes a significant amount of light with specific wavelengths. (2) Isobestic wave functions: There are significantly different in absorption among hemoglobin in the blood (HbO2) and reduction in hematocrit levels (Hb) other than at isosbestic light waves, which have been popularly used for PPG mild power source Gordy *et al.* [14] used measurements made at an isosbestic frequency range. The pulse should be relatively untouched by levels of blood oxygen substance. (3) Vascular surface depth: the density with which a given significance of optical radiation reaches the organisms is determined by the range of frequency. For transmission mode systems, PPG's catchment (study) volume can be 1 cm3 depending on the probe type. Through arterio-venous anastomosis shunt channels, PPG would offer data on tube nutritive and thermoregulatory blood flow.

1.2.3 Photoplethysmography's Early on and Modern Records

This section offers a quick outline of the untimely records of PPG and has been in use since the top-notch analysis piece of writing. In 1936, two research organizations, namely Molitor, Kniazuk under Merck Healing Organization, and Hanzlik *et al.* [22] from Stanford College of medication, defined comparable devices used to screen the blood quantity variations inside the ear of the rabbit subsequent venous occlusion along with the management of vasoactive capsules. Molitor and Kniazuk also disclosed capturing produced from human fingertips using a reflection mode PPG instrument. Hertzman, in the subdivision of body structure at St. Louis University faculty from medication, was an initiator of the PPG system launching.

Hertzman validated the PPG procedure in 1938 by visualizing the density of blood variations evaluated simultaneously with the help of automatic plethysmography with those detected simultaneously by PPG. Hertzman and Dillon [16] used separate electronic amplifiers to divide the AC and DC components and measured vasomotor activity. Hertzman [15] recognized several sources of error with the procedure, emphasizing the importance of good skin contact without applying excessive pressure that might cause blanching. He suggested that the measurement probe should not be moved against the skin. As a result of these observations, complex positioning devices were created. Another critical design consideration was identified as illumination. Hertzman mentioned a battery-operated torch bulb, which turned into much lesser than the excellent value due to its broad area, specifically within the infrared, which brought onto the typical warm-up cells, miscalculation of breathing dispersion outcomes and remarkable elucidation that blended pores and microvascular tissue flow of blood with more prominent vessel alerts. Additionally, maintaining a steady light power was not possible.

1.2.4 Building Blocks of Photoplethysmography

Modern photoplethysmography sensors use the technology of low-fee semiconductors, along with LEDs and matching photodetector models that work in the infrared wavelengths [2, 11, 23, 29] has produced an evaluation of visual sensor methodology for PPG and beats oximetry systems. Burke and Whelan [5], Naschitz *et al.* [24], and Ugnell and Oberg [28] emphasized the significant need for light source selection. LEDs are light-emitting diodes with a narrow single-bandwidth conversion, typically 50nm. The photodetector of optimal is also crucial [12, 30]. Its essential properties are as follows: they are selected from the matching light source color. The photodetector converts luminosity power into electro power. Those are extremely small, cheaper, sympathetic, and have high throughput.

Daytime filters can be used to protect near-infrared electronics. The photodetector is connected to electronic equipment with a slight noise, such as a trans-impedance amplifier and filtering circuits. The main DC component is reduced in size by a high pass filter, enabling the higher to the lower alternating current element to be brought up to a maximal marginal level of 1 V. To reduce undesirable noise in high bandwidth, electricity can take 50 Hz of electric power source, and filtering circuitry must be carefully chosen. Figure 1.2(a) depicts a design of an operational amplifier. In contrast, Figure 1.2(b) depicts different steps surrounding it, such as short surpass straining, elevated surpass straining, supplementary intensification, indicator inversion, and indicator boundary. This Model System Determines the PPG probe LED by a constant current driver stage.

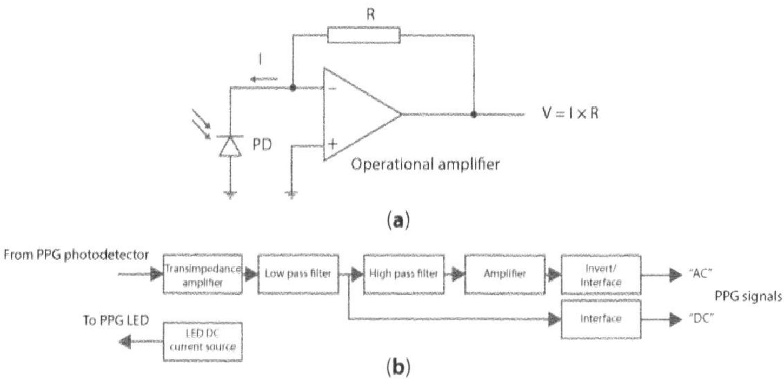

Figure 1.2 (a) Amplifier design; (b) Signal stages.

Transmission mode operation, where the tissue model is located among the starting place, detecting node, and indication form operating. The LED and detector are put side-by-side and are the two basic PPG operational setups. Transmission mode PPG has more constraints on the body areas that can be studied than reflection mode PPG. The PPG exploration can be made position safe to reduce probe-tissue movement artifact. Other causes of an artifact must be considered while using measurement technology. For example, ambient light interference can cause artifacts, which can be reduced in several ways, including combining valuable query addition to the cells using a dim Velcro twist hit, supplemental shadows in the research region and testing in low-light conditions, as well as digital filtration such as luminous attenuation filtration.

Other new technologies include PPG imaging expertise, digital consulting, and remote monitoring. Schultz *et al.* [17] and Huelsbusch *et al.* [27] used an exploratory liquid that evaporates close to the infrared PPG exploring device to study cutaneous blood circulation and associated syncopated anomalies. The goal of the technology was to learn more about maintaining vascular homeostasis permeability and diagnose complications associated with inflammatory processes and curative. Wieringa *et al.* [31] illustrated a contact-free several spectrum PPG measurement device for remote monitoring imaging primary breathing normalization (SpO2) dissemination. The arrangement will record films of matrices as two-dimensional topographically determined PPG sensory information at a few electromagnetic spectrums during differences in respiration values. An arterial oxygen picture may be helpful in a variety of diagnosing circumstances [10], including determining tissue viability.

PPG has much potential in telemedicine, including patient monitoring from afar or home. Miniaturization, usability, and robustness are essential to design considerations for such systems. This is demonstrated by the use of ring based finger system that uses PPG sensors to monitor heartbeat pulsations Rhee *et al.* [26]; Zheng *et al.* [34] and the necessary movement artifact drop, proper sensing location, and sensor calibration [26, 35]. The pulse, oxygen saturation, and respiration may all be detected, as well as hematocrit, which is obtained from optical properties at five variant bandwidths (569, 660, 805, 904, and 975 nm) in a PPG skin display and remote device monitoring entire house. In preliminary clinical testing, the hematocrit was within 10% of the standard gold value. Digital filtering techniques were used to retrieve respiratory data, and the standard ratio for red and near-infrared wavelengths was used to predict blood oxygen saturation (SpO2).

1.2.5 Protocol Measurement and Reproducibility

For example, in clinical physiological measurement, reproducibility is crucial to ensure the precision of detecting significant therapeutic effects. Elements that influence reproducibility include probe–tissue integration pressure, pulse oscillator throughput, motion artifact removal, relating body position, leisure, inhaling, consciousness, and weather conditions.

However, no internationally acknowledged standards for clinical PPG measurement exist. Published research is often based on studies that used widely disparate measurement technologies and methodologies, making it challenging to replicate PPG physiological results across research centers. Only a few studies have attempted to assess the continuity and reproduction of PPG capacity. Jago and Murray [18] conducted a significant investigation on the uncertainty in PPG measurements in a group of healthy adult participants. They looked at the consistency of PPG pulse transit time (PTT) measures taken from the ear, thumb, and toe locations during and between sessions. Both individual site measurements and both right-left side values were evaluated. The findings demonstrated the relevance of factoring in posture, ambient temperature, relaxation, and acclimatization. Bilateral assessments were more repeatable than individual site data because heartbeat, breathing, and blood pressure parameters are likely to influence both sides.

Many studies have also enumerated the complicated physical unpredictability of PPG waveforms collected at various body regions. The assessment of autonomic dysfunction and cardiovascular aging are two appliances that use the beat rate fluctuation in PPG parameters. However, obtaining an averaged heartbeat measurement to reflect an entity site can be valuable. To boost assurance in a particular period, amplitude, or form data retrieved using the beat rate from PPG, an averaging duration of at least 60 heartbeats has been advised [1].

1.3 Clinical Application – Heart Rate Monitoring

Heart rate is an essential physical metric to observe in various medical situations, medical centers, and the monitoring of patients. The AC aspect of the PPG heartbeats can calculate heart rate because it is correctly aligned with the chest. The data is commonly displayed next to the SpO2 level in oxygen therapy schemes. The core problem is that too much motion artifact or cardiovascular dysrhythmias can reduce the sense of trust in the rate factor. Machine tools were introduced that improve the effectiveness of beat rate diagnosis. Undemanding electronic filtration and zero-crossing diagnosis is used to fetch heartbeats and inhalation elements since the PPG in-ear communication [23]. The idea of PPG for heartbeat monitoring in emergency obstetric divisions was evaluated utilizing PPG, and Echocardiogram heart rate data was collected constantly and consistently over eight hours [19].

For 77% of the metrics, high-quality ECG transcripts were acquired. The PPG pulse rate was adequately documented, excluding those distorted by offset alteration of the signals (6%). There have been roughly 1% false negatives and 1% false positives in the PPG heartbeat. Quite enhanced methodologies, such as time-frequency methods based on the smoothed Wigner Ville dispersion, were used to derive pulse rate data from PPG sound waves [6, 32].

Necessary associations pulse from the study hand at rest and during limited gesture with indicators approximated from the posterolateral and office supplies source hands were used to estimate the validity of forecasting pulse rate. The moment methodology significantly outperforms two known algorithms for measuring object's weighted moving average (WMA) and fast Fourier transform (FFT). The mean and standard heart - rate irregularity was limited to 6bpm from 16bpm in WMA, and it was 11bpm in FFT. In correlation to different pulse pace data acquisition systems, Bland and Altman's assessment [4] discovered that pulse oximeter and radial piezoelectric pulses at the radial nerve have a high degree of similarity [13].

Yu *et al.* [33] presented a self-activating assessment of the trustworthiness of indication of heartbeat rates obtained since the patient's symptoms were monitored using ECG and PPG. They used a quality index for each reference heart rate to convey reliability. The support vector machine classifier (SVM) assessed the physiological waveforms. An adaptive peak identification technique was utilized to compute the heartbeat rate independently, filtering out movement caused noise. The method examined the usage of 158 randomly decided samples on 7-second facts examples from trauma patients accumulated throughout the helicopter transport. At least 92% of cases could be matched when the algorithm's results were compared to manual analysis performed by human professionals. The rules inferred a much less conservative sign of high quality in the remaining 8% of cases, primarily due to ambiguously labeled waveform samples. Sleep research has also benefited from automatic heart rate detection technologies. Foo and Wilson used a double measuring technique to improve PPG signals in poor perfusion circumstances, including an accelerator association detection and a filter with zero phases. A risk assessment matrix has been used to formulate a plan for instantaneously strengthening the PPG signal-to-noise ratio. A risk assessment matrix was used to determine the best approach for dynamically improving the PPG signal-to-noise proportion. While comparing to ECG pulse rate monitoring, the most significant error rate was less than 8%.

1.4 Summary

The generation of photoplethysmography has been brought on this evaluation, and its large capacity for usage in a wide variety of scientific checks has been hooked up. The assessment of the cardiovascular machine has been a first-rate awareness. The call for low-price, effortless, and handy tools is the number one concern, and the network primarily depends upon methodical surroundings. The condition of low-price and tiny semiconductor add-ons with the growth of computer-based beat wave evaluation strategies has contributed to a resurgence of hobby in the approach in current years. PPG-based era is utilized in a selection of commercially available clinical gadgets for figuring out oxygen dissemination, the level of blood pressure, and heartbeat rate, in addition to tracking autonomic features and figuring out the ailment of peripheral vascular. Although the features of the PPG waveform are not fully known, this success has been achieved. Equivalence of dimensions, enhancing ability to repeat, and presenting complete normal statistics levels for evaluation with sufferers and reading healing responses are demanding situations that the generation receives. Imaging using PPG, trouble-free endothelial tests to identify the dysfunction, and other measurement and analysis technologies are likely to advance in future studies.

References

1. Allen, J., The measurement and analysis of multi-site photoplethysmographic pulse waveforms in health and arterial disease. PhD Thesis, Newcastle University, Newcastle upon Tyne NE1 7RU, United Kingdom, 2002. https://ethos.bl.uk/OrderDetails.do?uin=uk.bl.ethos.247830

2. Barron, S.A., Rogowski, Z., Kanter, Y., Hemli, J., DC photoplethysmography in the evaluation of sympathetic vasomotor responses. *Clin. Physiol.*, 13, 561–72, 6, 1993.
3. Belcaro, G. *et al.*, Noninvasive investigations in vascular disease. *Angiology*, 49, 9, 673–706, 1998.
4. Bland, M., *An introduction to medical statistics*, 2nd edn, Oxford University Press, Oxford, 1995.
5. Burke, M.J. and Whelan, M.V., Photoplethysmography—Selecting optoelectronic components. *Med. Biol. Eng. Comput.*, 24, 647–50, 1986.
6. D. Chan, M. Hayes, P.R. Smith, Venous pulse oximetry. World Patent W,03/063697, 2003.
7. Challoner, A.V. and Ramsay, C.A., A photoelectric plethysmograph for the measurement of cutaneous blood flow. *Phys. Med. Biol.*, 19, 317–28, 1974.
8. Challoner, A.V.J., Photoelectric plethysmography for estimating cutaneous blood flow, in: *Non-invasive physiological measurements:*1, P. Rolfe (Ed.), pp. 125–151, Academic Press, London, 1979.
9. Cui, W.J., Ostrander, L.E., Lee, B.Y., In vivo reflectance of blood and tissue as a function of light wavelength. *IEEE Trans. BME*, 37, 632–9, 1990.
10. Dutch, J. and Redman, S., Psychological stress and arterial pulse transit time. *N. Z. Med. J.*, 96, 607–9, 1983.
11. Duck, F.A., *Physical properties of tissue*, Academic, London, 1990.
12. Fine, S. and Weinman, J., The use of photoconductive cells in photoplethysmography. *Med. Biol. Eng.*, 11, 455–63, 1973.
13. Foo, J.Y., Lim, C.S., Wang, P., Evaluation of blood pressure changes using vascular transit time. *Physiol. Meas.*, 27, 685–94, 2006.
14. Gordy, E. and Drabkin, D.L., Spectrophotometric studies. XVI Determination of the oxygen saturation of blood by a simplified technique, applicable to standard equipment. *J. Biol. Chem.*, 227, 285–99, 1957.
15. Hertzman, A.B., The blood supply of various skin areas as estimated by the photoelectric plethysmograph. *Am. J. Physiol.*, 124, 328–40, 1938.
16. Hertzman, A.B. and Dillon, J.B., Applications of photoelectric plethysmography in peripheral vascular disease. *Am. Heart J.*, 20, 750–61, 1940b.
17. Huelsbusch, M. and Blazek, V., Contactless mapping of rhythmical phenomena in tissue perfusion using PPGI. *Abstract Proc. SPIE: Medical Imaging: Physiology and Function from Multidimensional Images*, vol. 4683, Clough, A.V. and Chen, C.-T. (Eds.), pp. 110–7, 2002.
18. Jago, J.R. and Murray, A., Repeatability of peripheral pulse measurements on ears, fingers and toes using photoelectric plethysmography. *Clin. Phys. Physiol. Meas.*, 9, 319–30, 1988.
19. Johansson, A., Oberg, P.A., Sedin, G., Monitoring of heart and respiratory rates in newborn infants using a new photoplethysmographic technique. *J. Clin. Monit. Comput.*, 15, 461–7, 1999.
20. Jespersen, L.T. and Pedersen, O.L., The quantitative aspect of photoplethysmography revised. *Heart Vessels*, 2, 186–90, 1986.

21. Jones, D.P., Medical electro-optics: Measurements in the human microcirculation. *Phys. Technol.*, 18, 79–85, 1987.
22. Molitor, H. and Kniazuk, M., A new bloodless method for continuous recording of peripheral circulatory changes. *J. Pharmacol. Exp. Ther.*, 57, 1, 6–18, 1 May 1936.
23. Nakajima, K., Tamura, T., Miike, H., Monitoring of heart and respiratory rates by photoplethysmography using a digital filtering technique. *Med. Eng. Phys.*, 18, 365–72, 1996.
24. Naschitz, J.E. *et al.*, Pulse transit time by R-wave-gated infrared photoplethys-mography: Review of the literature and personal experience. *J. Clin. Monit. Comput.*, 18, 5–6, 333–42, 2004.
25. Naslund, J., Pettersson, J., Lundeberg, T., Linnarsson, D., Lindberg, L.G., Non-invasive continuous estimation of blood flow changes in human patellar bone. *Med. Biol. Eng. Comput.*, 44, 501–9, 2006.
26. Rhee, S., Yang, B.H., Asada, H.H., Artifact-resistant power-efficient design of fingering plethysmographic sensors. *IEEE Trans. Biomed. Eng.*, 48, 795–805, 2001.
27. Schultz Ehrenburg, U. and Blazek, V., Value of quantitative photoplethysmography for functional vascular diagnostics: Current status and prospects. *Skin Pharmacol. Appl. Skin Physiol.*, 14, 316–23, 2001.
28. Ugnell, H. and Oberg, P.A., The time-variable photoplethysmographic signal; dependence of the heart synchronous signal on wavelength and sample volume. *Med. Eng. Phys.*, 17, 571–8, 1995.
29. Webster, J.G., *Design of pulse oximeters*, Institute of Physics Publishing, Bristol, 1997.
30. Weinman, J. and Fine, S., Detectivities of photoconductive and silicon p-i-n light sensors in photoplethsymography. *T-I-T J. Life Sci.*, 2, 121–7, 1972.
31. Wieringa, F.P., Mastik, F., van der Steen, A.F., Contactless multiple wavelength photoplethysmographic imaging: A first step toward 'SpO2 camera' technology. *Ann. Biomed. Eng.*, 33, 1034–41, 2005.
32. Yan, Y.S., Poon, C.C., Zhang, Y.T., Reduction of motion artifact in pulseoximetry by smoothed pseudo Wigner–Ville distribution. *J. Neuroeng. Rehabil.*, 2, 3, 1–9, 2005.
33. Yu, C., Liu, Z., McKenna, T., Reisner, A.T., Reifman, J., A method for automatic identification of reliable heart rates calculated from ECG and PPG wave-forms. *J. Am. Med. Inform. Assoc.*, 13, 3, 309–20, 2006.
34. Zheng, D.C. and Zhang, Y.T., A ring-type device for the noninvasive measurement of arterial blood pressure. *Proc. 25th Annual International Conf. of the IEEE EMBC 4*, pp. 3184–7, 2003.
35. Zhang, X.Y. and Zhang, Y.T., The effect of local mild cold exposure on pulse transit time. *Physiol. Meas.*, 27, 649–60, 2006.

2

Adopting a Fusion Approach for Optical Amplification

E. Francy Irudaya Rani[1], T. Lurthu Pushparaj[2] and E. Fantin Irudaya Raj[3]*

[1]Department of Electronics and Communication Engineering, Francis Xavier Engineering College, Tamil Nadu, India
[2]MRI Research Lab, TDMNS College, Tamil Nadu, India
[3]Department of Electrical and Electronics Engineering, Dr. Sivanthi Aditanar College of Engineering, Tamil Nadu, India

Abstract

Transition and inner-transition metal-based semiconductors with 'd' and 'f' electron density demonstrate outstanding signal transmission with minimal signal attenuation throughout optical communication system interaction. These competencies are widely used in fiber-optic detection, health - care as well as industrial imaging, telecommunications equipment, Fiber-to-the-Premises (FTTP), communication infrastructure, defense systems, and High Definition (HD) and Conventional Scope (SD) Community Information Broadcast TV (CATV). However, the currently offered optical connection speed is limited in this scenario. Even the incidence of powerful reflective surfaces could be deleterious to the scheme in which such a platform is being used. To tackle the concerns of down signaling, transmission boosting or amplifier configurations that provide damned near polarization-independent characteristics, which are sometimes preferable, were established. The hybridization or mixing of far more productive semiconducting 'd' and 'f' block metals results in a much more compact arrangement with improved optical transmission transport. SOAs boozed with 'd' and 'f' electrons might be used in telecom systems as fiber-pigtailed constituents with narrow augmented stimulated emission (ASE). The high gain saturation in SOAs can also be used for nonlinear pattern recognition in telecommunications systems. Its configuration was far more versatile, with a small semiconductor chip containing electronics and fiber interconnection. A convenient housing will now provide great polarization-insensitive Faraday breakers at the input, output, or both ports.

**Corresponding author*: fantinraj@gmail.com

Chandra Singh, Rathishchandra R Gatti, K.V.S.S.S.S. Sairam and Ashish Singh (eds.) Modeling and Optimization of Optical Communication Networks, (11–34) © 2023 Scrivener Publishing LLC

Hybrid optical amplifiers (HOAs) play a critical role through wideband concert modulation and therefore are ubiquitously used in increased dense wavelength division multiplexed systems. The chapter summarizes the current state-of-the-art hybrid optical amplifier design, theoretical background, and various inline configurations. In HOAs, main concerns, including other achieved channel capacity, cross-talk, gain uniformity, and transitory consequences, have been discussed. Upon careful consideration, it has been determined that HOAs provide effectively gain flatness without the need for overpriced boost hollowing methodologies, but also a high precision of gain, signal to noise ratio, packet loss ratio, but instead vicissitudes.

***Keywords*:** Optical communication system, Semiconductor optical amplifier, Erbium-doped fiber amplifier, hybrid optical amplifiers, amplified spontaneous emission

2.1 Introduction

The present information era can be defined as high-bandwidth communication enabled by fiber communication systems. Signal deterioration occurs when signals are transmitted across thousands of kilometers long distances. Due to numerous passive components in the medium, the broadcast signals' strength gradually decreases as they travel along a communications platform. Indeed, the medium's attenuation continues to be a severe issue that influences light propagation over highly long distances via fiber optic cable. Signal deterioration must be avoided, which necessitates using the amplification procedure.

Additionally, a signal must have a minimum amount of baseline power for the information it transmits to be seen at the receiving end. Due to the limits imposed by transmission channels/systems, optical amplifiers with fiber optic and waveguides remain crucial in fiber transceivers. These limitations would result in fiber loss and dispersion, commonly handled by various amplifiers. Loss and dispersion are connected in nature [1], as seen by a pulse form that creates scattering or loss and vice versa.

Professor E. Snitzer built the first optical amplifiers in 1964, using neodymium and operating in a 1060nanometers spectral window, using the initial optical amplifier ideas established in the early 1960s. Researcher Snitzer also had the first erbium glassy laser on display. In 1970, more experiments at neodymium were done, and it was too early for broad application. Bell Labs developed the first mono fibers using these methods in late 1980. In 1985, erbium was utilized for magnification only at the University of Southampton and AT&T Bell Labs. Erbium's functioning at

1551 nm, the most critical wavelength in silica fibers, was a considerable gain. When contrasted to OEO regenerators, optical amplifiers are alluded to as all-optical. Optical amplification in the underwater sector is called "regenerators," which may be confounding to users from the conventional telecom sector.

Electro-optic signal boosters were used for amplifying in the past, in which the optical signal was transformed into a flow of electrons and afterward repeated using a receiver [2]. Nevertheless, resurfacing used to be a time-consuming and expensive procedure, mainly when multi optical systems were employed. As a result, an amplifier circuit has emerged as a viable approach for boosting transmitted signals throughout transmission. It is a mechanism that magnifies an optical power without converting it to an electrical signal, which is a property required in so-called repeaters.

2.2 The Mechanism Involved

Optical amplifiers magnify light beams via a procedure termed stimulated emission, which is analogous to the technique used in the functioning of lasers. Fiber optics are lasers without a feedback loop that gains an optic boost anytime the amplifier is pushed to induce population inversion [3]. The local incident light at any place within the device and the wavelength of the entering transmitted fiber determine the resulting optic gain. As a result, an amplifier's bandwidth is critical since it dictates the frequencies and intensity dependency of the optic yield (of an amplifier). Figure 2.1 depicts the overall shape of an optical amplifier.

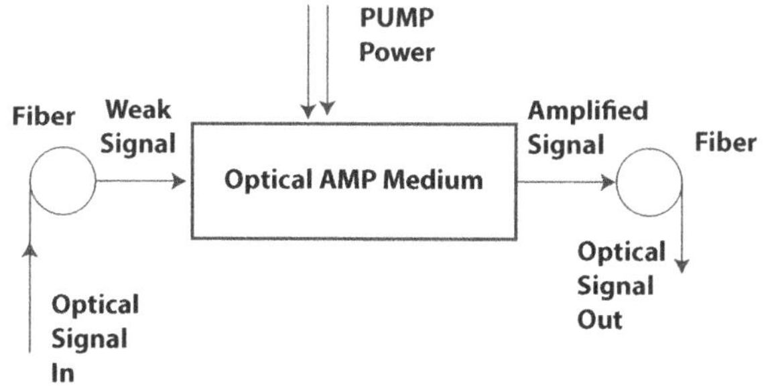

Figure 2.1 The overall silhouette of an optical amplifier.

2.3 Types of Amplifier

Fiber optics can be solid-state or fiber-based and available in various geometries and sizes [2]. Typically, an optical amplifier's functioning is based on feedback, generating adequate gain according to the specific frequency. Wave propagation amplifiers [4] are optical amplifiers that do not require feedback. In them, the magnified information only goes in one direction: forward. In solid-state amplifiers, the resonator is usually made of solid materials like silicon, and its form and size govern the amplification parameter. On the other hand, fiber-based amplifiers may rely on entirely elastic light scattering and specific doped silicon to convert the guide into an all-optical system to obtain the requisite gain.

2.3.1 Semiconductor Optical Amplifiers

Semiconductor fiber optics (Figure 2.2) is a solid-state amplifier that uses semiconducting lasers [5]. Different illumination reflections occur at the split aspects of the Fabry–Pérot type of interferometer, resulting in meaningful feedback. As a result, SOAs could be employed as amplifiers when pushed underneath the threshold. Although such amplifiers are inexpensive to construct, the optical transmission gain is particularly sensitive to heat and variations in the source light incidence. The interference induced by likeness again from terminal aspects is lessened when the SOA is of the signal propagation type.

Treating the aspects with an anti-reflective coating is a simple approach to lower the reflectance. For the SOA to act as a travel wave amplifier, one of several aspects' reflectance must be exceedingly low (0.1 percent). The

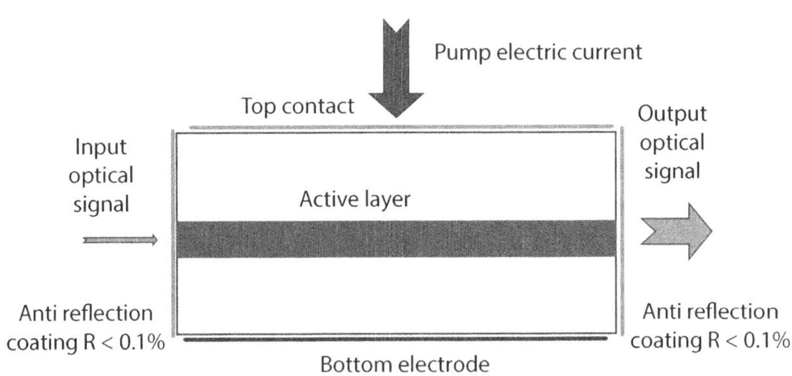

Figure 2.2 Construction of semiconductor optical amplifier.

power amplifier would determine the degree of minimal reflection. Low reflectance levels of features in a predictable form, on the other hand, remain very hard to achieve. One option in this situation would be to employ a laser with a slanted resonator cavity. The acute angle of such a laser system physically separates the specular reflection from the light flowing from the forward path. Related to the physiological features of light propagating in a leading channel, achieving a vanishing quantity of feedback is very difficult. An alternative to this layout is a display structure in which a translucent pane is added between the extremities of the active region and the sides. The signal spreads through the window area of such buildings, resulting in losses.

2.3.1.1 Various Phases and Progress of SOA

An SOA is a technique that magnifies a light beam and converts it to an electrical impulse in real-time [5]. Configurations are similar to silicon lasers in that they are made of semiconductor materials and function in the same way. They have a tiny footprint, straightforward construction, power efficiency, and long life span and are reasonably priced. As a result, they may readily be combined with another mass-producible optic, electrical circuits that can amplify and switch signals. SOAs are commonly used in optical technologies because of their high spectrum efficiency and energy transmission capabilities. SOAs are laser diodes that can magnify and transport light output from one end of a fiber to the other. They come in a tiny container and transfer data in both directions, minimizing device size.

Nevertheless, it has significant drawbacks, such as a high coupling loss, polarization dependency, and an excessive noise figure. Anti-reflective coatings of extremely high quality are required [6]. As a result, improving these restrictions will be a significant focus of research.

Even though optical signals are constantly attenuated for the period of transmission in optical networks, communications cannot be transmitted over thousands of kilometers with no optical communication technologies. Optical amplifiers are therefore crucial parts of long-distance transmission networks. Optical [2, 3], electronics [1], substance [7], telecommunication [8], and chemical modification [9] technologies are all relevant to SOAs. As a result, SOAs are a multidisciplinary research topic. Furthermore, with the advancement of 5G, wireless telephones, and light modern communications, many authorities now prioritize the potential growth of SOAs and feel that sufficient funding should be committed to helping this technology advance [5].

2.3.2 Fiber Raman Amplifiers

Any dielectric film's reaction to light becomes chaotic when exposed to a solid applied electric field. Asymmetric dispersion would happen due to such nonlinearity, and the wavelength of the light scattered would be decelerated, leading to a loss. As a result, photon scatter leads to a power outage at the incoming frequency. On the other hand, the scattering cross-sections stay extremely tiny at low input power levels, and the losses are virtually insignificant. However, with high input laser fields, the asymmetric process of stimulated Raman scattering (SRS) occurs, resulting in a significant loss. The quantity of dispersed light develops exponentially after the incoming optical power surpasses a threshold amount. SRS occurs when a powerful optical boost output propagates through silica fibers in fiber Raman amplifiers [10]. The intruding pump photon surrenders its electricity to generate a lesser photon with diminished energy. The medium transfers the leftover radiation as molecular vibrations, resulting in the generation of optical phonon scattering. To generate a gain, fiber Raman amplifiers are optically pumped. The energy disparity is referred to as the Stokes shift. Pumping and communication waves are pumped into a cable, and power is transmitted first from the focus point to the signaling beam through SRS as the two beams co-produce the fiber. The combined heat, power, and communication beams can be pumped into the fiber in a counter-propagating setup. After all, it relies on the pushing settings used to achieve the required gain with its distinctive range of benefits and drawbacks.

Fiber Raman amplifiers (Figure 2.3) have a wide bandwidth, which is beneficial for boosting many channels at once and brief optical impulses [11].

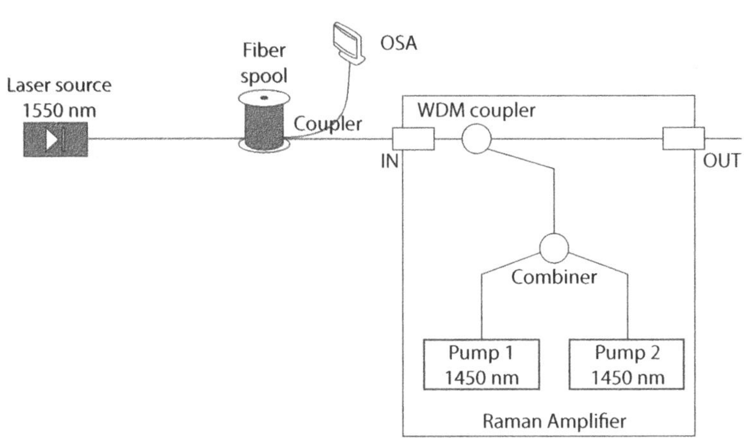

Figure 2.3 Fiber Raman amplifier scheme.

These amplifiers are also highly suggested for dispersed amplification since they can be used to offset fiber loss in predecessors' transceivers. However, these have the disadvantage of requiring high lasers for optically pumping, reducing the communication's cost-effectiveness.

2.3.3 Fiber Brillouin Amplifiers

Fiber Brillouin amplifiers work similarly to fiber Optical Amplifier, except that the output is generated by stimulated Brillouin scattering (SBS) rather than stimulated Raman scattering (SRS). When optically pumped amplifiers are utilized, SBS passes a portion of the power level to the signal [12–17]. The majority of the energy in each pump photon is utilized to make a signal photon, with the remainder being used to activate an acoustic phonon. As a result, the amplifier system depends on acoustic phonons rather than optical phonons, as in fiber Raman amplifiers [18, 19]. SBS is distinguished from SRS by the following characteristics:

- Amplification happens only when the information light proliferates in the reverse way of the focused beam in SBS. However, in SRS, both types of setups are used.
- SBS has a Stokes shift of about 10 GHz, or three orders of magnitude compared to SRS.
- Brillouin's boost spectra are restricted (less than 100 MHz). The fundamental disadvantage of fiber Brillouin amplifiers for magnifying optical signals in photonic communication systems is the poor gain-bandwidth product created due to the severely limited bandwidth. As a result, the Fiber Brillouin amplifier would be a better choice for boosting receiver sensitivity as a preamplifier. Such amplifiers also exhibit a high signal-to-noise ratio (over 15 dB).

2.3.4 Doped-Fiber Amplifiers

Clay minerals are used as a hollow resonator in dopant amplifiers. The parameters of these amplifiers are determined by the alloying elements rather than the silica fiber [2], which only serves as a host medium because these components are tainted with conventional silica filaments. The fiber amplifier may function in various wavelengths ranging from 0.5 to 3.5 m due to several types of dopants. The erbium-doped fiber amplifiers (EDFAs), for example, are particularly appealing since they operate at a 1.55 m wavelength, which corresponds to the lowest fiber loss [16, 17].

Erbium, a lanthanum element in the inorganic class, is the main component of EDFA Erbium was once considered a minor element. However, it is now expected that erbium will become what silicon is to semiconductor innovation in optoelectronics. A small amount of erbium loading in optical fibers, according to Emmanuel Desurvire [20], "allows the gain to be dispersed together with the fiber itself, decreasing the signal's power excursion." With this strategy, signal transfer from one fiber network to the next is nearly lossless."

If EDFAs are designed, the pump and signal beams can move with the unidirectional pumping arrangement. Bilateral boosting allows the amplifier to be forced into both directions simultaneously by using a highly doped laser at the two optical ends. Both sorts of arrangements have advantages and disadvantages. References [20, 21] have already reported on some of these pertinent setups. Figure 2.4 depicts the general EDFA amplification configuration.

The pumping approach and the numerous co-dopants inside the fiber core, such as Germania and aluminum, influence the gain properties of EDFAs. Because of the amorphous structure of silicon, the activity levels of Er3+ ions are broadened into Er3+ bands, allowing for a variety of potential crossings to pump the EDFA. Non - homogeneous widening of the EDFA gain profile is attributed to structural abnormalities, while homogeneous widening is caused by Stark splitting of multiple energy levels. The addition of aluminum to the core further broadens the gain spectrum.

Figure 2.5 demonstrates that EDFA pumping requires semiconductor beams with wavelengths of 0.97 to 1.45 m. Using silica fibers coated with aluminum and phosphate or fluorophosphate fibers can lower the pumping energy delivered.

The EDFA gain spectrum may differ between amplifiers over amplifiers within the context, even though the basic mixture is identical. This is because the EDFA grows proportional to the amplifier length, i.e., the length of the FP hollow resonator where scattering occurs. Gain is heavily influenced by the spectral properties of both the excitation and fluorescence

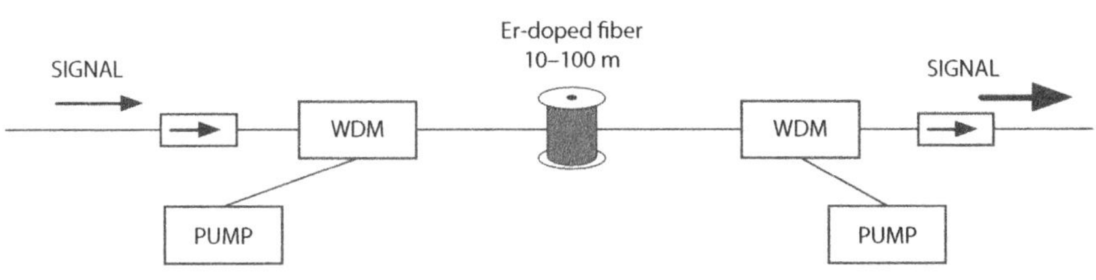

Figure 2.4 EDFA amplifier configuration.

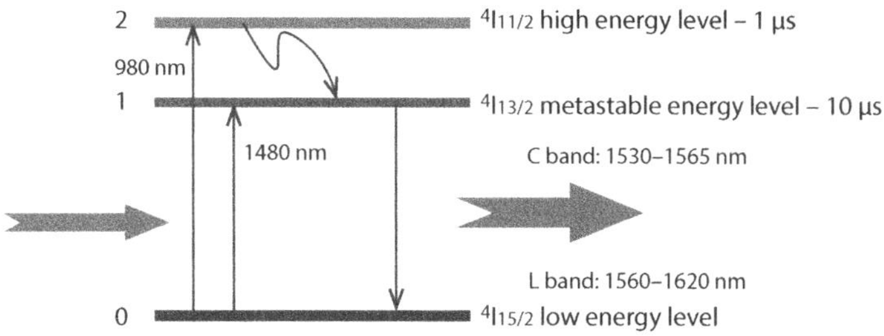

Figure 2.5 Erbium energy level schematic diagram.

cross-sections. Other device and operational characteristics, including Er3+ ionic strength, amplifier duration, inner diameter, and pumping power, are essential in determining its EDFA gain spectrum.

Because of their low noise levels, EDFAs are well suited for photonics communication devices. However, long-distance fiber-optic communication networks using many EDFAs suffer severe amplifier noise problems. The difficulties become significantly worse because the process functions in the anomalous dispersion zone of fiber. This is primarily due to modulation instability, a nonlinear phenomenon that boosts amplifier noise and lowers spectral characteristics.

As previously indicated, EDFAs are appropriate for light beam communications networks with wavelengths of about 1.55 m. However, the telecom network has a vast network of communication links designed for use at different wavelengths, including 1.3 m. Other types of amplifiers are required for signal enhancement in such communications infrastructure. Silica fibers doped with neodymium ions might be used to make fiber amplifiers that work in the 1.30–1.36 m wavelength range. However, unwanted phenomena, such as enthused absorption and hazardous transitions, restrict the technical specifications of such amplifiers. To address these concerns, many additional types of amplifiers were examined, including Nd3+ ion-doped fluoride fibers and ZABLAN (ZrF4-BaF2-LaF3-AlF3-NaF) fibers doped with praseodymium (Pr3+) ions, ytterbium-doped fibers [2]. The comparison of various optical amplifiers is in Table 2.1.

2.4 Hybrid Optical Amplifiers

Dense wavelength division multiplexing (DWDM) is a transmission mechanism that increases the records bandwidth of an optical fiber by simultaneously carrying various information and data signals at different

Table 2.1 Different amplifiers – evaluation.

Property	EDFA	RAMAM	SOA
Gain [dB]	achieve >40	achieve >30	achieve >30
Wavelength [nm]	1531–1626	1281–1651	1281–1651
Bandwidth (3 dB) [nm]	Between 30-60	Up to 100	Up to 61
Max. Saturation [dBm]	30	0.75 * pump power	19
Noise Figure [dB]	achieve >3.5	5	8
Pump power	achieve 25 dBm	achieve >30 dBm	achieve <400mA
Cost factor	Medium	High	Low

wavelengths over the same fiber. A unique wavelength is assigned to each message signal in a DWDM transmission line from a pool of wavelengths. In a DWDM system, the channel separation between two channels can be as small as 1 nm. An optical amplifier is an essential part of a DWDM transmission system. A skinny range of optic bearers must be configured in an optical DWDM transmission line so that several data streams may be delivered utilizing the very same frequency. A high-speed optical amplifier with a smooth gain reaction for a band of frequencies is required to provide identical magnification for every stream of a DWDM distribution grid. Hybrid amplifiers are a vital technology for multi-terabit DWDM transmission systems development [21, 22].

Different optical amplifiers with different gain sides of the spectral bands are linked together in various configurations in hybridization amplifier innovation to provide desired gain bandwidth as required by the transmission system. Hybrid amplifiers are made up of many optical communication systems coupled in a series or parallel manner. At first, a de-multiplexer splits the DWDM message signal into distinct frequency bands, as in a simultaneous arrangement of mixed optical amplifiers. The spectral divisions are then boosted by different amplifiers with amplitudes in the relevant wavelength range and finally concatenated using an optical coupler. The parallel arrangement of hybrid optical amplifiers is reasonably easy to create and may be applied to any optical amplifier. The occurrence of unused bandwidth between those output areas of independent amplifiers due to a frequency component in the optical coupler is the fundamental restriction of the simultaneous construction of mixed optical amplifiers. In addition, the system's noise figure worsens owing to losses produced by optical interconnects placed in front of amplifiers. Optical interconnects are not needed inside the series arrangement of hybrids transceivers, and the boosted bandwidth is rather broad.

2.4.1 EDFA and SOA Hybrid

To increase the gain spectra of an EDFA, a combination of EDFA and an SOA is used. An SOA and a couple of EDFA with a gain range of 1,551–1,600 nm form a three-stage hybrid fiber amplifier in L-Band [23]. Multiple wavelengths and combination optical amplifiers created a sixteen-station DWDM rail line with a 10 Gbps data communication rate. Their efficiency was assessed using the chromatic dispersion coefficient and data transfer range. Study findings of EDFA and SOA have been conducted. Their effectiveness has been evaluated on other optical amplifiers such as EDFA–Raman fiber amplifier (RFA) and RFA–SOA. At a range of 100km, the combination arrangement of RFA-EDFA (12.16 dBm and 12.88 dBm) was found to have the maximum transmitted optical power (12.16 dBm and 12.88 dBm) as well as the lowest BER (10–40 and 10–18) for 2.1 ps/nm/km and 4.1ps/nm/km monochromatic dispersion assessment scales, correspondingly.

Likewise, when combined against EDFA–RFA and SOA–RFA, EDFA–SOA generates significantly higher amplified photoelectric effect (ASE) disturbances [24]. Hence the aggregate performance of the network diminishes as just a result of such limitations established on transmission augmentation by that of the presence of intermodulation distortion therein optical distribution transformer. That ASE formed by EDFA–SOA diminishes this recipient's sensitivity to achieve a specific bit error rate (BER). As a result, EDFA–SOA is not frequently used in long-haul optically DWDM communication systems.

2.4.2 EDFA and FRA Hybrid

RFA's achieve band may be changed by changing the system pressure and pumping frequencies. This RFA feature may be utilized to increase an EDFA's gain bandwidth. Furthermore, RFA has a far lower noise figure than other optical communication systems. Thus, an EDFA fragment established prior and now after dissemination reimbursement fiber (DCF) has so far been utilized to capture more significant gain in sequence utilizing EDFA compartments deployed previously and now after DCF, and

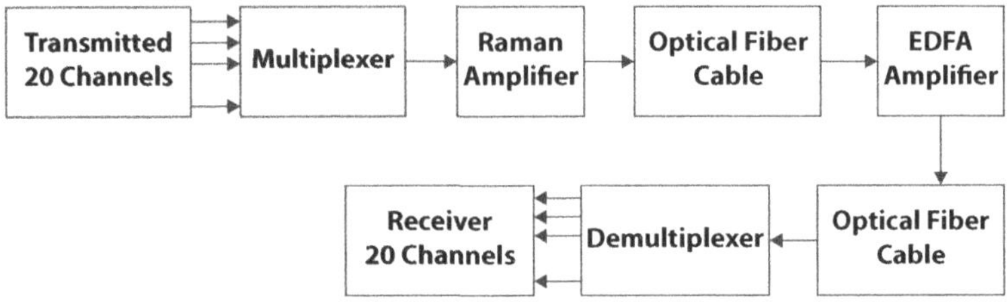

Figure 2.6 EDFA-RAMAN hybrid configuration.

perhaps sufficient merit of BER, acquire, wide bandwidth, as well as Q factor [1]. An ideal design of an EDFA- RFA mixture optical amplifier has been built with various configuration options, and the significance of fiber nonlinearities has also been examined [2]. An EDFA-RFA hybrid amplifier (Figure 2.6) was used for smooth gain attributes to examine a DWDM transmission system with a 25 GHz channel capacity. The variance in gain throughout the radio band appears to rise as the power of data transmission increases. For a signal transmission of 3 mW in the frequency region 187–190.975 THz, a uniform gain of 10 dB is attained with a gain fluctuation of little less than 4.5 dB [3].

2.4.3 RFA and SOA Hybrid

These quasi features generated with SOA can sometimes be reduced via RFA. RFAs have a more comprehensive intensification range, narrow bandwidth aberrations, and narrow bandwidth rate levels. RFAs might give a suitable vast magnification bandwidth and minimal instabilities with enhanced impulses, including reduced noise statistic quantities. In contrast, SOAs appear to be somehow highly portable and also have the possibility of synchronizing multiple tasks such as coherent detection with electro-optic revitalization. RFA, on the other hand, necessitates a powerful pumping system. RFA can be used to provide broad bandwidth with a narrow bandwidth rate and then a wide enhanced frequency. For a 200-kilometer circuit, several CWDM systems with a throughput of around 60 nanometers were created utilizing an SOA-RAMAN unique optical accelerator setup. A RAMAN amplifier boosts the system's gain while minimizing the SOA amplitude skew [25]. Numerous combinations of genetic recombination optical amplifiers, which include EDFA-SOA-RAMAN, EDFA-SOA, but also SOA-RAMAN, have already been constructed in a DWDM scheme with both a diverse variety of channels, as well as measurable Q factor, BER, as well as SNR have also been published [26].

2.4.4 Combination of EYDWA as well as SOA

Due to the increased doping level, Er–Yb co-doped wavelength amplification (EYDWA) can achieve the highest gain and flat gain reaction. SOA can deliver immense output power with low fluctuation. As a result, a combination EYDWA–SOA architecture may be used in urban networks to produce more significant benefits with fewer gain fluctuations [27]. Together with the establishment of many an EYDWA–SOA mixture optical accelerator for quite a 100-Gbps DWDM distribution network with something like a network lifetime of 0.2 nanometers and any yield smoothness technique, a smooth boost of 14 decibels despite fluctuation inside this yield of 0.75 decibels has been observed [27].

2.4.5 EDFA–EYCDFA Hybrid

Due to doping, the fundamental restriction of EDFA is that various values obtained exist at varying frequencies. By employing a specific pump power, the Er–Yr co-doped fiber amplifier (EDFA) can prevent the density quenched effect and obtain high gain with gain smoothness [28]. This high-performance configuration using erbium-doped fiber accelerator (EDFA) & EYCDFA has been evaluated for a 14meters in length EDFA pushed at 980nm using Spectrophotometer, and just a smooth yielding of 36 decibels besides a signal strength equal 4.2 decibels have been validated [28].

2.4.6 TDFA Along with RFA Hybrid

RFA can enhance for either range of frequency, but its main disadvantage would be double Rayleigh dispersion, thus causing amplified signals to be distorted. A TDFA–RFA unique optical amplifier setup can be used to provide around RFA's limitations. A TDFA–RFA hybrid amplifier can encompass a broad span of something like the small frequency band enabling amplified. A widespread excellent optical amplifier with both the cumulative manner with TDFA–RFA has been built within 1,460–1,525 nanometer range of wavelength, with only an accrue oscillation of 24 decibels detected [29].

2.4.7 EDFA and TDFA Hybrid

In today's environment, amplifiers' broad amplitude spectra and noise-canceling figure levels are critical [28–30]. The primary purpose is to increase

Fabry–Perot amplifiers' functioning to gather a large frequency band spacing in a DWDM distribution line. While EDFA is committed to delivering enough gain flattening, its gain band is quite limited, decreasing the number of lines in a DWDM power system. As a result, the EDFA enhancement frequency must be increased. Using TDFA in a hybrid arrangement with EDFA can help to overcome EDFA's shortcomings. In the 1,480–1,555nm wavelength region, a 96-channel DWDM communication scheme with 0.8 nm channel spacing was developed with TDFA-EDFA and EDFA–TDFA unique optical amplifier design, and satisfactory Q factor and BER values were reported [31]. For a wavelength range of 1,460–1,540nanometers, a TDFA–EDFA combination optical amplifier was examined in gain properties. The findings showed a boost of 20 dB and a signal-to-noise ratio of less than that 7 dB [32]. A comparison of the different hybrid optical amplifiers is shown in Table 2.2.

2.5 Applications

Optical amplifiers can be used for a variety of purposes. The employment of SOAs as preamplifiers improves optical detection sensitivity. The information is optically magnified before it reaches the receiver in this type of application. This rectifier circuit boosts the information to a threshold where even the reception's frequency proportion is improved. In order to make up for signal transmission loss, such boosters are also employed in local area networks (LANs). SOAs can also be utilized to augment signal power as power amplifiers.

Furthermore, a power amplifier can boost up to 100 kilometers of optical transmission distances. Nevertheless, most of this is determined by the function generator and transmission attenuation. Finally, enhancements are used in transmission links to maximize data transmission effectiveness [33].

Having reviewed the various types of SOA implementations, it is crucial to remember that they have several shortcomings, such as polarization responsiveness, open-ended multi-dimensional, and multi, including considerable interaction loss. Each of these prohibits your usage as inline accelerators. Fiber amplifiers, on the other hand, do not have these problems and may be used successfully for signal amplification in 1.55 m-based communication lines. However, because fiber amplifiers do not operate much at that frequency, SOAs remain a superior option for 1.3 m-based photonics systems. In a spectral network, SOAs may also be utilized as a wavelength transformer and a rapid switch for spectrum routing [34].

Table 2.2 Comparison between the optical amplifiers.

A mixed arrangement of optical amplifier	Achieved as well as NF (decibel)	Achieved uniformity	The number of features with their proximity	Obtaining frequency band
RAMAN with DCF and EDFA	achieve >10	4.5 decibel	160 streams 25 GHz width	3.97 Terahertz
SOA with EYDWA	achieve >14	0.75 decibel	100 streams 0.2 nm width	20 Nanometer
TDFA with EDFA	achieve >20	achieve dissimilarity proportion <0.4	-	80 Nanometer
RAMAN with EDFA	achieve >12	1.2 decibel	35 streams 1-nm spacing	23.5 Nanometer
EDFA with EYDWA	Gain >36	Flat	-	-

2.5.1 Telecom Infrastructure Optical Power Amplifier

Optical amplifiers are a crucial part of every optical transmission network, and they are not just critical long-haul applications like undersea. [35–37] constitute excellent texts on optical signals and lay a foundation for the following sections. With 'One Gigabyte' per second and 'Ten' Gigabyte transponders, the maximum tensile bandwidth is typical 'Eighty' kilometers; however, some signal repeaters potentially exceed 120 kilometers. While 'Eighty' kilometers may very well be traversed without ever using countermeasures, further distances require the use of such transmitting error detection and correction gear. In 2008, the development of integrated devices [38] changed the landscape. In 2019, synchronized technologies with the most extensive transmission capacity of 200 Gigabyte were widely accessible, as are technologies with a frequency of 400 Gigabyte; unfortunately, the above technologies are costly, and effective optical reaching is limited to some few hundred kilometers. In comparison to the current scenario, a newer generation of semiconductors circuits known as digital processing applications (DSP) would be able to boost this rate to 600 Gigabytes, theoretically lengthening the visual distance to 400 kilometers.

Communication systems have the advantage of being able to amplify many optical signals. In comparison, OEO regenerators should only be used for one transmission at the moment and need expensive multiplexing and transmission of message processes. Optical amplifiers use high spectral efficiency, the exact mechanism used in laser beams to boost the transmitted signal. Non-feedback laser beams are another name for optical amplifiers. To induce frequency imbalance of the perovskite materials, approved devices are pushed (supplied with energy) physically or electronically. In optical amplifiers, type reversal describes how some system components – photons – have tremendous energy or quantum effects than would otherwise be conceivable. The above electronic electrons are unsustainable and eventually result in increased conditions during population relaxes lengths of 1.1 ns to 1.1 milliseconds (further restrictions are possible and therefore are investigated in far more focused research on optical communication systems) [39]. Numerous optical amplifier configurations used in practical systems are depicted in Figure 2.5. An induction type is regularly utilized for relatively short trips of 150 kilometers (Figure 2.7a).

Usually use a combination with little more than an instrumentation amplifier how we want to avoid the extra optical strengths provided by enhancers; nevertheless, we almost always have to incorporate an optical filtration system to limit interference throughout this setup (see Figure 2.7b). Whenever frequencies exceed 250 kilometers, a program that

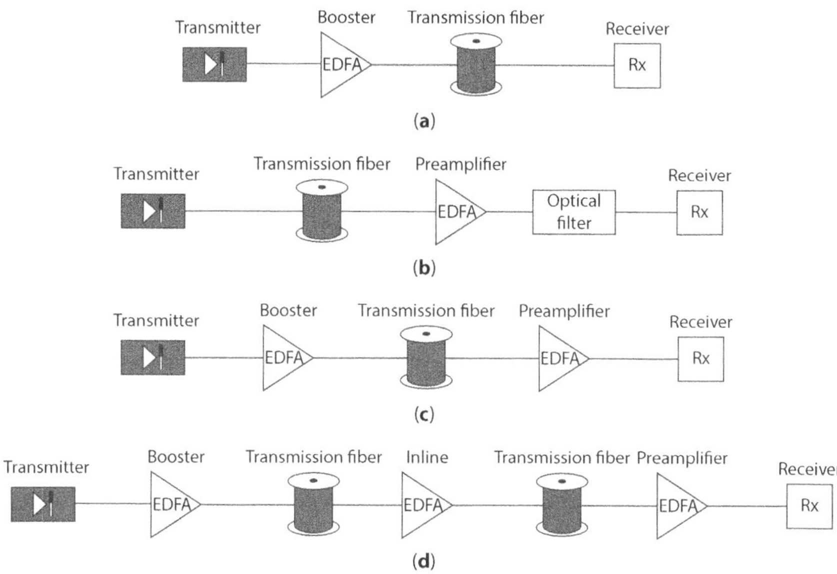

Figure 2.7 (a) For shorter distances (100–150 km) use a booster-only setup, (b) for longer trips, use a preamplifier-only arrangement. (c) An amplifier and processor arrangement is used for up to 200 km range. (d) Configuration of a booster, inline, and preamplifier for long travels (threads that cascade).

includes either a boosting or a threshold is recommended (see Figure 2.7c). Incorporated amplifiers must always be employed to produce more significant spiraling optical gaps (see Figure 2.7d). Optical filtering may be necessary, including all systems [40–55] with preamplifiers to reduce noise, although they are usually not necessary for boosted and combined amplifiers. This same finished product used Raman cranking to achieve a frequency of 350 kilometers; however, due to the unstable Raman effect in crystalline silicon, Raman pumping necessitates ultrahigh powers (up to 1 Watt), which might also necessitate extensive eye protection. It is good to remember that perhaps the durations shown are only estimates and are significantly dependent on existing network technologies (The collecting and analyzing data seems to be the most critical metric.)

2.6 Current Scenario

The researchers have proposed several optical amplifiers that may be used for various applications. The scope of this introductory chapter is too limited to provide for all of them simply. Organic semiconductor lasers, to mention a few, have made substantial contributions to the field of natural light-emitting switching devices (LESD). These transistors absorb much light and have a wide spectral range. Furthermore, because they operate in the visible spectrum, they are highly suitable for various purposes. These amplifiers are optically pumped, and the boosting scheme has improved slowly but surely to the point where small sources, such as semiconductor lasers [56], may now be used with excellent efficiency. Because significant absorbing provides a big gain, the gain width product for such solid-state amplifiers becomes very large [57]. Furthermore, these amplifiers are compatible with optical fibers made of polymer.

Wideband technologies are typically used in communications and multimedia, and as a result, optical amplifiers are intended to amplify all routes with varying wavelengths simultaneously. As such a response, fiber optics with outstanding optical non-linearities, bidirectional interference, boost flattening, and a large transmit power relationship, among several other features, will still be in demand. If certain constraints are satisfied, the amplifier would only be suitable for dense-WDM circuits. Hybridization fiber optics (HFO) appear promising because they are cost-effectively suitable for high data transmission rates [58]. HOA refers to more than one optical amplifier in any arrangement. Significant gain across a wide bandwidth combined with large channel capacity and lower nonlinear losses are prospective benefits of implementing such a method. Intermodulation, distortion, and nonlinear losses are all problems with them. When high gain and gain bandwidth with minimal variance are needed in DWDM systems, HOA can be employed. However, the relative virtues and drawbacks of various designs have prompted researchers to develop new concepts for designing such amplifiers with improved efficiency [59, 60].

2.7 Discussion

The current mission of the optical research methodology is to establish an optical communication network that does not require the use of a piece of software that replicates an optical communication line in real-world settings before it is built. Despite how difficult it may appear, we can build a new technique that meets our needs. This optical amplifier predicated on semiconductors gain medium in which the output reflectivities are still not affected is known as just a semiconductors optical amplifier. A semiconductor single-mode broadband having a fixed transverse dimension (1–2 m) and a specified length (0.5–2 mm) transmits the signaling light. The broadband pattern seems to have many similarities with both the active site, which is driven with or without an electrical charge like a fiber-coupled laser source with anti-reflection layers mostly on termination mirrors. The injecting current increases the carrier density within the valance band, permitting photonic transition first from conducting to the valence bands. Photon frequencies considerably well above the bandgap produce the most gain.

SOAs exhibit significantly stronger quadratic deformities in the measure of self-manipulation and four-wave blending at intermediary power ratings. A resistance value semiconductors amplifier compares against high-power fiber amplifiers in terms of output strength and maintenance strength. Furthermore, their boost frequency seems comparable to that of a high-power fiber amplifier, and it functions in a variety of wavelength ranges. The gain in signal strength is picoseconds because the upper-state duration of energy input is so low. SOAs can be used for optical fiber communication systems depending on nonlinearities or cross-phase manipulation and signal amplification. The specific phenomenon should help with bandwidth interpretation in frequency-division spanning systems, modulate regeneration, clocking regeneration, communication revitalization, and data modeling. Many optical communication systems have routinely used SOAs due to their ease of production, cost-effectiveness, and small size. Since of their rapid gain recovery, low tolerance current, and temperature-insensitive functioning, quantum dots-SOAs are among the most appealing forms of SOAs. Furthermore, the output gain characteristic of the QDs may be easily controlled by adjusting the radii of the QDs. There seem to be numerous devices in optoelectronic devices and photonics design that could be implemented utilizing a broadband optical amplifier.

DWDM in optoelectronics has occurred due to the use of a hybrid optical communication system. Optical processors can indeed be employed in various hybrid topologies to improve optical signal quality while minimizing the constraints of conventional amplifiers. Hybrid optical amplifiers have been shown to improve the communication range of DWDM systems while also mitigating nonlinear characteristics and fiber impairments. In the case of many data channels, the problem encountered by composite optical communication systems should be to provide and sustain high gain with enhanced bandwidth. The optisystem simulation tool can also analyze mm-wave massive MIMO accelerators with DWDM transmission networks, allowing several topologies of hybrid optical communication systems to be thoroughly investigated. For different hybrid configurations of optical amplifiers, several vital parameters, including BER, Q factor, SNR, and ultimate control of receiver end, can indeed be investigated.

2.8 Conclusions

It is impossible to achieve longer distances without any need for amplifiers or transmitters. Thus, the article outlines the fundamentals of repeating and amplification and the optical fiber amplifiers individually. Optical amplifiers are critical components of today's remote communication systems, which extensively use fiber-based networks using the WDM concept. Indeed, there are advantages and disadvantages to using various configurations, determined mainly by the operational requirements. The BER and Q-factor values of something like the transistors mentioned above were used to assess performance. According to tests, this Raman amplifier's pumping transistors may magnify the transmitted information, albeit with comparatively little power (300 mW). The BER should be less than 1012 for fully optical communication systems. The hybrid, as well as modified optical amplifiers, has a lower error rate than the SOA, according to the investigation. The EDFA is probably one of the most widely utilized fiber amplifiers. However, due to the global economy, SOAs are being widely implemented. According to the explanation under the topics above, 'd' and 'f' block metals boozed SOA fiber amplifiers are viable candidate technologies for optical transmission multiplication in high-speed, cost-effective fiber optic communication networks. Except for broadband applications, SOAs can be used in networking applications previously thought to be unsuitable for such powerful optical technology. Intra bands quadruple wave ambiguity occurs when SOA-based wavelengths processors are used, necessitating relatively high boost outputs.

References

1. Ahmad, H., Azmy, N.F., Azmi, A.N., Zulkifli, M.Z., Ismail, M.F., Configurable triple wavelength semiconductor optical amplifier fiber laser using multiple broadband mirrors. *Microwave Opt. Technol. Lett.*, 62, 46–52, 2020.
2. Lu, Y., Wang, F., Gu, J., Shi, L., Peng, M., Huang, J., Kang, W., Photonic generation of the broadly tunable radio-frequency signal using a reflective semiconductor optical amplifier. *Opt. Appl.*, 49, 27–36, 2019.
3. Yang, X., Lindberg, R., Margulis, W., Fröjdh, K., Laurell, F., Continuously tunable, narrow-linewidth laser based on a semiconductor optical amplifier and a linearly chirped fiber Bragg grating. *Opt. Express*, 27, 14213–14220, 2019.
4. Marketandresearch.biz., Global semiconductor optical amplifier (SOA) market research report 2019, Marketandresearch.biz, New York, NY, USA, 2019.

5. Imran, M., Collier, M., Landais, P., Performance analysis of semiconductor optical amplifier as a gate switch. *AIP Conf. Proc.*, 2146, 020016, 2019.
6. Transparency Market Research, Optical amplifiers market—Global industry analysis, size, share, growth, trends and forecast 2015-2023, Transparency Market Research, Albany, NY, USA, 2020.
7. Wang, W., Allaart, K., Lenstra, D., Semiconductor optical amplifier gain anisotropy: Confinement factor against the material gain. *Electron. Lett.*, 40, 1602–1603, 2004.
8. Yang, S., Zhang, Y., Li, Q., Zhu, X., Bergman, K., Magill, P., Baehr-Jones, T., Hochberg, M., Quantum dot semiconductor optical amplifier/silicon external cavity laser for O-band high-speed optical communications. *Opt. Eng.*, 54, 1–5, 2015.
9. Soto, H., García, E., Marquez, H., Valles, N., A simple experimental technique to obtain an optimum anti-reflection coating in semiconductor optical amplifiers. *Microw. Opt. Technol. Lett.*, 27, 223–225, 2000.
10. Aoki, Y., Properties of fiber Raman amplifiers and their applicability to digital optical communication systems. *J. Lightwave Technol.*, 6, 1225–1239, 1988.
11. Han, B., Zhang, X., Zhang, G., Lu, Z., Yang, G., Composite broadband fiber Raman amplifiers using incoherent pumping. *Opt. Express*, 13, 6023–6032, 2005.
12. Mohd Nasir, M.N., Al-Mansoori, M.H., Abdul Rashid, H.A. *et al.*, Multiwavelength Brillouin-erbium fiber laser incorporating a fiber Bragg grating filter. *Laser Phys.*, 18, 446, 2008.
13. Mohd Nasir, M.N. *et al.*, Broadly tunable multi-wavelength Brillouin-erbium fiber laser in a Fabry-Perot cavity. *Laser Phys. Lett.*, 5, 812–816, 2008.
14. Mohd Nasir, M.N. *et al.*, Low threshold and efficient multi-wavelength Brillouin-erbium fiber laser incorporating a fiber Bragg grating filter with intra-cavity pre-amplified Brillouin pump. *Laser Phys. Lett.*, 6, 54–58, 2009.
15. Mohd Nasir, M.N. *et al.*, Widely tunable multi-wavelength Brillouin-erbium fiber laser utilizing low SBS threshold photonic crystal fiber. *Opt. Express*, 7, 12829–12834, 2009.
16. Johari, M.I. *et al.*, Ring cavity multi-wavelength Brillouin-erbium fiber laser with a partially reflective fiber Bragg grating. *J. Opt. Soc. Am. B*, 26, 1675–1678, 2009.
17. Mohd Nasir, M.N. *et al.*, On the preamplified linear cavity multi-wavelength Brillion-erbium fiber laser with low SBS threshold, highly nonlinear photonic crystal fiber. *Laser Phys.*, 19, 2027–2030, 2009.
18. Song, K.Y. and Hotate, K., 25 GHz bandwidth Brillouin slow light in optical fibers. *Opt. Lett.*, 32, 217–219, 2007.
19. Xing, L., Zhan, L., Luo, S., Xia, Y., High-power low-noise fiber Brillouin amplifier for tunable slow-light delay buffer. *IEEE J. Quantum Electron.*, 44, 1133–1138, 2008.

20. Desurvire, E., Giles, C.R., Simpson, J.R., Zyskind, J.L., Efficient erbium-doped fiber amplifier at a 1.53 μm wavelength with a high output saturation power. *Opt. Lett.*, 14, 1266–1268, 1989.
21. Chaugule, S. and More, A., WDM, and optical amplifier. *2nd International Conference on Mechanical and Electronics Engineering (ICMEE 2010)*, Kyoto, Japan, 1–3 Aug 2010, pp. 232–236.
22. Singh, S., Sharma, M.L., Kaur, R., 32 × 10 and 64 × 10 Gb/s transmission using hybrid Raman-Erbium doped optical amplifiers. *Int. J. Adv. Comput. Sci. Appl. Spec. Issue Wirel. Mob. Netw.*, 2011, 76–80, 2011.
23. Yeh, C.H., Lai, K.H., Huang, Y.J., Lee, C.C., Chi, S., Hybrid L-band optical fiber amplifier module with erbium-doped fiber amplifiers and semiconductor optical amplifier. *Jpn. J. Appl. Phys.*, 43, 5357–8, 2004.
24. Kaler, R.S., Simulation of 16 × 10 Gb/s WDM system based on optical amplifiers at different transmission distance and dispersion. *Optik*, 123, 1654–1658, 2012.
25. Iannone, P.P., Reichmann, K.C., Zhou, X., Frigo, N.J., 200 km CWDM transmission using a hybrid amplifier. *Optical Fiber Communication Conference*, California, USA, 6–11 Mar 2005, vol. 4, pp. 76–80.
26. Bhaskar, S., Sharma, M.L., Kaur, R., Performance comparison of different hybrid amplifiers for different numbers of channels. *Int. J. Adv. Comput. Sci. Appl. Spec. Issue Wirel. Mob. Netw.*, 2011, 19–25, 2011.
27. Singh, S. and Kaler, R.S., Novel optical flat-gain hybrid amplifier for dense wavelength division multiplexed system. *IEEE Photon. Technol. Lett.*, 26, 173–6, Jan 15 2014.
28. Vashi, R.R., Desai, A.H., Choksi, A.H., Modeling of gain flattening using EDFA-EYCDFA in cascading mode. *Int. J. Emerging Trends Technol. Comput. Sci.*, 2, 135, July–Aug 2013.
29. Emami, S.D., Hajireza, P., Abd-Rahman, F., Abdul-Rashid, H.A., Ahmad, H., Harun, S.W., Wideband hybrid amplifier operating in S-band region. *Prog. Electromagn. Res.*, 102, 301–313, 2010.
30. Iannone, P.P. and Reichmann, K.C., Hybrid SOA-Raman amplifiers for fiber-to-the-home and metro networks. *National Fiber Optic Engineers Conference 2008*, San Diego, California, US, 24–28 Feb 2008, pp. 1–6.
31. Kaur, I. and Gupta, N., Comparative analysis of hybrid TDFA-EDFA and hybrid EDFATDFA configurations for 96 channels DWDM system for S + C bands. *2014 IEEE 16th International Conference on Transparent optical networks*, Graz Austria, 6–10 July 2014, pp. 1–4.
32. Sakamoto, T., Aozasa, S.I., Yamada, M., Shimizu, M., Hybrid fiber amplifiers consisting of cascaded TDFA and EDFA for WDM signals. *J. Light. Technol.*, 24, 2287–93, June 2006.
33. Agrawal, G., *Fiber-optic communication systems*, Academic Press, New York, 2002.
34. Mohammed, K.A. and Younis, B.M.K., Comparative performance of optical amplifiers: Raman and EDFA. *TELKOMNIKA Telecommun. Comput. Electron. Control*, 18, 1701–1707, 2020.

35. Agrawal, G.P., *Fiber-optic communication systems*, 4th ed., Wiley, New York, NY, USA, 2010.
36. Senior, J.M. and Jamro, M.Y., *Optical fiber communications*, 3rd ed., Financial Times/Prentice Hall, New York, NY, USA, 2009.
37. Digonnet, M.J.F., *Rare-earth-doped fiber lasers and amplifiers*, Revised and Expanded 2nd ed., Marcel Dekker, New York, NY, USA, 2001.
38. Munster, P., Vojtech, J., Sysel, P., Sifta, R., Novotny, V., Horvath, T., Sima, S., Filka, M., F-OTDR signal amplification, in: *Proceedings of the Optical Sensors 2015*, 13–16 April 2015, vol. 9506, International Society for Optics and Photonics, SPIE, Prague, Czech Republic, pp. 28–36, 2015.
39. Pakarzadeh, H., Golabi, R., Peucheret, C., Two-pump fiber optical parametric amplifiers: Beyond the 6-wave model. *Opt. Fiber Technol.*, 45, 223–30, 2018.
40. Tiwari, U., Rajan, K., Thyagarajan, K., Multi-channel gain and noise figure evaluation of Raman/EDFA hybrid amplifiers. *Opt. Commun.*, 281, 1593-1597, 2008.
41. Raj, E.F.I., and Kamaraj, V., Neural network based control for switched reluctance motor drive, in: *2013 IEEE International Conference on Emerging Trends in Computing, Communication and Nanotechnology (ICECCN)*, 2013, March, IEEE, pp. 678–682.
42. Sijini, A.C., Fantin, E., Ranjit, L.P., Switched reluctance motor for hybrid electric vehicle. *Middle-East J. Sci. Res.*, *24*, 3, 734–739, 2016.
43. Raj, E.F.I., and Balaji, M., Analysis and classification of faults in switched reluctance motors using deep learning neural networks. *Arab. J. Sci. Eng.*, *46*, 2, 1313–1332, 2021.
44. Francy Irudaya Rani, E., Niranjana, R., Shah Mohammed Ilyas, S., Suresh, A., Udhaya Prakash, S., Students bus tracker (SBT) enabled GPS device for regular monitoring of heavy vehicles through android application. *Proceedings of the 5th International Conference on Trends in Electronics and Informatics*, ICOEI, pp. 363–369, 2021.
45. Rani, E.F.I., and N.R., Novel engineering of smart electronic wheelchair with physiotherapy treatment compatibility. *2019 Third International Conference on I-SMAC (IoT in Social, Mobile, Analytics and Cloud) (I-SMAC)*, pp. 1–4, 2019.
46. Aiysha Farzana, F.M. and Rani, E.F.I., Technically improved image and video enhancement using adaptive gamma correction with weighting distribution based contrast enhancement techniques. *Asian J. Appl. Sci. Technol.*, 3, 1, 50–57, 2019.
47. Vedhapriyavadhana, R., Rani, E.F.I., Theepa, M., Simulation and performance analysis of security issue using floodlight controller in software defined network. *2018 International Conference on Emerging Trends and Innovations in Engineering and Technological Research (ICETIETR)*, pp. 1–6, 2018.
48. Gampala, V., Sunil Kumar, M., Sushama, C., Fantin Irudaya Raj, E., Deep learning based image processing approaches for image deblurring. *Mater. Today: Proc.*, 2020, doi:10.1016/j.matpr.2020.11.076.

49. Fantin Irudaya Raj, E. and Appadurai, M., Minimization of torque ripple and incremental of power factor in switched reluctance motor drive, in: *Recent Trends in Communication and Intelligent Systems: Proceedings of ICRTCIS 2020*, Springer Singapore, pp. 125–133, 2021.
50. Ch, G., Jana, S., Majji, S., Kuncha, P., Tigadi, A., Diagnosis of COVID-19 using 3D CT scans and vaccination for COVID-19. *World J. Eng.*, 19, 2, 2021.
51. Jain, D., Sangale, D.M.D., Raj, E., A pilot survey of machine learning techniques in smart grid operations of power systems. *Eur. J. Mol. Clin. Med., 7*, 7, 203–210, 2020.
52. Agarwal, P., Ch, M.A., Kharate, D.S., Raj, E.F.I., Balamuralitharan, S., Parameter estimation of COVID-19 second wave BHRP transmission model by using principle component analysis. *Ann. Romanian Soc. Cell Biol.*, 25, 5, 446–457, 2021.
53. Deivakani, M., Kumar, S.S., Kumar, N.U., Raj, E.F.I., Ramakrishna, V., VLSI implementation of discrete cosine transform approximation recursive algorithm. *J. Phys.: Conf. Ser.*, 1817, 1, 012017, 2021, March. IOP Publishing.
54. Priyadarsini, K., Raj, E.F.I., Begum, A.Y., Shanmugasundaram, V., Comparing DevOps procedures from the context of a systems engineer. *Mater. Today: Proc.*, 2020. doi:10.1016/j.matpr.2020.09.624.
55. Raj, E.F.I., and Appadurai, M., The hybrid electric vehicle (HEV)—An overview. *Emerging Solutions for e-Mobility and Smart Grids*, pp. 25–36, 2021.
56. Singh, S. and Kaler, R.S., Investigation of hybrid optical amplifiers for dense wavelength division multiplexed system with reduced spacing at higher bit rates. *Fiber Integr. Opt.*, 31, 208–220, 2012.
57. Rocha, A.M. and Nogueria, R.N., Flexible single pump hybrid fiber amplifier for the S+C bands. *Opt. Commun.*, 320, 105–108, 2014.
58. Singh, S. and Kaler, R.S., Novel optical flat gain hybrid amplifier for dense wavelength division multiplexed system. *IEEE Photonics Technol. Lett.*, 26, 173–176, 2014.
59. Lalam, N., Ng, W.P., Dai, X., Wu, Q., Fu, Y., Sensing range improvement of Brillouin optical time-domain reflectometry (BOTDR) using inline erbium-doped fiber amplifier, in: *Proceedings of the 2017 IEEE SENSORS*, Glasgow, UK, 29 October–1 November 2017, pp. 1–3.
60. Ji, W., Li, X., Kang, Z., Xue, X., Design of WDM-RoF-PON based on improved OFDM mechanism and optical coherent technology. *J. Opt. Commun. Netw.*, 7, 74–82, 2015.

3

Optical Sensors

M. Shanthi[1]*, R. Niraimathi[2], V. Chamundeeswari[3] and Mahaboob Subahani Akbarali[4]

[1]Department of Electronics and Communication Engineering, University College of Engineering Ramanathapuram, Ramanathapuram, India
[2]Department of Electrical and Electronics Engineering Mohamed Sathak Engineering College, Kilakarai, India
[3]Department of Electrical and Electronics Engineering, St. Joseph's College of Engineering College, Chennai, India
[4]Department of Electrical and Electronics Engineering, National Institute of Technology, Tiruchirappalli, India

Abstract

The field of optical fiber has made great development in the previous 25 years [16]. During the 1960s, researchers became increasingly interested in the potential of fiber optic cables for data transmission and other applications. When compared to microwave and other technologies, laser systems can transport a large volume of data. The first experiment that used a laser was to transmit a laser beam freely across the air. Aside from the benefits, fiber optical sensing has sparked attention due to cost reduction and development. As a result, fiber optic sensors develop when fiber optic telecommunications are coupled with optoelectronic devices. Numerous studies have been done with fiber optic sensors in the previous decades using various methodologies [11].

Keywords: Optical fiber, data communication, optical sensors and low attenuation

3.1 Introduction

Figure 3.1 shows a fiber-optic sensor system with a fiber-optic cable, remote sensor, and amplifier.

**Corresponding author*: mlshanthi_2199@yahoo.co.in

Chandra Singh, Rathishchandra R Gatti, K.V.S.S.S.S. Sairam and Ashish Singh (eds.) Modeling and Optimization of Optical Communication Networks, (35–52) © 2023 Scrivener Publishing LLC

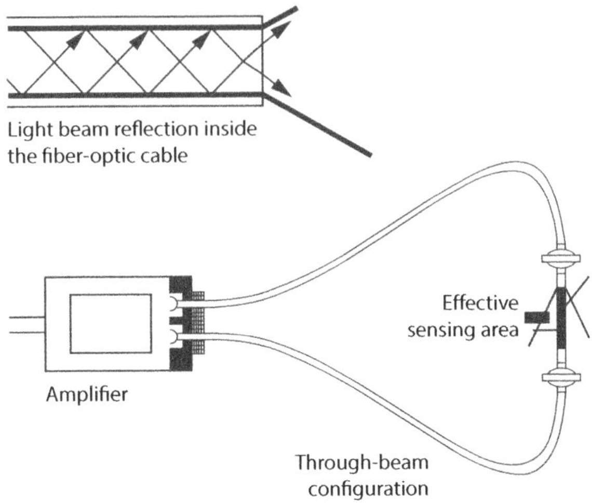

Figure 3.1 Schematic representation of fiber optic sensor system.

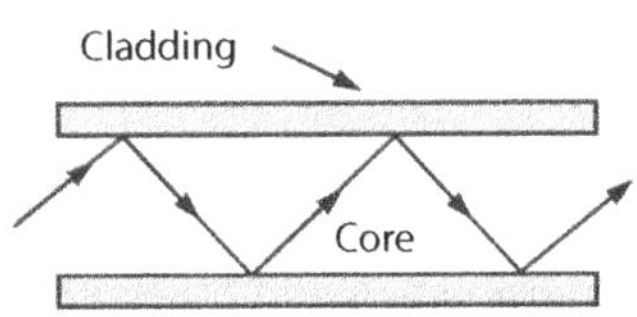

Figure 3.2 Fiber core surrounded by cladding material.

The transmitter and receiver of a fiber-optic sensor system share a single housing. The sensor can reach places that are inaccessible to ordinary photoelectric sensors thanks to the fiber-optic connection linked to the amplifier.

The lights are moved into and out of regions using a wire that connects back to the sensor. As seen in Figure 3.2, that causes it to behave in this manner [4].

3.2 Glass Fibers

To protect cladded fibers, glass optical fibers have very small glass support with a diameter of around 0.0511mm. A stainless-steel armored sheath or a polyvinyl chloride jacket can be used to clad fibers. Bundles made of glass fiber (GF) can endure temperatures of up to 450°F. Special order cables having a temperature rating of up to 1200°F are utilized above 450°F.

3.3 Plastic Fibers

Plastic fiber optic cable contains a single strand with a diameter of around 0.254–1.52 mm and is utilized in extremely confined spaces where bending is required often. In terms of cores, multi-core high flex plastic fiber differs from the regular plastic fiber. When subjected to strong chemicals and solvents, the bending radius might be as small as 1 mm. Teflon or nylon can be used to protect plastic fibers from the elements. Three elements impact the attenuation of light energy: the fiber material, the distance traveled in the fiber, and the wavelength of the light. At all wavelengths, glass fibers operate reliably. However, plastic fibers absorb light from infrared LEDs. Red LEDs, for example, have less attenuation in plastic optical fiber and are thus more widely used [4].

Optical fibers made of glass swiftly rose to prominence as the preferred medium for transmitting light. In the beginning, optical fibers were not allowed to be replaced with coaxial cables because of the high losses in fibers. As a result of their high losses, early fibers were unsuited for communication [11]. Imperfections in the fiber material are what cause optical fiber transmission loss, as many scientists discovered in 1969. It wasn't until 1970 that Corning Glass Works created a multimode fiber with a loss of under 20 dB/km. As early as 1972, the same company created a multimode fiber, in which Optoelectronic components have improved and cost reductions have led to the introduction of new product lines. Products from fiber optics and optoelectronics were combined in the last revolution to make fiber optic sensors.. A decrease in material loss and a rise in the detection sensitivity for losses meant that changes in phase, wavelength, and intensity from external disturbances on the fiber itself could be detected. Due to [11], the development of fiber optic sensing was made. This article provides an overview of fiber optic sensors and their many uses.

3.4 Optical Fiber Sensors Advantages Over Traditional Sensors

In comparison to traditional sensors, optical fiber sensors are more dependable and robust in harsh and challenging settings [1, 10]. The following is a list of benefits.

→ Their modest size and cylindrical shape make them easy to integrate into a variety of applications.

- → Electrical current cannot flow because of a lack of conductivity.
- → Compact, with minimal attenuation and power consumption. Multifunctional sensing capabilities.
- → Extremely durable and impervious to the effects of severe conditions.
- → Significant degree of sensibility.
- → Multiplexing and remote sensing capabilities are included [11].
  - → environmental ruggedness and wide bandwidth
  - → Optical sensors provide the extra benefit of long-range coverage. Nowadays, sensor networks are utilized, which removes the need for separate photonics and electronics conversions at each sensing location, lowering costs and increasing flexibility [18].

3.5 Fiber Optic Sensor Principles

Figure 3.3 depicts the overall layout of an optical fiber sensor system. To process electronics and analyze optical spectrum, it comprises of optical sources such as lasers, LEDs, and laser diodes, optical fiber that can be single or multimode, sensor or modulator element, optical detector, and actuation circuitry. Amplitude, phase, color, and polarization state may all be manipulated in Fiber optic sensor devices [18].

3.6 Classification of Fiber Optic Sensors

The fiber optic sensors are grouped into two categories depending on where the sensor is located on the fiber optic cable.:

(i) Intrinsic Fiber-Optic Sensor
(ii) Extrinsic Fiber-Optic Sensor

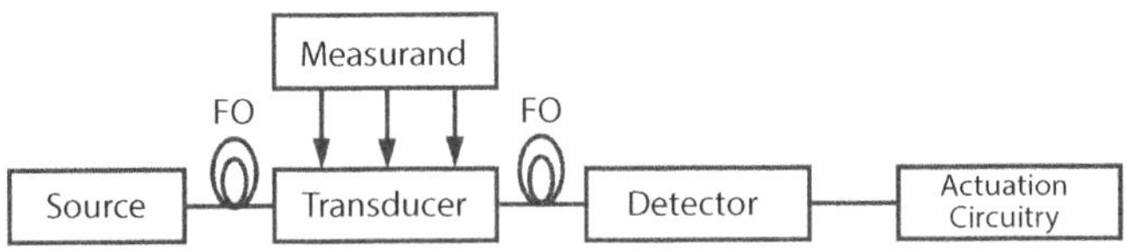

Figure 3.3 The general structure of fiber optic sensor.

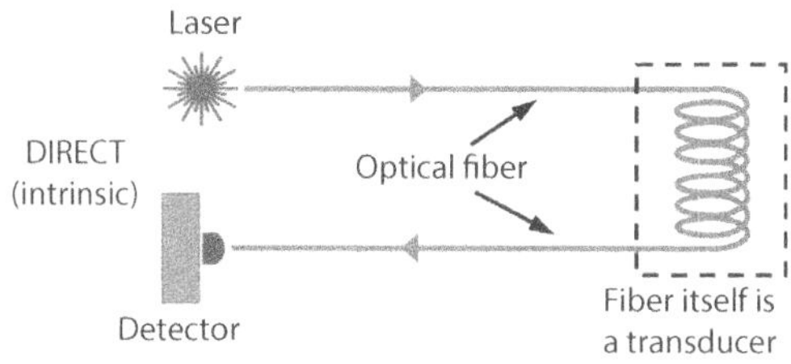

Figure 3.4 Basic concept of intrinsic fiber sensor.

3.6.1 Intrinsic Fiber Optic Sensor

Fiber optic sensors with intrinsic sensing are those in which the sensing takes place within the fiber themselves. They depend on the optical fiber's unique properties to alter the path of the incident light as it passes through the fiber. It is possible that one of the physical characteristics of the light signal in this sensor may be frequency, phase, polarization, or intensity. Distributed sensing over a long distance is the ultimate property of the intrinsic fiber optic sensor [5]. Intrinsic fiber optic sensors are seen in Figure 3.4.

3.6.2 Extrinsic Fiber Optic Sensor

Extrinsic sensors need light to leave the fiber and travel to the outside sensing zone before returning to the fiber. In this case, the fiber acts as a conduit for transmitting light to the detecting location [6]. Extrinsic type fiber optic sensors, represented by a black box, may use the fiber as information carriers. Based on the data received from the black box, it generates a signal of light. All kinds of optical signal-generating devices may be used to make the black box. Measurements include rotational displacement, twisting and torque as well as the velocity of vibrations and acceleration. These sensors' primary advantage is their capacity to enter areas that would otherwise be out of bounds for human beings. Figure 3.5 depicts the fundamental concept of an extrinsic fiber optic sensor.

Aircraft jet engines are an excellent illustration of this sensor in action. Fiber optics are used to transmit radiation to a radiation pyrometer outside of the engine. Transformers' internal temperatures may be monitored using a sensor of this kind [20]. These sensors have a high degree of noise immunity.

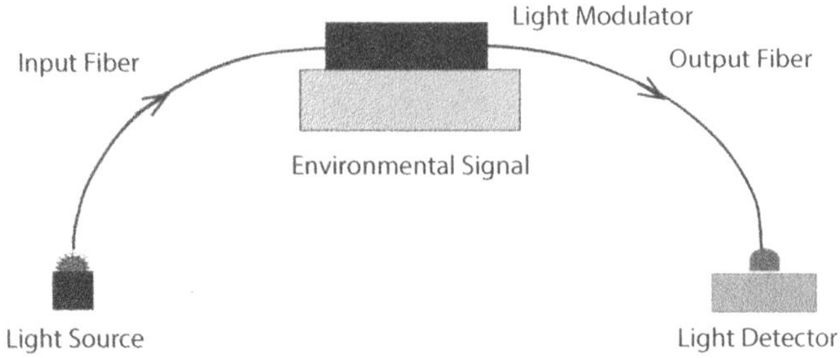

Figure 3.5 Extrinsic type fiber optic sensors.

There are three kinds of fiber optic sensors depending on their functioning principles:

(i) Intensity
(ii) Phase
(iii) Polarization

3.6.3 Intensity-Modulated Sensors

These sensors take advantage of large-core multimode fibers [20]. Unnerving circumstances may be detected by sensors by detecting changes in light intensity. Intensity modulation is linked with principles like reflection, transmission and micro bending. A transmissive or reflecting target may be included in the fiber to achieve it. Additional processes, including absorption, polarization, fluorescence, and scattering, may be used independently of these three basic concepts [16]. Phase-modulated sensors need less light to function, while intensity-modulated sensors require lighter to operate [20].

Figure 3.6 shows two optical fibers held closely together as the basis for this sensor. Injecting light into an optical fiber causes the light to form a cone whose angle is determined by the light's diffraction coefficient. In this way, vibration amplitude may be measured intelligently and accurately. Many restrictions in the system do not exist in the environment due to the sensors' varying losses. Micro- and macro-bending losses are included in the variable losses. Splices and connections at joints are also included in the list of losses. Intensity-based fiber optic sensors use multiple wavelengths to compensate for the losses. One benefit of fiber optic sensors is that they are inexpensive and work well when used in a dispersed fashion. They are also easy to implement and may be multiplexed. On the downside, there are issues with light intensity and relative measures, for example [20].

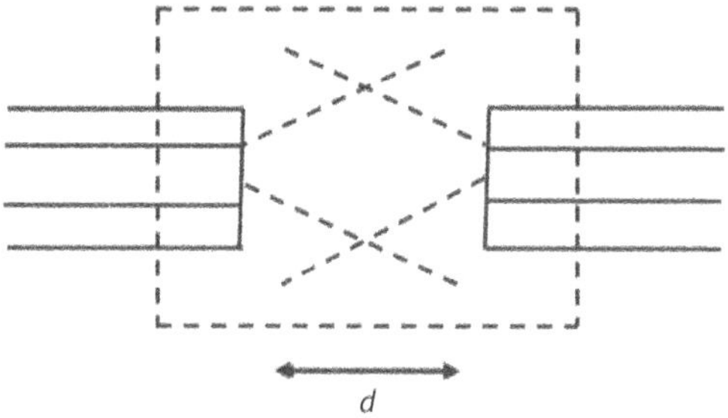

Figure 3.6 Sensor for vibration in fiber optic cables.

Micro bend fiber optic sensors are types of intensity-based sensors, such as evanescent-wave sensors.

3.6.3.1 *Intensity Type Fiber Optic Sensor Using Evanescent Wave Coupling*

The evanescent wave phenomenon occurs when light travels through a single-mode fiber and spreads into the surrounding cladding area as well. Depending on how far apart the two fiber cores are, you'll get different levels of coupling strength. It propagates to the second fiber core that is nearby when light is launched into the first fiber. This causes the evanescent wave of light to pass through both fiber cores. As a result, an evanescent wave may be coupled. Changes in the surroundings may be detected by monitoring the second fiber's intensity change [20]. Chemical sensors make extensive use of this feature. In contrast to other sensor types, this one has the drawback of being sensitive only under certain circumstances.

3.6.3.2 *Intensity Type Fiber Optic Sensor Using Microbend Sensor*

Optical fiber microbending causes transmission loss, which is where this sensor gets its power, and it measures that loss (Figure 3.7). Bending a

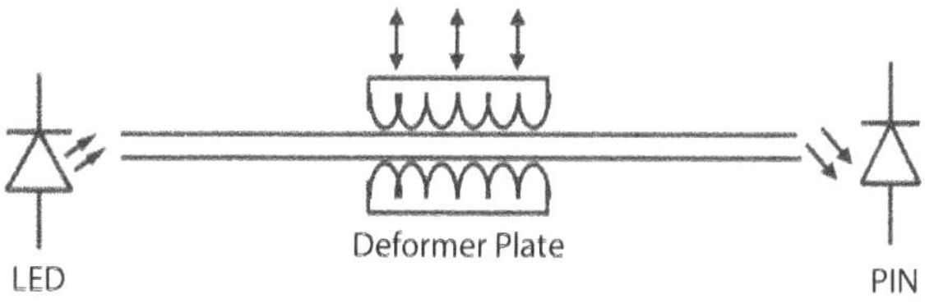

Figure 3.7 Schematic illustration of optical fiber sensor with micro-bending.

multimode fiber waveguide redistributes light power among the fiber's many modes. A steep bend reduces the propagation of fiber light because more light is linked to radiation modes and thus less is propagated [7, 17]. As the fiber deforms, the critical angle of the incoming light in the core is exceeded (microbends). As a result, energy is redistributed between the core and the cladding phases. When the directed higher-order core modes are linked to the cladding modes, light transmission in the fiber is reduced. Mode coupling may be achieved between a series of corrugated plates, particularly deformer plates, by compressing the fiber. As a result, microbending reduces the light intensity by allowing light to seep into the cladding. The decrease of light intensity may be monitored and correlated to create various microbend sensors [2]. Microbend sensors have the benefit of being simpler to install and less expensive than other kinds of fiber optic sensors [3].

3.6.4 Phase Modulated Fiber Optic Sensors

An external physical event might cause a change in the phase shift between two cohering propagating beams that have taken divergent paths. Phase shifts are detected by comparing signal fibers with a reference fiber's phase shifts. One of the beams in an interferometer is subjected to the detecting environment and suffers a phase shift, while the other is isolated and may be used as a reference point. The beams become interfering whenever they are recombined. Most common interferometers are those from Michelson and Mach-Zehnder through Sagnac and Fabry-Perot to the diffraction interferometer [17]. The two interferometers represented in the figure below were built by Mach-Zehnder and Michelson.

Figure 3.8 shows a Michelson-type interferometer, which is a folded Mach Zehnder interferometer. The Michelson interferometer may be

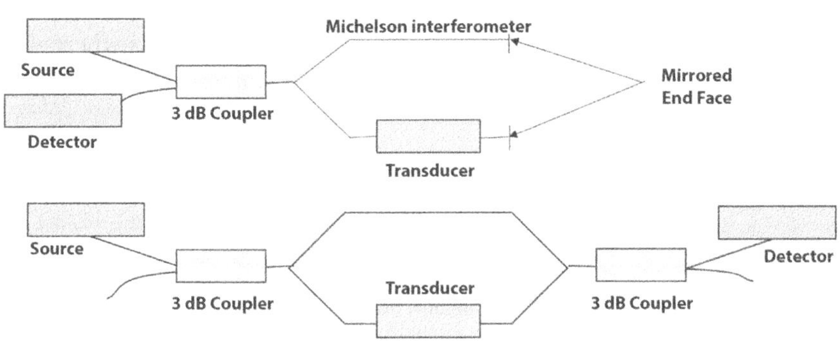

Figure 3.8 Schematic of a fiber optic phase sensor.

set up with just one optical fiber coupler and no further components. Light travels between the sensor and reference fibers twice, resulting in a twofold increase in optical phase shift. Consequently, Michelson may be more sensitive. For sensor interrogation, only one fiber between the source and detector module is required when using Michelson-type Interferometers. However, a high-quality reflection mirror is required for the Michelson interferometer [20]. In contrast to intensity-modulated sensors, phase-modulated sensors are considered highly accurate and errors free. Polarization is a function of strain birefringence, while color change is directly proportional to modifications in the optical signal's absorptivity, reflectivity, transmission, or luminescence.

3.6.4.1 Fiber Optic Gyroscope

The 'Fiber Gyroscope' is a widely used single-mode fiber-optic interferometric sensor. Sagnac rings are used in this interferometer because of their ancestry in classical astronomy. There's a phase shift produced when two light beams traveling in opposing directions go around a fiber coil that's spinning perpendicular to the coil's plane [20], which is called the Sagnac effect.

3.6.4.2 Fiber-Optic Current Sensor

To determine the direct current, fiber optic current sensors have been used. An optical fiber with a single end that takes advantage of the magneto-optic effect (also known as the Faraday Effect) wraps around the conductor to transport the current. Current up to 6000A may be monitored in either direction with an accuracy of 1 %.

3.6.5 Polarization Modulated Fiber Optic Sensors

The direction of the electric field component of the light field is referred to as the polarization state of the light field. The polarization kinds of light fields include elliptical, linear, and circular. Electric field direction does not change throughout the propagation of light for an elliptical polarization state, but the electric field direction does change and the end of the vector produces an ellipse shape, hence "elliptical polarized light" [13]. Pressure or strain induces an effect on the fiber's phase difference between different modes. This process is known as the photoelastic effect or induced refractive index. The optical fiber operates as a linear retarder when subjected to

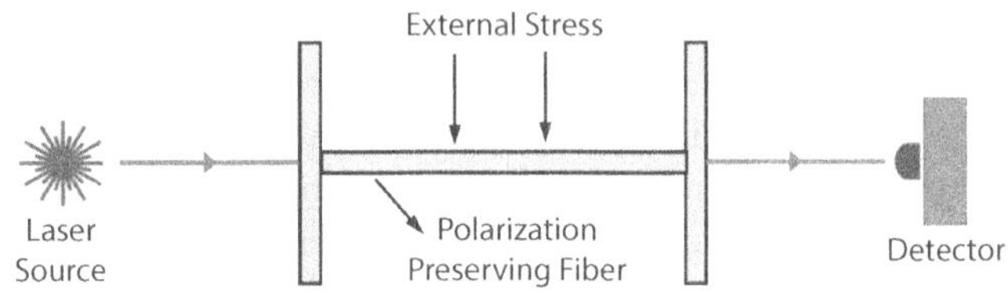

Figure 3.9 An illustration of a fiber optic sensor based on polarization.

external perturbation. As a result, output polarization fluctuations may be used to detect external disturbances [5].

A fiber-optic polarization sensor is shown in an optical configuration in Figure 3.9. A polarizer shapes light coming from a source by polarizing it. A length of birefringent polarization-protecting fiber is used to hold the polarized light at a 45° angle to the fiber's chosen axis. Sensing fiber is the term used to describe this segment. The phase difference between the two polarization states changes when subjected to stress or strain. The output polarization changes as a result of external perturbations. The output polarization state at the fiber's other end may be used to identify external disturbances [20].

Based on their intended use, fiber optic sensors may be divided into three categories:

(i) Physical sensor
(ii) Chemical sensor
(iii) Bio-Medical sensor

3.6.6 Physical Sensor

It is a device that transforms a physical quantity into a signal that an observer or instrument can read. Unlike chemical or biological sensors, they take measurements of temperature and other factors that govern the flow of mass or energy. Temperature sensors, resistance sensors, conduction sensors, heat sensors, and capacitance sensors utilize these sensors.

3.6.6.1 Temperature Sensors

These sensors collect temperature data from a source and display it in a way that other devices or people can comprehend. A mercury thermometer in glass serves as an example of a temperature sensor. The mercury within the glass expands and shrinks as the temperature changes. The outdoor temperature serves as the measurement's source element.

The viewer may determine the temperature by looking at the mercury's location [12].

3.6.6.2 Proximity Sensor

In the absence of any point of touch, a proximity sensor detects things that are close by. Because the sensors do not come into direct touch with the item being felt, they have a long service life and excellent dependability. Closeness sensors come in a variety of forms: inductive, capacitive, ultrasonic, photoelectric, hall-effect, and others. There are many distinct kinds of proximity sensors. The proximity sensor sends out an EM beam and watches for changes in the environment. As the name implies, the item that the sensor detects is known as the proximity sensor's target. A wide variety of sensors of this kind are utilized throughout the manufacturing and automation processes. A shaft's distance from its support bearing may be calculated using vibration monitoring technology [12].

3.6.6.3 Depth/Pressure Sensor

Lake and other freshwater resources' depth or pressure may be measured using water depth. Water depth, for example, is crucial for understanding water resources and how they are being used. A simple marker pole may be used to measure depth, but in several automated applications, a pressure sensor implies the depth of the water. Water pressure may be used to estimate depth. This kind of sensor makes use of a submersible pressure transducer to detect pressure changes and produce an electrical current that can be automatically read and interpreted. In addition to lakes, streams, and rivers, pressure transducers may be utilized in a wide variety of different situations [21].

3.6.7 Chemical Sensor

A chemical sensor is an analyzer that measures and detects chemical qualities in an analyte. Chemical data are converted into electronics that can be used in applications like medical, automotive appliances, nano, micro, and thermal technology. All these share two main components namely receptors and transducers. The overview of the chemical sensor is shown in Figure 3.10.

Analyzers that measure and detect the chemical properties of an analyte use chemical sensors as part of their analysis. Chem data is transformed into electronic data, which may be utilized in a variety of applications, including medical devices and automobile electronics. Receptors and

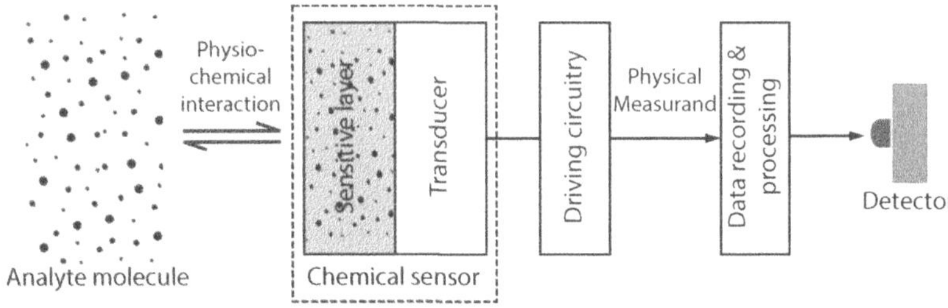

Figure 3.10 Diagram of a chemical sensor (ResearchGate).

transducers are common to all of them. Figure 3.10 depicts a high-level overview of the chemical sensor. It is necessary to utilize transducers to take in chemical data and transform it into electrical data before sending it to another mechanical device, such as a computer. Transducers, which display data on a screen by increasing or decreasing resistance, may be used.

Using a chemical sensor such as a breathalyzer, a human's blood alcohol content (BAC) may be determined. There is a direct correlation between the quantity drank and the number of alcohol molecules exhaled. When determining whether or not a person is fit to drive, this test is utilized. When receptors engage with alcohol molecules, they come into contact with sulfuric acid, potassium dichromate, and water, all of which set off the chain reaction. When a chemical difference is detected between two chambers, an electric signal is generated as an output signal shown by a needle or screed. This helps to raise the specter of BAC.

Fiber-optic chemical sensors have many benefits over conventional sensing methods because of the unique properties of optical fibers. However, the most significant benefit of these sensors is that they are completely impervious to adverse environmental conditions including electromagnetic interference, very high temperatures, and extremely alkaline pH levels. These qualities allowed for the development of a wide range of fiber-optic chemical sensors in the environmental, medicinal, and industrial sectors. Fiber-optic chemical sensors, on the other hand, collect data from a variety of different sensors, which results in a far more complex signal. Fibers have the ability to transfer massive amounts of data. Chemical sensors that can perform real-time measurements on many analytes using fiber optics are expected to be developed soon [6].

3.6.8 Bio-Medical Sensor

One definition of a biosensor is: a device that detects the presence of an unknown chemical or biochemical substance by translating the result of

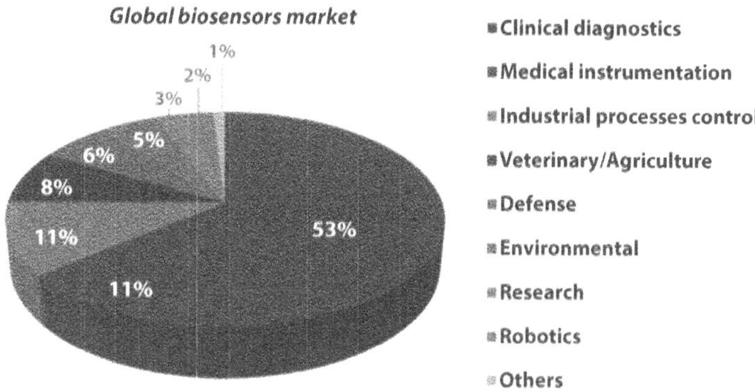

Figure 3.11 Graph showing the global biosensors market share by category.

a biological interaction into a measurable change in some other signal Substrate, biological interface, and biological receptors make up the biosensor structure. There are a variety of bio receptors that may be used in research. All of them have been developed to look for additional molecules that scientists refer to as "analytes" or "biomarkers."

Figure 3.11 shows the range of biosensor applications, which include not just healthcare but also industrial processes, agricultural testing, veterinary, defense, and robotics.

To further process the substrate converts the biochemical interactions into measurable variables, which are then converted back into electrical signals for use elsewhere. Using nanotechnology-based methods, the intermediary layer, referred to as the biofunctionalization interface, connects the bio receptors to the substrate after that. Well-known methods for developing biosensors include electrochemistry-based approaches, micro(Opto)-electromechanical system (MEMS/MOMES) approaches, and optical approaches. Mechanical vibrations are detected using MEMS as a working principle. A small structure known as a microcantilever senses mechanical vibrations as a result of the interaction between bio receptors and the target biomolecules. Because of their high production costs, even though they represent a sensing technology with enormous potential, these devices remain a work in progress. MEMS/MOEMS contributions have remained flat over the last decade, according to (Figure 3.12a).

Electrochemical biosensors, on the other hand, rely on the electrons produced during redox reactions to function. Analytes and bio receptors are reacting, and this is the result of that interaction. They include those that use analyte–bioreceptor interaction to detect electrons produced by other

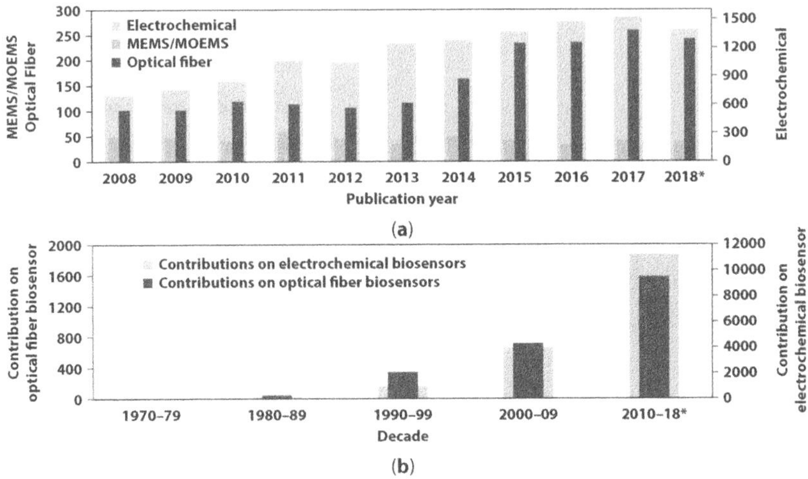

Figure 3.12 Involvement of sensing technologies, such as electrochemical, EMS, or MOEMS, or optical fiber AND biosensor (a Scoptis search for these terms will turn up relevant results). (a) The development of MEMS/MOEMS between 2008 and 2018. (b) Comparison of the contributions made by electrochemical and optical fiber-based biosensors.

substances, as well as those that use the electrons emitted by the analyte–bioreceptor contact alone.

Functionalizing electrodes created with bio-detection in mind is a cinch. In terms of lowering analyte detection limits, this technique has shown to be the most appealing and effective, and it is also the most cost-effective. As a result, the only technology now available to meet the biosensors market's needs is electrochemical-based. In terms of research, they've shown a consistent rise over the past decade, indicating that this technology is steadily improving (Figure 3.12a).

Optical biosensors are the second most widely studied class. Photonic crystals, optical resonators, integrated optics, and optical fibers are all used in some of the system's configurations. Since 2013, optical fiber's impact has grown dramatically (Figure 3.12a and b). Endoscopy, varicose vein eradication, and the prevention of benign prostatic hyperplasia are all examples of optical fiber uses that are now commonplace (BPH). Optical fibers' suitability for medical applications is proven by several characteristics, including:

(i) There are no known negative effects when biological substances contact optical fibers since they are mainly composed of silica and plastic.

(ii) Working in hazardous media: it has shown excellent working results in nuclear settings and the ocean. As a result, this technology's performance in complicated biological matrices must be optimized.
(iii) Electrochemical sensors are harmed by electromagnetic interference (EMI), while optical fiber sensors aren't. In comparison to the light frequencies that pass through the fibers, the electromagnetic fields around them have a far smaller range of frequency variation.
(iv) Multiparameter sensing: It is capable of multiplexing and integrating data from various sensors in optical networks, whether in the time or wavelength regimes.
(v) Low cost: Reduced manufacturing costs have been achieved as a consequence of optical fiber-based communications technology's continuous development. It's now cheaper to create medical gadgets using this technology.
(vi) Light propagation configuration flexibility: while optical fibers guide light when used as cables, the light flow may also be shaped by changing the configurations of the fiber structure.
(vii) It is possible to insert fibers within the organism to highlight its interior cavities, and physical variables may be monitored using a catheter because of their tiny size, flexibility, and lightweight. Open surgery may be avoided because of this technology.
(viii) The phrase "lab on hre" was created by the researchers to describe any kind of modification, printing process, or deposition that may be carried out in and around optical fibers to allow bio-detection [19].

3.7 Optical Fiber Sensing Applications

Fiber optic sensors have a wide range of uses. Optical sensing is a great alternative to conventional electrical sensors, especially in inclement weather or when trying to detect anything a long way away. Electrically passive, resistant to EMI and produced noise, and nonconductive are just a few of the benefits of Fiber Bragg Grating (FBG) optical sensing. In outdoor and industrial settings with hazardous gases or voltages, the fiber's nonconductive and noncorrosive characteristics make it an excellent choice. Due to its immunity to EMI, it is also a cost-effective measuring method since signal conditioning is not required. Medicine, agriculture, civil infrastructure monitoring, and traffic monitoring are just a few of the industries that utilize FBG optical sensing [14].

3.7.1 Application in the Medicinal Field

Researchers across the globe are working on optical-chemical and biochemical sensing, and they're finding more and more uses for it in industries including environmental monitoring, healthcare, and chemical analysis. Absorption and fluorescence are key optical chemical sensing physical phenomena. Even though chemical luminescence, Plasmon resonance, and Raman scattering are all in use, health care seems to have the greatest future growth potential [8]. Optically-based biosensors are finding ever-increasing use in medicine across a wide range of specialties [9].

3.7.2 Application in the Agriculture Field

Agribusiness and food production When it comes to determining crop maturity, determining chlorophyll levels, and evaluating reflectance in flowers and leaves to assess plant health, optical sensing is very useful. Quality control is also critical for ensuring that food items are of the highest possible quality. Classification of incoming feed, online quality control, and analysis of oxygen content within sealed packaging all utilize optical sensors [15].

3.7.3 Application in Civil Infrastructure

The environmental difficulties that structural health monitoring faces are common. Wire installation, a lightning protection system, external assessment and possible maintenance are all part of an electrical monitoring setup. These drawbacks may be overcome by using an optical sensor system. The system's weight and complexity may be drastically reduced by daisy-chaining numerous sensors. Since copper wire corrodes and conducts electricity, optical fiber is less susceptible to damage from lightning. The use of optical sensors significantly reduces the amount of upkeep needed on the system. Applications for civil FOS devices include monitoring of huge construction sites, monitoring of roads and bridges and dams, and monitoring of airport runway loads [16].

Fiber optic sensors are widely employed in the transportation industry to monitoring transportation systems including shipping, trains, airplanes, automobiles, and more. Dimensions, harshness, and weight are the primary difficulties in electrical monitoring systems that may be alleviated with the assistance of optical sensors that can be successfully installed for decades without the need for maintenance. Long-term railway and ship hull surveillance are particularly well-suited to this technology. The use of numerous sensors on a single, extremely thin fiber lowers the overall weight of the monitoring system, which has particular advantages in aircraft applications [21, 22].

3.8 Conclusion

Fiber optics development and application have advanced at a fast pace in recent years. It's intentionally honed to pick up on outside disturbances. The major optical fiber sensors have been covered in this article. Numerous benefits of optical fiber sensors make them extensively used in various fields, including lightweight, easy launch light, downsized size and low ISI. They are also highly resistant to electromagnetic interference and have high sensitivities. They have broad bandwidth. All of the optical fiber's unique properties make it a superior sensor, and also offer up a slew of new research possibilities.

References

1. Kashyap, R., Photosensitive optical fibers: Devices and applications. *Op. Fiber Technol.*, 1, 1, 17–34, 1994.
2. Berthold, J.W., Historical review of microbend fiber-optic sensors. *J. Lightwave Technol.*, 13, 1193–1199, 1995.
3. Luo, F., Liu, J., Ma, N., Morse, T.F., A fiber optic microbend sensor for distributed sensing application in the structural strain monitoring. *Sens. Actuators*, 75, 41–44, 1999.
4. Yu, F.T.S. and Shizhuo, Y., *Fiber optic sensors*, Marcel Decker, Inc., Newyork, 2002.
5. Wolfbeis, O.S. and Weidgans, B.M., Fiber optic chemical sensors and biosensors: A view back. *Optical Chemical Sensors, NATO Sci. Ser. II*, vol. 224, Springer, Dordrecht (NL), pp. 17–44, 2006, chapter 2, ISBN 1-4020-4609-X.
6. Krohn, D., Overview of fiber optic sensors, in: *KMI's 26th annual Newport Conference*, October, 2003, 2003.
7. Baldini, F., Falai, A., De Gaudio, A.R., Landi, D., Lueger, A., Mencaglia, A., Scherr, D., Trettnak, W., Continuous monitoring of gastric carbon dioxide with optical fibres. *Sensors Actuators*, 90, 132–138, 2003.

8. Baldini, F. and Giannetti, A., Optical chemical and biochemical sensors: New trends. *Proc. SPIE*, 5826, 485–499, 2005.
9. Yu, F.T.S., Yin, S., Ruffin, P.B., *Fiber optic sensors*, CRC Press Taylor & Francis Group ISBN 9781420053654, United State of America, 2008.
10. Fidanboylu, K. and Efendioglu, H.S., Fiber optic sensors and their applications. *5th International Advanced Technologies Symposium (IATS'09)*, Karabuk, Turkey, May 13-15, 2009.
11. Bhatt, A., Sensors: Different types of sensors, Engineer Garage, An EE world online resource, February 2011. https://www.engineersgarage.com/sensors-different-types-of-sensors/.
12. Vayalil, R.S., Bisen, A.Y., Kherde, S.J., Fiber optics and its types for sensing applications in various fields. *Int. J. Eng. Res. Technol. (IJERT)*, 1, 7, 1–7, September - 2012.
13. Sabri, N., Aljunid, S.A., Salim, M.S., Fouad, S., Fiber optic sensors: Short review and applications. *Springer Ser. Mater. Sci.*, 204, 299–311, 2015.
14. Presti, D. L. and Cimini, S., Plant wearable sensors based on FBG technology for growthand microclimate monitoring. *Sensors*, 21, 19, 6327, 2021, https://doi.org/10.3390/s21196327.
15. Sabri, N., Aljunid, S.A., Salim, M.S., Ahmad, R.B., Kamaruddin, R., Toward optical sensors: Review and applications. *J. Phys. Conf. Ser.*, 1–7, 423, 012064, 2013.
16. Khandelwal, P., Optical fiber sensors: Classification & applications. *IJLTEMAS*, II, VII, 22–25, July 2013.
17. Sabri, N., Aljunid, S.A., Salim, M.S., Fouad, S., Fiber optic sensors: Short review and applications, in: *Springer Series in Materials Science*, December 2015.
18. Socorro-Lerànoz, A.B., Santano, D., Del Villar, I., Matias, I.R., Trends in the design of wavelength-based optical fibre biosensor (2008—2018). *Biosens. Bioelectron.*, 1, 100015, 2008-2018, June 2019, https://doi.org/10.1016/j.biosx.2019.100015.
19. Fidanboylu, K. and Efendioğlu, H. S., Fiber optic sensors and their applications, in: *International Advanced Technologies Symposium (IATS'09)*, Karabuk, Turke, May 13-15, 2009.
20. Physical Sensors, Lake scientists, the online source for lake science and technology, https://www.lakescientist.com/physical-sensors/.
21. Singh, C., Sairam, K. V. S. S. S. S., MB, H., Global fairness model estimation implementation in logical layer by using optical network survivability techniques, in: *International Conference on Intelligent Data Communication Technologies and Internet of Things*, 2018, August, Springer, Cham, pp. 655–659.
22. Singh, C., Implementation of fiber network survivability by using PL approach, in: *Communication and Computing Systems*, Ist Eidtion, pp. 112–116, CRC Press, 2019. https://doi.org/10.1201/9780429444272

4

Defective and Failure Sensor Detection and Removal in a Wireless Sensor Network

Prasannavenkatesan Theerthagiri

Department of Computer Science and Engineering, GITAM School of Technology, GITAM University, Bengaluru, India

Abstract
In a Wireless Sensor Network, sensor failure is a serious problem since it separates the network into discontinuous parts. The sensor network's functions are affected by these partitions. The primary research issues include tracking the current condition of sensors and eliminating malfunctioning sensors. The Scanning Algorithm for Cut Tracking (SCT) and the Elimination of Faulty Sensor (EFS) algorithms are used in this proposed chapter. The SCT Algorithm will be used to track the state of all sensors in this chapter. This approach is very scalable because the workload does not rise as the number of sensors grows. The EFS Algorithm removes faulty sensors from the network. The tree structure is used in the EFS technique. Highly powerful sensors have been set to the highest level in the tree form. Leaf placement is set with low-power sensors. The chosen sensor is removed from the network. As can be seen, the results show that the proposed methodology will undoubtedly aid in network recovery.

Keywords: WSN, tracking, elimination, failure sensors, network recovery

4.1 Introduction

Wireless Sensor Networks are used in a variety of domains, including military [12], environmental monitoring, and tracking [26–30]. WSNs face several challenges when transporting data from a base sensor to a sink sensor, including secure data transfer, power consumption, and fault tolerance.

Email: vprasann@gitam.edu

Chandra Singh, Rathishchandra R Gatti, K.V.S.S.S.S. Sairam and Ashish Singh (eds.) Modeling and Optimization of Optical Communication Networks, (53–74) © 2023 Scrivener Publishing LLC

The majority of the terminated sensors lose their connections due to a physical failure in the network. This leads to disruptions in the information delivery process [13–15]. If a problem occurs in a sensor, all procedures associated with that sensor will fail. The entire treatment will fail if any process is carried out with that malfunctioning sensor [7]. As a result, if a sensor loses its membership with the base sensor, it is considered faulty.

Sensor failure leads the net to fragment, resulting in failure nodes in the network. This divides the purpose of the internet connection. As a result, the overall WSN's QOS should be reduced [31-37]. The fault has an impact on the dialogue as well as the system's QoS [22–24]. The tracking of malfunctioning sensors is required for a well-organized dialogue [16–21]. To achieve the best results, the sensors in the network must work together [25]. The following diagram shows a graphical representation of cut S (V, E). Let 1 be the starting point, and 10 be the final destination. Vertices 2, 3, 4, 5, 6, 7, 8, and 9 are energetic vertices.

Dark borders show the attachment between energetic vertices. Blue vertices have become faulty because of a cold climate and short battery life. The energetic vertices have separated from the blue vertices. As a result, their affiliation has been missed, and dashed lines show these detachments. Due to node failure, blue vertices have separated from energetic vertices. As a result, blue vertices are faulty vertices. As a result, the graph (V, E)'s malfunctioned vertices are 11, 12, and 13. Because of the cuts, the network has been divided into several portions. As a result, facts transfer becomes more complicated due to cuts, which halt facts transfer. Figure 4.1 Illustrates the graph with malfunctioning nodes.

The SCT and EFS procedures are discussed in detail in this article. The status of each sensor was tracked using the scanning approach in the SCT procedure. The three criteria used in this scanning approach are 1. CF, 2. Activation Status, and 3. Initial power level. The primary goal is to keep track of the status of all sensors. A visual representation of the state has

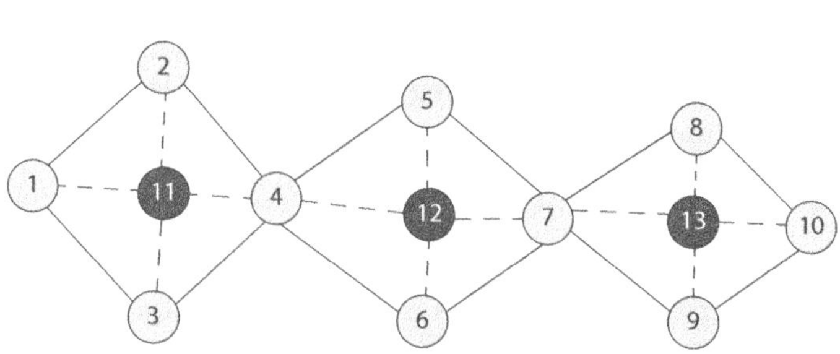

Figure 4.1 Graph illustrating malfunctioned nodes.

been created. This process visualized the total number of malfunctioning sensors. This page also includes a performance evaluation of this procedure. The malfunctioning sensors were removed from the system using the EFS technique. The tree structure is used in the EFS method to visualize the WSN. The basic sensor serves as the root node in this tree. The node's N ID and Power Level have been assigned statically in the tree.

The incredibly powerful sensors have been tuned to a higher level of localization. Low-power sensors pinpoint the location of a leaf in the tree. The terminal level's lowest power level sensor is being clicked. The selected sensor is removed from the tree by selecting the REMOVE SENSOR button in the frame. The notification of removed sensors, as well as their N IDs, has been shown. This process of elimination continues until the system is free of all low-power level sensors. The two techniques implemented tracked the problematic sensors and removed them from the system, making our system fault-free. As a result, these processes eliminate the shortcomings of CVD [16], CAM [20], DFS [18], DDFS [19], and other power-saving procedures. These processes laid up a clear path for the system to recover.

4.2 Related Works

Using the net's condition information could be highly beneficial for many apps and phenomena now growing for MANET [1]. Except that it is not limited to the tracking of net splitting. The appearance of a net fault is net splitting. A single linked net topology collapses into many net topologies separated from one another. Both approaches reliably tracked divisions, according to the simulations.

The writers of article [2] used a broadcast tree to preserve the sensor's power. Sensors with the lowest power level are put into sleeping mode. The sink sensor acts as a root node, while the sensors with the highest power level act as branch nodes. Sensors with low power levels are used as terminal nodes. Terminal nodes are put to sleep, but branch node remnants are kept active. The tree is rebuilt regularly to maintain consistent power in each sensor.

The article's authors [3] created the ABAD protocol in the NS2 apparatus. They demonstrated the superiority of this treatment to ARP. It becomes more secure in terms of sensor power utilization. The PDR has increased by 36.44 percent, and the PLR dropped by 46.52 percent. The prevalence of Alzheimer's disease has dropped by 30.25 percent. Throughput has increased by 36.44 percent, but RE has remained unchanged by 5.88 percent.

Security techniques for de-perimeterized apps were not expanded, and flaws in 3-factor authentication were not correctly recognized.

The authors provided a professional power approach in the article [4]. The process includes two steps: parent discovery and data forwarding. The Parent Discovery Step begins after the net has been configured. This step employs BFS and the Shortest Path algorithm, and it secures the parent and uncle nodes in each sensor's memory. After tracking the failing nodes in the Data Forwarding Step, the Center recursively applied the Parent Discovery Step. Suppose a defective node cannot reach its subsequent hop parent node. In that case, the Data Forwarding Step sends records and the parent node's lone position to an uncle node, alerting the Center that cares about the net recovery and re-executing the Parent Discovery Step. The article's writers [5] chose a central sensor and built it to transmit an ALIVE message over the internet. When the ALIVE note from the center sensor is missed more than a predetermined number of times, Boundary Sensors detect a malfunction.

4.3 Proposed Detection and Elimination Approach

We have divided our procedure into two steps: status tracking and faulty sensors elimination as given in Figure 4.2.

The state of each sensor was tracked in this section using the SCT algorithm and the scanning method.

Elimination of Defective Sensors: The EFS algorithm was used to eliminate faulty sensors in this section.

Status tracking: We used the swing java API to implement this task. Sensors of small size (5-10) were used to test their performance. The technique is divided into two sections: 1. sensor Initialization and 2. Cut Tracking.

Sensor Initialization: All of the sensors' parameters are set up in this section. We have constructed a framework. We have created a dynamic table in

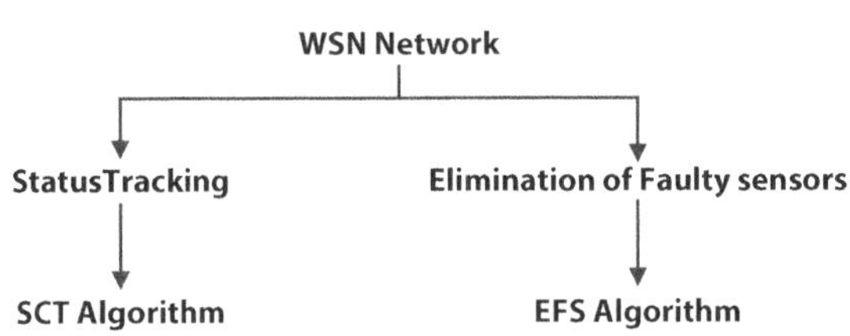

Figure 4.2 Categorization of procedure into two steps.

this frame. The rows in the table represent the number of sensors. The sensor's parameters are shown via columns. The following parameters [6] were used:

N ID: N ID stands for node Id, used to identify sensors exclusively.
QL stands for the sensor's distance from the basis sensor.
CF: The sensor's conversation factor is abbreviated as CF. It expresses the degree to which the sensors are conversing with one another.
Initial Power: The quantity of power required by the sensor to complete various activities is known as initial power.

The activation state is a scalar status object value that each sensor uses. Values are typed in text boxes and added to the table by clicking the ADD Button in the frame during the values assigning phase. This method will be repeated until the entire value in the table has been inserted. Figure 4.3 shows the process of initialization.

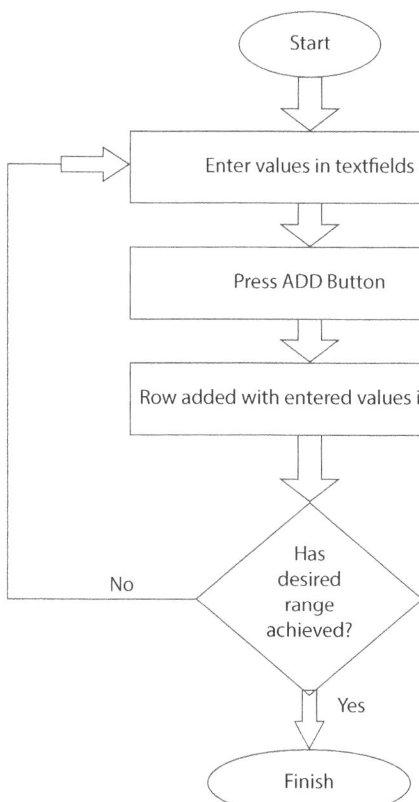

Figure 4.3 Flowchart showing the process of initialization.

Cut Tracking: The first step is to click the table's failure nodes tracking button. The location of the malfunction and damaged sensors will be tracked in this step. Initially, a dialogue box appears, displaying the N_ID and QL in the database and describing the sensor positioning operation.

Cut Tracking by Initial Power Level Scheme
The tracking of failure nodes is accomplished using a scanning approach that monitors the sensors' initial power levels. The DCD algorithm [8–11] is the primary source of inspiration for this technique. We have decided on a tiny network. All of the assigned power level values for each sensor will be scanned. The entered values will be compared to the value set as the threshold. If the entered value is less than the threshold, the sensor with that lower value is considered faulty. As a result, these sensors cannot send data due to a lack of power. There will be no data given. The delivery of data is halted. Those sensors with low power levels will be easily discovered by scanning the power level. In the frame, the lowest power level will be indicated. The value 5J was chosen as the threshold. Sensors with an energy level less than 5J will be considered faulty.

Cut Tracking via Conversation Level
This process begins by scanning the status of all sensors' conversations. A sensor with a 0 conversation status value will be unable to communicate with other sensors and base sensors. As a result, it will be unable to function during the allocation of resources procedure and other required operations. As a result, 0 conversation status sensors will be considered defective. In the frame, the most negligible conversational value will be visible. The N IDs of faulty sensors will be visible. In the Dialogue box, the number of malfunctioning sensors will be displayed. If the conversation values are more than 0, this method will not reveal a defective sensor.

Cut Tracking via Activation Status
The first step in this process is to provide status values to all sensors. The implicit value was used to

Create a value contrast to find the lowest status value. Sensors with low state values will be considered faulty. If the status values are equal to 1, no sensor is faulty; otherwise, if the value is 0, failure nodes are present in the network, and N ID plus the sensor's site is displayed. The number of malfunctioning sensors is shown. Finally, the addition of faulty sensors will be seen by adding the total number of faults obtained from the three

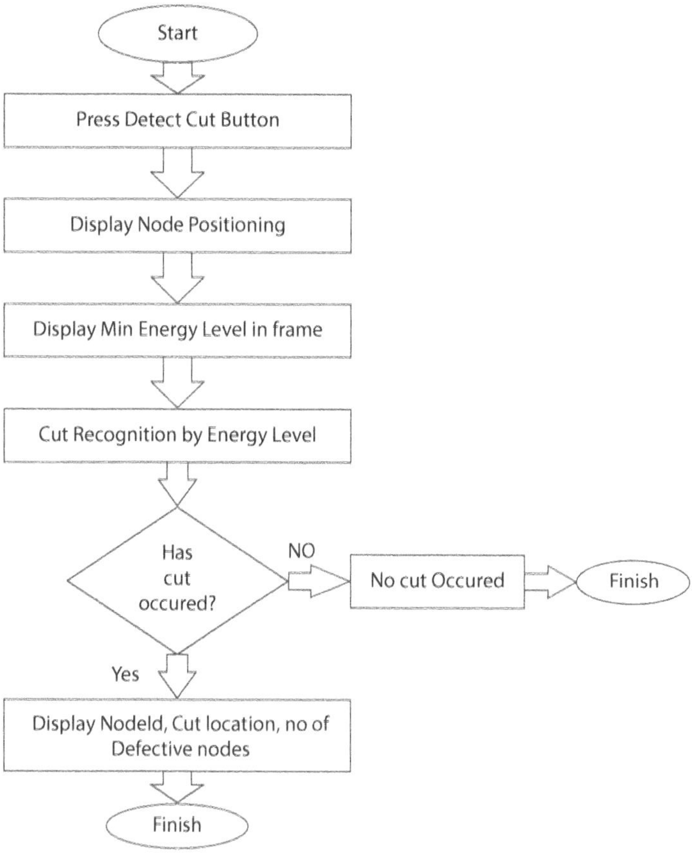

Figure 4.4 Flowchart showing scanning process by power level.

components. Figures 4.4–4.6 provide pictographic representations of failure node tracking using three parameters and flowcharts.

This diagram can be explained in the following steps:

1. The user adds dynamic values to the table.
2. The scanning process begins by assessing the initial energy level to identify failure nodes.
3. When comparing the entered energy level to the predetermined expected value, if it meets the criterion, the sensor node is energetic; otherwise, it is malfunctioning.
4. By checking the contact level, the scanning recognizes failure nodes once more.

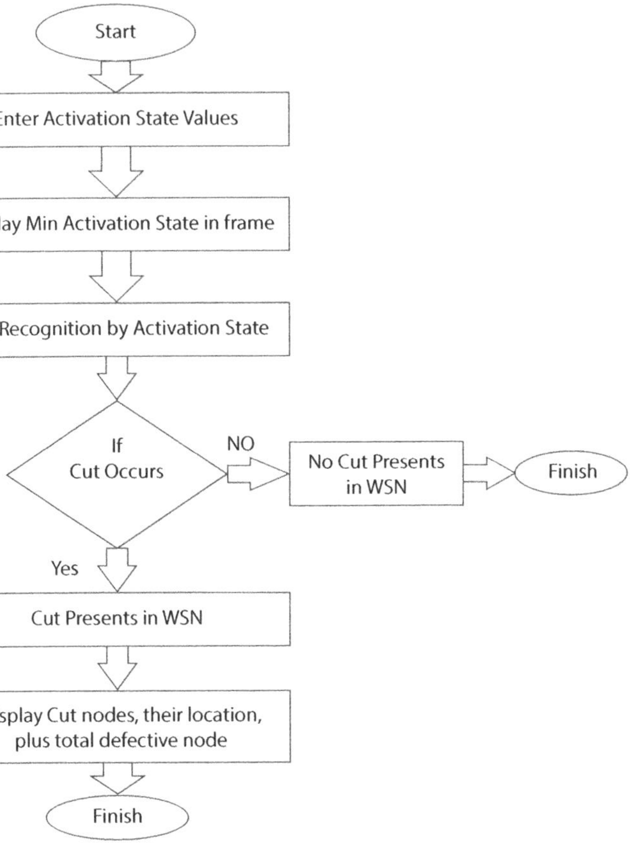

Figure 4.5 Flowchart showing scanning process by conversation level.

5. If the entered contact level meets the criterion, the sensor node is active; otherwise, it is malfunctioning.
6. The scanning mechanism then identifies failed nodes by looking at their activation state.
7. If the entered activation state satisfies the condition, the sensor node is energetic; otherwise, it is malfunctioning.
8. Each sensor node's scanning procedure results are entered into the record database.
9. Finally, a record table appears, displaying the results of the scanning procedure and the status of all nodes according to each factor.

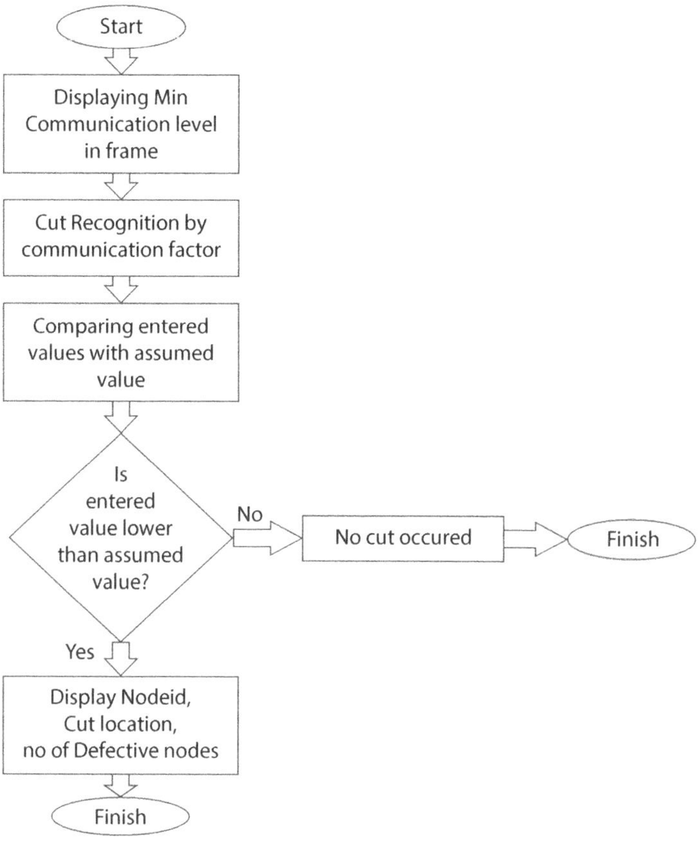

Figure 4.6 Flowchart showing scanning process by activation state.

We have devised a straightforward approach. The user assigns values to each sensor field in the frame in this operation. The smallest values of Power Level plus the CF field have been visualized in the frame. Then a dialogue box appears, displaying the sensor's position in a table with the fields N ID and QL. We worked on the sensor's factor records that had been entered. The dialogue box is used to visualize the outcomes. The Threshold values were compared to the records entered. The ideal value required for accurate performance is the threshold value. If the data are less than the threshold, the screen will display an alert notice dialogue box with the N_Id plus the index of the malfunctioning sensor. The Power Level Threshold value is expected to be 5. The CF field's threshold value has been set to 1. The Activation Status threshold value has been set to 1. We are using values for CF and Activation Status (0 or 1). This concept was primarily influenced by

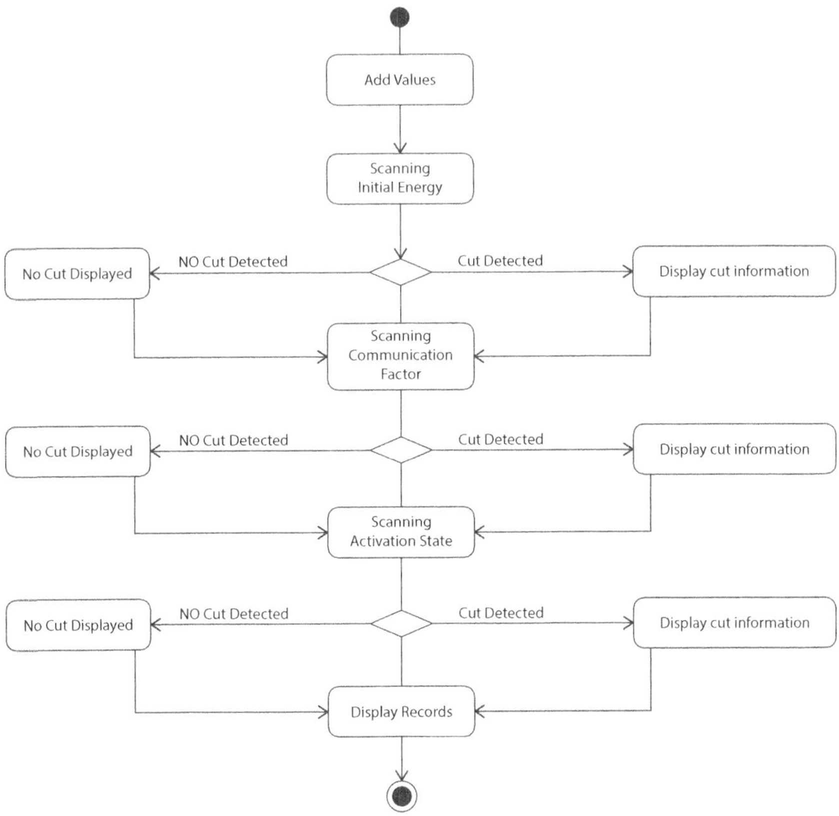

Figure 4.7 Activity diagram showing the overall failure nodes recognition process.

the article [9]. Figure 4.7 shows the activity diagram of overall failure nodes recognition process. Finally, each sensor's data will be visualized in a table in the dialogue box. Finally, erroneous sensor addition will be computed. They are calculated using the following formula:

$$\text{Sum} = \text{sum} + \text{commfailure nodes} + \text{iacnode}$$

Where the sum is the total number of defective sensors, the number of defective sensors determined using Power Level Scanning is the sum. No. of defective sensors estimated using Contact Level Scanning = Commfailure nodes Iacnode = Activation Status Scanning computed the number of defective sensors

Finally, the installation of malfunctioning sensors will be visible on the screen via a dialogue box. The following is the scanning algorithm for tracking failed nodes:

4.3.1 Scanning Algorithm for Cut Tracking (SCT)

```
1.  Assign values to fields.
2.  Scanning power levels of each sensor.
3.  IF (Power Level < Threshold value){
4.        THEN fault occurred.
5.              SHOW index and N_ID of sensor}
6.  Scanning Contact Levels of each sensor.
7.  IF (Contact Level < Threshold value){
8.        THEN fault occurred.
9.              SHOW index and N_ID of sensor}
10. Scanning Activation Status of each sensor.
11. IF( Activation Status < Threshold value){
12.           THEN fault occurred.
13.                    SHOW N_ID and position of each sensor}
14. SHOW status record of each sensor.
15. Compute SUM of faulty sensors obtained via three factors
16. SHOW the SUM of faulty sensors.
```

The significant benefit of this procedure is that there will be no effect on the operation in case of an increasingly large amount of sensors.

Eliminating Defective Sensors from Network

In this section, we hierarchically visualize WSN in a frame where the base sensor serves as the root node. To represent the WSN, we use the JTree data structure. In JTree, each node's N ID and Power Level have been statically allocated. Each row on the tree is highlighted with a different color, symbolizing the level's active power. The extremely powerful sensors have been placed on the tree's higher level. The terminal level is where the less powerful sensors are positioned.

The Logic Behind Locating Sensors

The main reason for putting the low-power sensors in the leaf area is that if the parent level is deleted, the entire subtree will be eliminated. Low-power sensors are thus isolated at the leaf level to preserve the elimination of other active sensors.

The malfunctioning sensors can be removed using the REMOVE SENSOR button on the frame. The action event occurs when this button is clicked, and the code associated with the REMOVE SENSOR button is executed. As a result, the sensors in question will be removed from the system. The cmd screen will display notifications of sensors that have been removed in sequential order. Each sensor's N_ID will be displayed on the cmd screen. The elimination procedure will continue until all defective sensors have been removed from the system. The list of defective sensors will be visualized in the cmd screen once all of the problematic sensors have been eliminated.

As a result, we have met our second goal and removed the faulty sensor from the network. The technique for Eliminate Faulty Sensor Algorithm is as follows, based on the preceding explanation:

4.3.2 Eliminate Faulty Sensor Algorithm (EFS)

1. Assigning N_ID + Power Level to each node statically.
2. Setting node's location.
3. If (Power Level<Threshold value)
4. Then Set node to upper location of JTree.
5. Else
6. Set the node to the leaf location of JTree.
7. Highlight each level of JTree with different colors.
8. Visualize WSN in tree form.
9. Select the lowest power sensor.
10. Click the REMOVE SENSOR button.
11. Elimination of Faulty sensor.
12. Visualization of notification of eliminated sensor via cmd screen.
13. If (All faulty sensors have been eliminated).
14. Then End the process.
15. Else go to step 9.

The result of code execution has been given in the result section of this research article. This process is elaborated via the flowchart given in Figure 4.8.

Description of flowchart: The preceding flowchart effectively displays the complete EFS method in sequential order. First, JTree assigns N ID + Power Level to each node statically. The condition will then determine the proper placement. If the power level is less than the threshold, the sensor will be set to the leaf location. It will be set to the upper location. Then we receive a JTree in the frame colored at different levels. In the frame, there is a REMOVE SENSOR button. The sensor with the lowest power level is chosen. Then the REMOVE SENSOR button is pressed. The sensor has been deactivated in the system. The notification of a removed sensor is sent via the cmd screen, which displays the removed sensor's N_ID. This procedure will be repeated until the system is free of all malfunctioning sensors.

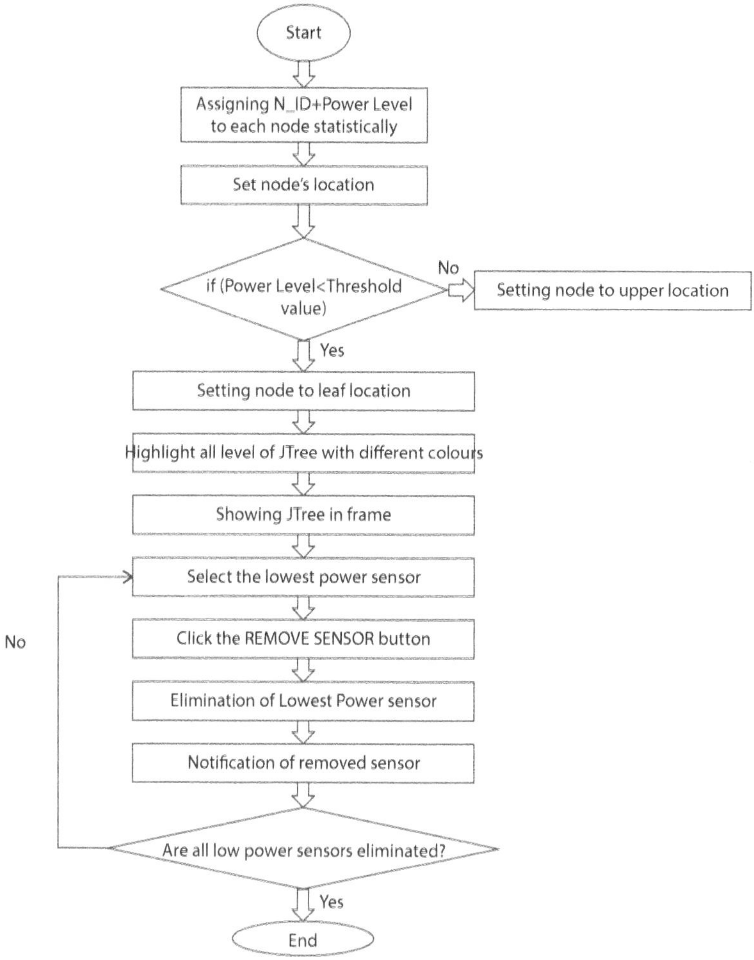

Figure 4.8 Flowchart visualizing the process of faulty sensor elimination.

As a result, we removed the problematic sensors from the system and built a defect-free system. As a result, this approach assists us in moving towards the net recovery process. Placing specific objects in the locations of eliminated sensors and those who perform the same functions as those eliminated sensors is part of our future work.

Advantages of EFS Procedure

It prevents the elimination of the subtree from the system by preserving the removal of other active sensors via the logic of setting sensors to the leaf position. This process keeps the entire system's power. It reduces the

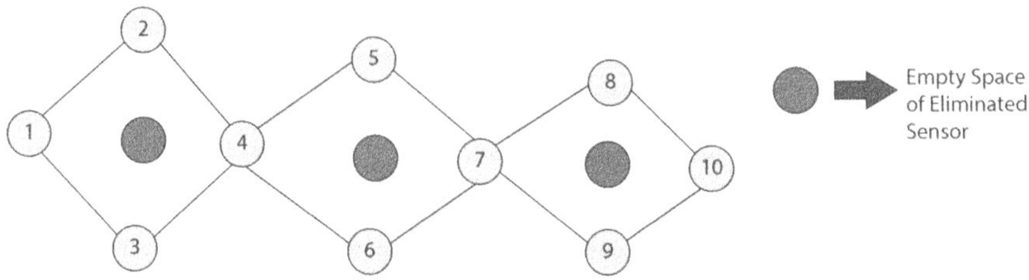

Figure 4.9 Graph visualizing the EFS procedure.

amount of time spent waiting. This method is highly sequential as well as cyclic. This method is very scalable because adding many sensors does not influence work.

Graphical Illustration of Eliminated Faulty Sensor Procedure
The operation of our EFS technique is depicted in the graph below. We used the same graph as in the introduction to show the difference between the graph with fault sensors and without fault sensors. The malfunctioning sensors were removed from the system using the EFS technique. There are vacant areas in their location where an object that serves the same function as the deleted sensor is required. Empty gaps are represented in the graph by circles that are highlighted in red. This method eliminates the possibility of defective sensors in our system. As a result, there will be no issues with data transport being blocked in the system. All difficulties relating to the chat have been resolved. The net restoration technique was given a new path thanks to this procedure. Our future study will include the placement of things that function similarly to those sensors. The effective operation of our technique is depicted in Figure 4.9.

4.4 Results and Discussion

The results of the suggested EFS and SCT algorithms are discussed in this section. Figure 4.10 depicts a screenshot of the malfunctioning sensor. The record depicted in Figure 4.11 shows the state of each sensor. The total number of malfunctioning sensors in WSN is depicted in Figure 4.12.

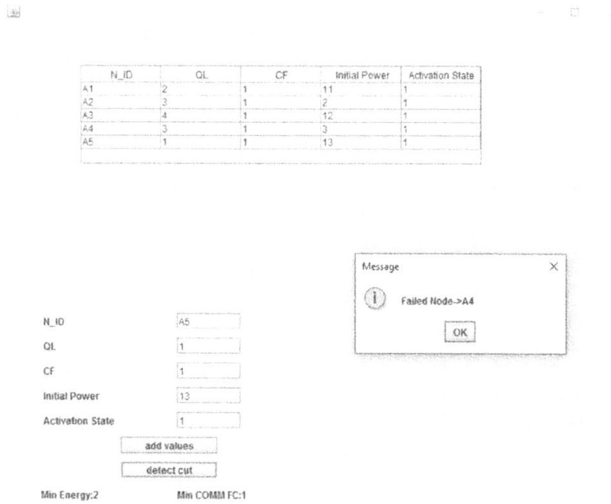

Figure 4.10 Snapshot visualizing the defective sensor.

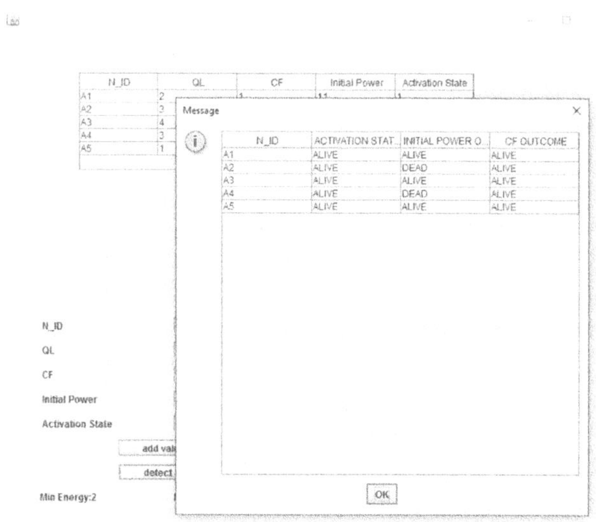

Figure 4.11 Snapshot visualizing the record showing status of each sensor.

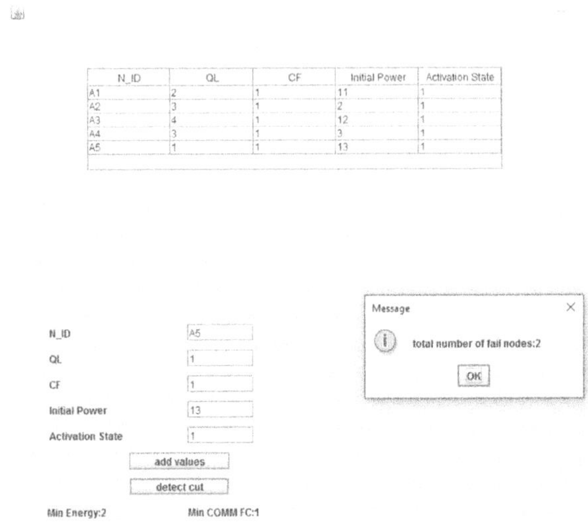

Figure 4.12 Snapshot visualizing the sum of faulty sensors.

4.5 Performance Evaluation

We have done a procedure analysis in this section. MS-Excel is used to carry out this analysis technique. The two graphs below depict the variables used and their values entered during the value assigning process. The N IDs of sensors are displayed on the X-axis. The Y-axis displays the values entered by the user for three variables. The three factors have been shown in the graphs as a series. 1. Activation Status, 2. CF, and 3. Power Levels are the three criteria. To visualize these records, we created graphs. Two variables, CF and Activation Status values, stay constant in the 2-D graph shown in Figure 4.13. The values of both variables line up and create a straight line. A single line has been displayed for these two variables in the graph. Both graphs effectively depict the factor's assigned values for each sensor.

The graph in Figure 4.14 shows the data from the malfunctioning sensor at the time of power level scanning during the execution phase. The N IDs of sensors are represented on the X-axis in this graph. The Power level of each sensor is represented on the Y-axis. The assigned values were compared to the threshold value during the scanning step. The value 5J was chosen as the threshold. Sensors with a Power level below a certain threshold are classified as faulty sensors. The malfunctioning sensors A3 and A4

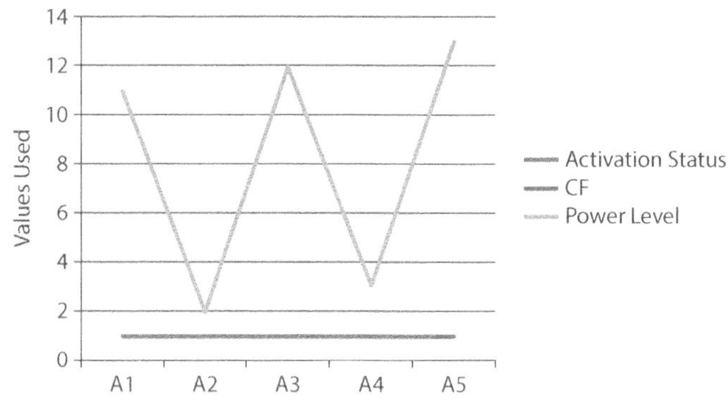

Figure 4.13 2-D Graph visualizing entered variables.

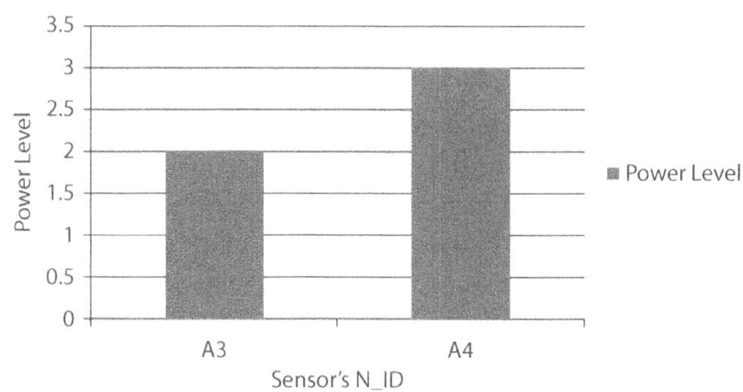

Figure 4.14 Graph visualizing the defective sensor's record.

are shown in Figure 4.14, together with their related Power levels 2J and 3J obtained during the execution phase.

The performance of our approach is depicted in Figure 4.15. The X-axis displays the running time in minutes. The number of sensors used is represented on the Y-axis. Our approach found that increasing the number of sensors does not affect workload. This graph shows that the execution time does not alter as the number of sensors in the network grows. With more sensors, the execution time remains constant. This depicts the scalability of our technique. It performs well when the number of sensors is increased.

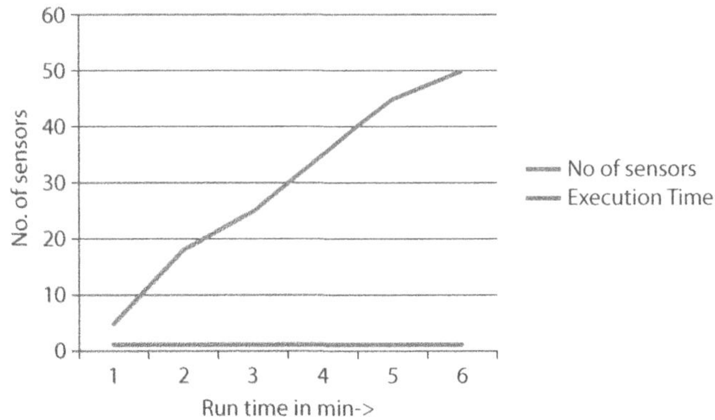

Figure 4.15 Graph visualizing the work's performance.

4.6 Conclusion

The primary study field examined in this paper is detecting problems in sensor networks. In our research, we describe how to develop a technique that is energy efficient and takes less time to detect failing nodes. The SCT technique was developed as a research project to track network failure nodes. The linear search approach is used to identify failure nodes in this procedure. Scanning takes place on the values of each node in each factor. During the scanning process, each result is compared to the assumed value. If the entered value meets the criterion, the node will be certified as active; otherwise, it will be inactive. Our technique is sequential as the scanning process moves from one factor to the next. To reduce instant delay, we have simplified our technique. Finding failure nodes increases the likelihood of restoring network connectivity to its former state. Another process we provide is the EFS algorithm, which eliminates malfunctioning sensors. The WSN is presented using the tree form in this process.

Each node in this tree has been allocated an N ID and a Power Level. The incredibly powerful sensors have been placed at the top of the tree. Low-power sensors have been assigned to the tree's leaf location. The sensor for the lowest power level is being removed. The notifications for the sensors that have been withdrawn have been delivered in consecutive orders. The EFS technique prevents additional active sensors from being removed from the system by using the logic of placing low-power sensors at the tree's terminal point. When compared to alternative processes, this approach saves time. It is also highly scalable, as it performs well when the number of sensors is increased. We were able to locate and remove the problematic sensors from the network in this manner. This procedure paved the way for the following step in the network recovery process. Our subsequent work will include comparing our technique to existing procedures and developing a solution to the network recovery challenge.

Furthermore, our SCT process is performed sequentially. The time delay in the SCT technique can be significantly reduced by converting from a series to a parallel operation. Future work will include locating active objects in vacant areas left by deleted sensors.

References

1. Hauspie, M., Carle, J., Simplot, D., Partition detection system in mobile ad-hoc networks. *Proc. Second Mediterranean Workshop Ad-Hoc Networks*, pp. 25–27, 2003.
2. Parmar, S., Shevale, K., Wakhare, K., Rudrakanthwar, N., Cut detection and prevention in wireless sensor network. *Int. J. Eng. Tech. Res. (IJETR)*, 3, 2, 266–269, 2015.
3. Sudhaman, K., Anand, M., Raja, B., Malathy, V., Angle based adversary node detection in wireless sensor networks. *J. Crit. Rev.*, 7, 6, 542–553, 2020.
4. Venkatesan, P., FUCEM: Futuristic cooperation evaluation model using markov process for evaluating node reliability and link stability in mobile ad hoc network. *Wireless Netw.*, Springer, 26, 4173–4188, 2020.
5. Mitra, P., Roy, S., Chandra, S., Saha, I., Saha, B., Energy-efficient approach on cut detection in wireless sensor networks. *Int. J. Comput. Sci. Inf. Technol. (IJCSIT)*, 6, 1282–1286, 2015.
6. Ritter, H., Winter, R., Schiller, J., A partition detection system for mobile ad-hoc networks. *Proceedings of IEEE Conference on Sensor and Ad-Hoc Communications and Networks (SECON)*, pp. 489–497, 2004.
7. Mathew, R. and Annadevi, E., Artificial routing protocol for cut detection of cut vertices. *IOSR J. Comput. Eng. (IOSR-JCE)*, 9, 8–12, 2012.
8. Arun Sathya, M. and Nellai Nayaki, V., FNDRA: Failure or dead node detection and recovery algorithm for wireless sensor networks. *Int. J. Adv. Res. Comput. Commun. Eng.*, 4, 9, 495–499, 2015.
9. Barooah, P., Chenji, H., Stoleru, R., Kalmar-Nagy, T., Cut detection in wireless sensor networks. *IEEE Trans. Parallel Distrib. Syst.*, 99, PrePrints, 483–490, 2011.
10. Menon, V., Mhase, P., Kanekar, S., Nikram, N., Cut detection in wireless sensor network. *Int. J. Comput. Sci. Technol. (IJCST)*, 3, 2, 121–126, 2015.
11. Theerthagiri, P., COFEE: Context-aware futuristic energy estimation model for sensor nodes using Markov model and auto-regression. *Int. J. Commun. Syst.*, e4248, 2019.

12. Anna Devi, E. and Martin Leo Manickam, J., Distributed cut detection in wireless sensor network. *Res. J. Pharm. Biol. Chem. Sci. (RJPBCS)*, 8, 2, 1279–1284, 2017, ISSN: 0975-8585.
13. Rewagad, P.P. and Patil, H.P., DCD algorithm in wireless sensor networks with enhance security mechanism. *Int. J. Latest Trends Eng. Technol. (IJLTET)*, 3, 2, 246–250, 2013.
14. He, T., Krishnamurthy, S., Luo, L., Yan, T., Gu, L., Stoleru, R., Zhou, G., Cao, Q., Vicaire, P., Stankovic, J.A., Abdelzaher, T.F., Hui, J., Krogh, B., VigilNet: An integrated sensor network system foe energy efficient surveillance. *ACM Trans. Sens. Netw.*, 2, 1, 1–38, 2006.
15. Won, M. and Stoleru, R., Destination-based cut detection in wireless sensor networks. *Proceedings of IEEE/IFIP International Conference on Embedded and Ubiquitous Computing (EUC)*, 2011.
16. Won, M. and Stoleru, R., Towards robustness and power efficiency of cut detection in wireless sensor networks. Elsevier *Ad-Hoc Netw.*, 9, 3, 249–264, 2011.
17. Senturk, I.F., Akkaya, K., Yilmaz, S., Distributed relay node positioning for connectivity restoration in partitioned wireless sensor network. *IEEE Explore Digital Library*, 2012.
18. Theerthagiri, P. and Menakadevi, T., Futuristic speed prediction using auto-regression and neural networks for mobile ad hoc networks. *Int. J. Commun. Syst.*, 32, 9, e3951, 2019.
19. Xiong, S. and Jianzhong, An efficient algorithm for cut vertex detection in wireless sensor network. *International Conference on Distributed Computing System*, 2010.
20. Pimple, J. and Pandey, Y., Distributed source separation detection algorithm. *Int. J. Sci. Res. Publ.*, 2, 12, 1–4, 2012.
21. Cormen, T.H., Leiserson, C.E., Rivest, R.L., Stein, C., *Introduction to algorithms*, Second Edition, MIT Press, London, 2002.
22. West, D.B., *Introduction to graph theory*, Second Edition, Prentice-Hall, Upper Saddle River, 2001.
23. Liu, X., Xiao, L., Kreling, A., Liu, Y., Optimizing overlay topology by reducing cut vertices. *ACM Workshop on Network and Operating System Support for Digital Audio and Video (NOSSDAV)*, 2006.
24. Milic, B. and Malek, M., Adaptation of the breadth-first search algorithm for cut-edge detection in wireless multihop networks. *ACM International Workshop on Modeling Analysis and Simulation of Wireless and Mobile Systems (MSWiM)*, 2007.
25. Prasanna Venkatesan, T., Imposing packet relaying for mobile AdHoc networks using genetic algorithm. *Int. J. Sens. Wirel. Commun. Control*, 10, 617–624, 2020.
26. Duthie, R.N. and Sarwade, N.P., Sensor node failure detection based on round trip delay and paths in WSNs. *IEEE Sens. J.*, 14, 2, 2014.

27. Patel, P.S., Bohara, M.H., Chahwala, B.D., Sensor fault detection in wireless sensor networks and avoiding the path failure nodes. *Proceedings of 2nd IRF International Conference*, 2014.
28. Abbasi, A.A., Younis, M.F., Baroudi, U.A., Recovering from a node failure in wireless sensor-actor networks with minimal topology changes. *IEEE Trans. Veh. Technol.*, 62, 1, 256–271, 2013.
29. Prasannavenkatesan, and Menakadevi, T., Resource-based routing protocol for mobile AdHoc networks. *Songklanakarin J. Sci. Technol.*, 42, 4, 889–896, 2020.
30. Anuradha, M.P., Swetha, A.M., Doraipandian, M., Fault node detection and connectivity with mobile relay node in wireless sensor networks. *J. Comput. Sci. (JCSSP)*, 16, 551–558, 2020.
31. Akyildiz, I.F., Su, W., Sankarasubramaniam, Y., Cayirci, E., Wireless sensor networks: A survey. Elsevier *Comput. Networks*, 38, 393–422, 2002.
32. Culler, D., Estrin, D., Srivastava, M., Overview of sensor networks. *IEEE Comput.*, 37, 41–49, 2004.
33. Estrin, D., Govindan, R., Heidemann, J., Kumar, S., Next century challenges: Scalable coordination in sensor networks. *Proc. of 5th Annual Mobicom, ACM*, 1999.
34. Pottie, G.J. and Kaiser, W.J., Wireless integrated network sensors. *Commun. ACM*, 43, 51–58, 2000.
35. Theerthagiri, P. and Menakadevi, Significance of scalability for on-demand routing protocols in MANETs. *Proceedings of IEEE Conference on Emerging Devices & Smart Systems*, Namakkal, March 4-5, pp. 76–82, 2016.
36. Zhao, F., Liu, J., Guibas, L., Reich, J., Collaborative signal and information processing: An information directed approach. *Proc. IEEE*, 91, 8, 1199–1209, 2003.
37. Prasannavenkatesan, T., Raja, R., Ganeshkumar, P., PDA-misbehaving node detection & prevention for MANETs, *Proceedings of IEEE Int. Cof. Communication and Signal Processing*, Melmaruvathur, April, 1808–1812, 2014.

5

Optical Fiber and Prime Optical Devices for Optical Communication

Srividya P.

R.V. College of Engineering, Bangalore, India

Abstract

In recent times, global businesses demands for faster, secure and high capacity communication systems for all their network operations. This demand is increasing at a faster pace. Fiber optic communication plays a major role in fulfilling these demands. Data transmission from one destination to another by transmitting infrared light pulses via optical fiber is known as fiber optic communication. Developments in fiber optic communication have brought about a drastic change in telecommunication industry. Fiber optic communication is an optimum choice meant for transmitting data in gigabits and more. It can be utilized to transmit video, audio, telemetry and data over long distances by means of local area networks or using network of computers [9]. Light-wave technology is used in fiber optic communication to transfer data over optical fiber by converting electrical signals into light signals. An optical fiber is a cylindrical waveguide made out of silica or plastic. They are extremely thin (only barely thicker than a single strand of human hair) and flexible. Optic Fiber cable which is a pack of optical fibers protected by a flexible and transparent sheath is most suited for higher data rate transmissions over longer distances without electromagnetic interference. Since it uses light pulses for communication, it offers boundless performance capabilities. In addition, this technology does not involve the usage of electromagnetic field. Hence, the data cannot be intercepted or jumbled with other signals.

Usage of fiber optic cable for optical communication has paved way in establishing links over long distances with higher data rates, light weight, higher security and with low transmission loss. Safety, security and speed of the optical fibers come at a lower cost compared to the increasing cost of copper, which is used

Email: srividyap@rvce.edu.in

Chandra Singh, Rathishchandra R Gatti, K.V.S.S.S.S. Sairam and Ashish Singh (eds.) Modeling and Optimization of Optical Communication Networks, (75–98) © 2023 Scrivener Publishing LLC

in other cable technology. The advantages of optical fiber communication have attracted its usage for wide range of applications like Ethernet systems, broadband distributions and even data networking. 5G, cloud computing, IoT, Augmented reality/Virtual reality, health care systems, Massive machine communications, smart cities have become some of the driving forces in development of optical technologies.

Fiber optic transmission system consists of four different blocks-transmitter, fiber optic cable, optical repeater and receiver. These blocks are explained in detail in this chapter. The chapter also deals with the optical sources, materials used to make the fiber, benefits and drawbacks of Optical fiber. Further an insight into the recent advancements in fiber technology is also highlighted.

Keywords: Optical fibers, LED, laser, multi-core fibers, WDM, SDM, DWDM, all optical network

5.1 Introduction

Optical fiber communication (OFC) has brought about an innovative change in telecommunication and data networking domains. It is a technique used to transmit audio, video or telemetry data signals as light pulses using optical fiber as the transmitting medium in both terrain and underwater installations. The signals in electrical form are converted into light pulses at the transmitter end and are converted back into electrical pulses at the receiver end using optical sources and optical detectors respectively. Usage of optical fiber cable provides long distance communication with lower losses and with higher data rates. When OFC was in the early stages of invention, only light source, optical fiber as transmitting medium and photodetectors were used. At present, a number of active and passive optical devices are also being used along with the optical link to perform various network functions like signal reinstatement, signal routing and signal switching in optical domain.

The technologies used in optical communication are also radically changing due to the demand for high speed connectivity. The leading technological advancements in 5G, smart cities, Internet of Things, connected vehicles, cloud computing, augmented and virtual realities have become the driving force in the recent developments of optical technologies.

5.2 Optic Fiber Systems Development

Communication, which is one of the basic needs of people, is the motivation behind the remarkable development in the communication system. The parameters demanding up gradation are improvement in transmission fidelity, higher data rate to enable more data transmission and aid in increasing the transmission distance between relaying stations. During the initial stages of telecommunication, a single line wire was used in data transmission. Later with the demand to transmit huge data at faster rates, coaxial cables were introduced. The bandwidth limitations of these devices paved way for optical communication systems. OFC was possible with the discovery of lasers in 1960s and optical fibers in 1970s [7, 8]. The first generation link was operational at a wave length of 850 nm. It used GaAs based sources, Silicon photo detectors and optical fibers. Fiber loss and intermodal dispersions were found to be more in the system. The transmission wavelength was then shifted from 850 nm to 1300 nm and 1500 nm with the developments in optical sources and photo detectors. This shift helped in reducing the drawbacks of 850 nm system and also increased the repeater less transmission range. The 1500 nm system provided lowest attenuation with data transfer rate of about 2.5 Gbps over 90 km repeater less range [1]. This resulted in the signal dispersion by silica fibers to be more at 1500 nm wavelength. The transmission rates of more than 10 Tbps is now possible with the advent of higher quality lasers, optical amplifiers and the usage of wavelength division multiplexing. Recent developments include usage of Graphene material and two dimensional (2D) nano-materials in building optoelectronic devices [5]. Graphene is one of the strongest material and approximately 200 times stronger than steel and lighter than a paper. Most predominantly studied 2D nano-materials include transition metal dichalcogenide like molybdenum disulfide and black phosphorus. Although these materials exhibit superior electrical and optical properties, their sensitivity to external conditions poses a challenge in mass production of optoelectronic devices.

5.3 Optical Fiber Transmission Link

The major components of optical fiber communication link are the transmitter circuitry comprising of a light source with a driver circuitry, signal regeneration circuitry and a receiver circuitry comprising of photo detector, amplifier and a signal restorer circuit all connected through optical fibers as shown in Figure 5.1. Apart from these components, the link also has optical connectors, splice, coupler and amplifiers. The different blocks of the system varies based on the application.

Input to the transmitter circuitry can either be in analog or in digital forms. Light sources of suitable dimension are made use of to launch the optical signal into the fiber. The light source generates optical stream that is modulated to transmit the data. Presence of light denotes "1" and the nonexistence of light denotes "0". Light emitting diodes (LEDs) and laser diodes are most important source of light because of their feature of modulating their optical signal output by varying the bias current. In case of high data rate transmission of about 2 Gbps, direct modulation leads to signal distortion. This is avoided by using an external modulator to vary the light amplitude from the laser diode source. For shorter wavelengths (800 nm

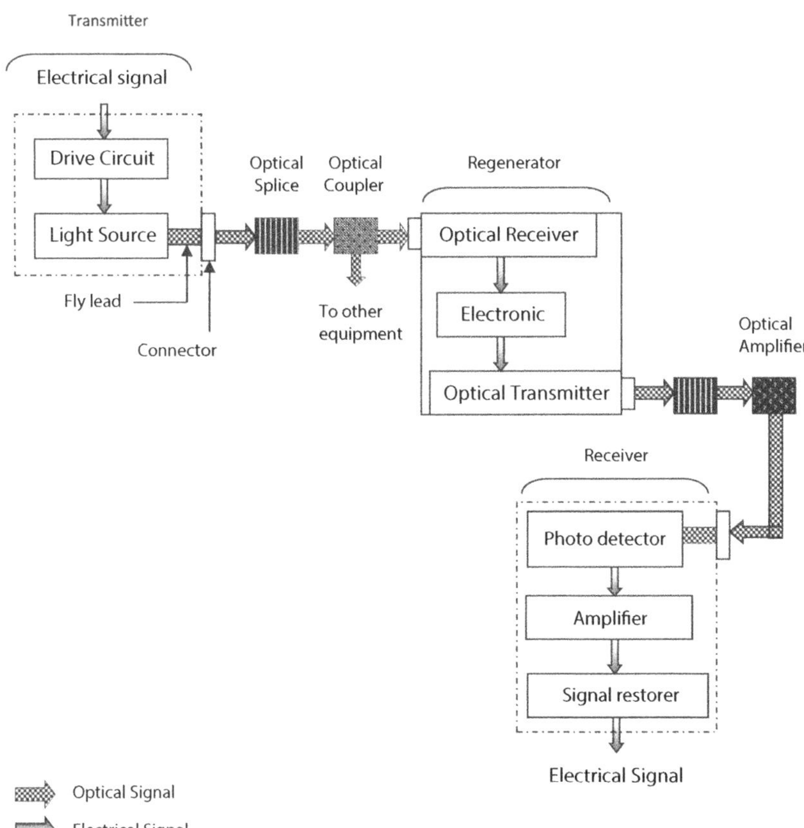

Figure 5.1 Components of optical fiber communication link [1].

to 900 nm), alloys of GaAlAs is used to make light sources and for longer wavelength (1100 nm to 1600 nm) alloys of InGaAsP is used.

The light signal is transmitted through an optical fiber to the detector that converts light into electrical pulses. The cable attenuation and light signal distortion due to absorption, scattering and dispersal mechanisms in glass material along the cable limits the maximum distance over which the signals can be transmitted through an optical fiber. This limitation can be overcome by using Opto-electric repeaters or regenerators and amplifiers in the transmission path. Optical fiber can be installed aerially, under water or buried underground. Due to cable weight, manufacturing and installation limitations, a single cable length is restricted to few kilometers. The cables are then connected together using optical splices to establish long distance links.

The receiver consists of a photo diode that converts the week optical signal received by it into electrical signals. For shorter wavelengths (800–900 nm), silicon is used to make photo diode and for longer wavelength (1100–1600 nm) alloys of InGaAs is used.

Light signals of different wavelengths in the range of 1300–1600 nm can be transmitted simultaneously over an optical fiber by using wavelength division multiplexing technique. Light pulses (precisely laser light) are preferred in OFC because laser light provides a single wavelength as opposed to other sources like bulb or sunlight that gives light of many wavelengths.

5.4 Optical Sources Suited for Optical Fiber Communication

Optical source suitable for optical fiber communication must have the following characteristics:

- Physical dimension must suit the geometry of the optical fiber.
- Must possess narrow radiation pattern.
- Must possess the capacity to modulate directly by varying the driving current.
- Must have quick response time.
- Must be able to generate ample output power to couple into the fiber.
- Must be stable, efficient, reliable and cost effective.

Table 5.1 Difference between laser and LED optical sources.

Laser diodes	LEDs
Optical resonant cavity is used to produce optical energy. Hence forms a coherent source.	Does not use optical resonant cavity to produce optical energy. Hence forms an incoherent source.
The output beam is monochromatic and is highly directional.	The output radiation from LED has broad spectral width.
As it needs thermal and optical circuits for stabilization, it involves complex drive circuitry.	Requires lesser complex drive circuitry than Laser diodes.

Since Injection laser diodes (ILDs) and Light emitting diodes (LEDs) possess most of the above characteristics, they are the principle luminescent or light sources utilized in optical communication. Comparison between LEDs and Laser diodes sources are listed in Table 5.1.

The PN junction of both LED and laser diodes are constructed with III-V semiconductor materials which have direct band gap. The junction is called recombination region due to forward biasing the junction, injection of the carriers (electrons and holes) into respective p and n regions where they are either recombined radiatively or non-radiatively. If they recombine radiatively, a photon of energy hv is emitted and if they recombine non-radiatively, heat is dissipated.

Light emission in the form of a photon can either be spontaneous or could be stimulated by any other photon with precise energy level. In both the process, electrons must be raised to higher energy state from lower energy state by supply of energy.

5.5 LED as Optical Source

LED is a PN junction diode that converts electrical current into light output. This process is called electroluminescence. LED is an elementary transmitting device for low cost applications as it is cheaper compared to other light sources. LEDs find its use in applications where data rate ranges between 10 to 100 Mbps and transmitting devices are few kilometers (e.g. Local area network). The usage of LEDs offers only a small portion of input power that enters the optical fiber. This reduces the transmission efficiency. This can be overcome by using high power drivers to facilitate long distance transmission.

On forward biasing the PN junction, the electrons from N region gets accelerated and recombines with the holes present in the P region. Since the electrons occupy the conduction band and holes occupy the valence band there exists a dissimilarity in the energy levels between the two. If the electrons and holes recombine, a fraction of the energy gets dissipated in the form of heat and light.

LEDs must acquire the below mentioned characteristics if it has to be used as an optical source:

- Higher quantum efficiency – This is associated with the portion of injected electron hole pairs that are recombined radiatively. It is achieved by confining the emission of electrical input pulses to PN junction's active region where radiative recombination occurs.
- Rapid emission response time – Response time is a measure of lag between the application of electrical pulses to the LED and arrival of optical pulses from it. This factor limits the bandwidth by which the source modulation can be achieved by varying the applied pulses.
- Output radiance must be higher – Radiance indicates the amount of optical power emitted into a unit solid angle for each unit area of the emitting surface. This is measured in Watts. The measure must be higher to ensure that higher power gets coupled into the optical fiber. This is achieved by confining the optical emissions within the active region of the PN junction.

The quantity of radiative recombination and the area of emission are dependent on the semiconductor material used and on the device fabrication. Only direct band gap materials exhibits radiative recombination. Hence, the semiconductor used for active layer of the optical source has to be made of direct band gap. None of the solitary element semiconductors exhibit direct bandgap when compared with binary compounds made from group-III (e.g. Al, In, Ga) and group-V (e.g. P, Sb, As) materials which exhibit it. This has motivated in forming many ternary and quaternary combinations of these elements in forming direct band gap materials appropriate for optical sources. GaAlAs, a ternary alloy is used for operation at 800–900 nm and for operation at higher frequencies, quaternary alloy InGaAsP is used. The main advantage of using these combinations of ternary and quaternary alloy is the possibility in matching the lattice parameters of the hetero structure by appropriate amalgamation of binary, ternary and quaternary materials. A heterostructure is formed by combining two different semiconductors with varying band gap energies. The various combinations of materials used in LED construction and their characteristic are shown in Table 5.2.

The two configurations of LED being used in optic fiber are surface emitters and edge emitters. Surface emitters have simple structure, offering low or average output power and operating speed. Although they are reasonably inexpensive, the emitting area is larger than edge emitters. They emit light in all the directions and hence are Lambertian emitters as shown in Figure 5.2. Thus only a fraction of light enters the fiber resulting in poor coupling.

Table 5.2 LED material and its characteristic.

Material	Energy gap	Wavelength	Applications
Gallium Phosphide (GaP)	2.24 eV	550 nm	To make emitters in the visible spectrum
Aluminum Arsenide (AlAs)	2.09 eV	590 nm	
Gallium Arsenide (GaAs)	1.42 eV	870 nm	To make emitters in the near infrared portion spectrum (first window)
Indium Phosphide (InP)	1.33 eV	930 nm	
Gallium-Aluminum Arsenide (GaAlAs)	1.42–1.61 eV	800–900 nm	
Indium-Gallium-Arsenide-Phosphide (InGaAsP)	0.74–1.13 eV	1100–1600 nm	To make emitters in the near infrared portion spectrum (second window)

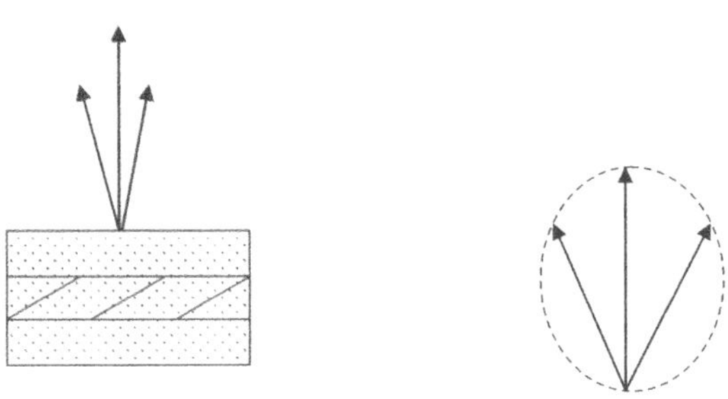

Figure 5.2 Surface emitting LED and its radiation pattern.

Edge emitters have more complex structure, high operating speed and offers high output power level since the emitting region is very narrow (30 to 50 μm). They are relatively expensive devices and offers high-speed performance. The emission spectrum of edge emitter is very narrow as shown in Figure 5.3. Hence, it offers good coupling capacity.

Super luminescent diode (SLED or SLD) is a variant of edge emitters. It is a blend of LED and laser, has narrow emission spectra, and offers high power density and internal gain like laser. However, the optical output still remains incoherent unlike lasers. It finds application in optic gyroscopes. Its emission optical bandwidth ranges between 5 and 750 nm.

SLED has a PN junction with active region between the two, a waveguide and facets. When the device is forward biased, device becomes optically active and electric current flows from P side to N side. Light is emitted by spontaneous emission over wide wavelength range and due to random recombination of holes and electrons. The light is then amplified as it travels along the waveguide. The material composition of the device and the injected current level decides the intensity and peak wavelength of the emitted light. SLED only has one-pass amplification for spontaneous emission along the waveguide. The feedback is insufficient to generate lasing action and can be achieved by using a tilted waveguide and facets with anti-reflection coating. Thus, SLED is a laser that cannot generate stimulated emission due to lack of optical feedback. Figure 5.4 shows the comparison of LED, Laser and SLED.

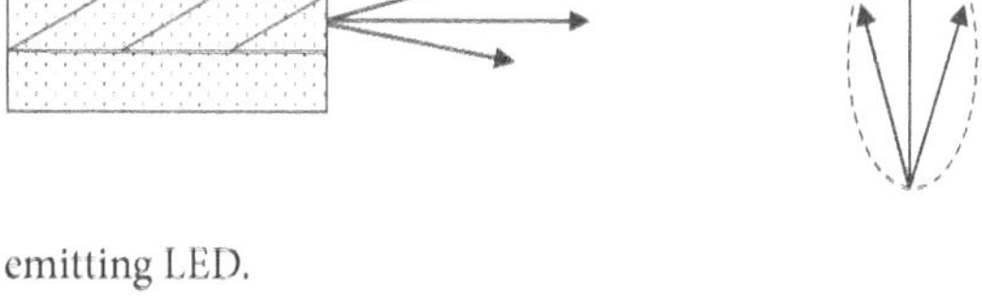

Figure 5.3 Edge emitting LED.

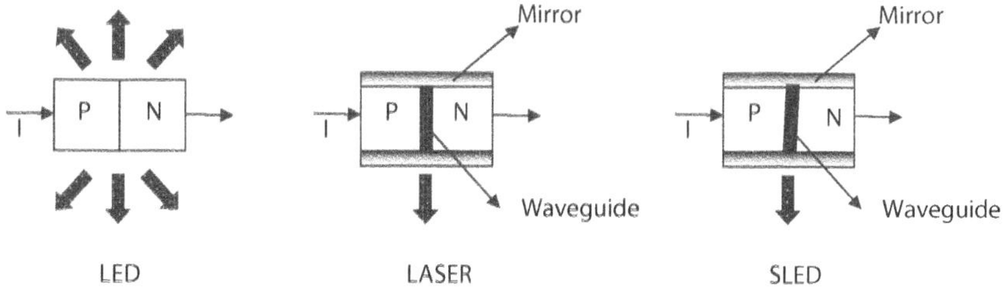

Figure 5.4 LED, Laser, and SLED comparison.

As the light is predominantly emitted in a single direction, SLED has high optical power density. Thus, it has high spatial coherence similar to laser. As SLED device does not have a cavity which is wavelength selective, it can emit broad spectrum. This makes the device low temporal coherent. This makes SLED behave as LED and the low temporal coherence avoids the problem of speckle noise.

5.6 Laser as Light Source

Light amplification by stimulated emission of radiation is abbreviated as Laser. Lasers are coherent light sources that produce narrow beam of light. The need for high speed/high performance transceivers has been the driving force in the active research in the field of laser. The following features of laser diodes make it suitable for optical fiber communication:

- Response time is lesser than 1 ns.
- Optical bandwidth lesser than 2 nm.
- Capacity to couple several tens of mill watts of optical power even into optical fiber with small core diameter.

Lasers come in varying measurements ranging from the size of a salt grain up to the size of a room. Lasing medium could either be a crystal, gas, liquid or a semiconductor. Fundamental lasing operation is divided into three steps – Absorption, Stimulated emission, Spontaneous emission. These three lasing operations are illustrated in Figure 5.5 where ground energy state and the excited energy state are represented as E0 and E1 respectively. The transition between these states comprises either emission or absorption of photon of energy given by Planck's law, hv = (E1 - E0).

When the system is excited by a photon of energy hv, system absorbs it and electron in E0 state jumps to E1 state. Since E1 is in an unstable state, the electron cannot stay there for longer duration and returns back to E0 by emitting the absorbed energy hv called spontaneous emission. The electron transition from E1 to E0 can also be made by applying a photon of energy hv when the electron still remains in E1 state. This is termed as stimulated emission. This gives the photon of energy 2 hv. Usually the densities of electrons at lower energy level (E0) will be higher than electrons at higher energy level (E1). If the reverse condition exists, stimulated emission will increase and this is called population inversion.

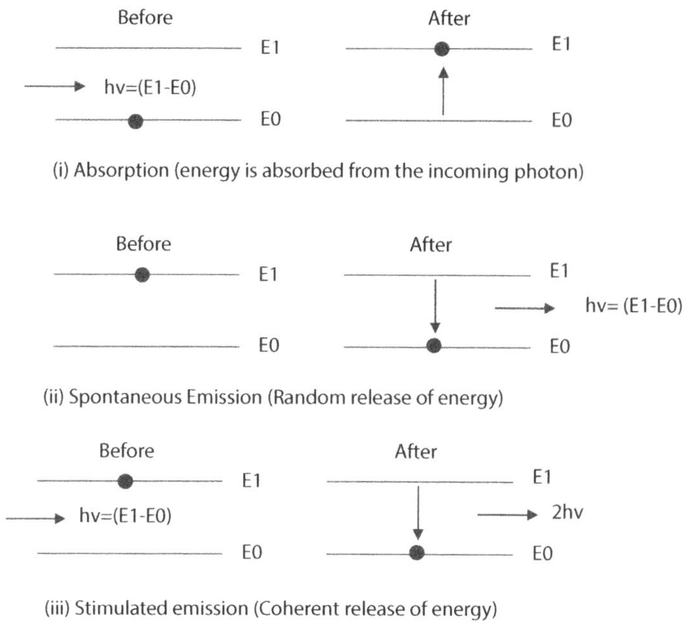

Figure 5.5 Energy level diagrams showing the lasing operations.

Lasers used for fiber optic transmitters are classified into three types: Vertical Cavity Surface Emitting laser (VCSE), Fabry Perot (FP) laser and Distributed Feedback (DFB) laser.

Vertical Cavity Surface Emitting Laser: In VCSEL, laser beams are emitted from the top perpendicular to the surface. VCSEL can be tested on wafer before they are inserted into individual device. Whereas, edge emitting semiconductor laser emits in parallel from the laser cavities made at the middle of the chip and can be tested only after the production. VCSELs are cheaper in construction and can be used for distances of about 500 m or less. VCSELs have lower power consumption and higher coupling efficiency with optical fiber.

Fabry Perot Laser: It is a type of laser that supports many longitudinal modes. The radiation is generated in a cavity called Fabry-Perot resonator cavity. The length of the cavity or the active region is roughly between 250–500 μm, width is 5 μm to 15 μm and thickness is 0.1–0.2 μm. The cavity is enclosed by a pair of partially reflecting mirrors that provides optical feedback in longitudinal direction. It makes the device work like an oscillator with higher gain. There exists many resonant frequencies and the device will oscillate at those frequencies for which the gain is higher to overcome the losses in the cavity. This device offers a large output power, has smaller angle of divergence, narrow spectrum and higher modulation rate. They can be used for longer distance than VCSEL.

Distributed Feedback Laser: It incorporates a grating into one of the cladding layers around the active layer of a laser diode. Insertion of grating reflects back a specific wavelength into the cavity and allows others to pass through. Thus the desired wavelength is fed back into the cavity. Since the feedback is distributed, it is called distributed feedback laser. DFBs are more expensive than other lasers but are ideal for long distance transmission.

At present, high repetition rate (HRR) pulse fiber lasers have drawn much attention owing to its superior capacity in fields like communication systems, astronomical spectrograms and high speed electric-optic sampling. Cavity mode spacing usually decides the pulse repetition rate in a fiber. Hence in HRR pulse fiber laser, the repetition rate is increased by increasing the cavity mode spacing and decreasing the cavity length. However, owing to the restraint on the cavity length, pulse of Ghz pulse repetition rate is achieved. As an alternate to this, number of mode locked pulses in the fiber can be increased. These pulses distribute equally along the laser cavity.

5.7 Optical Fiber

It is a dielectric waveguide through which electromagnetic energy travels as light rays in the direction parallel to the axis. It is cylindrical in shape and operates at optical frequencies. A set of directed electromagnetic waves that travel along the fiber is termed modes of waveguide. The guided mode is a repeated pattern of electric and magnetic field distribution at equal intervals of time all along the fiber. Only certain modes that satisfy homogeneous wave equation and the boundary condition at surface of the waveguide can transmit through the guide [10].

The most commonly used structure of optical fiber consists of dielectric cylinder called core surrounded by another dielectric cylinder called cladding. The radius of core and cladding are 'a' and 'b' and the refractive indices are $\eta 1$ and $\eta 2$ ($\eta 2 < \eta 1$) respectively. Since the cladding has lower refractive index than core, the light traversing through the core experiences total internal reflection on reaching the core-cladding interface, thereby the core of optical fiber is able to confine the light rays within itself. Cladding does not serve the purpose of light propagation. Rather it helps in providing mechanical support to the fiber, avoids contamination of the core with which it might come in contact with and also reduces scattering loss that occurs when there are dielectric discontinuities at the surface of the core. The core cladding structure is then encapsulated into an elastic, plastic material that further enhances the mechanical strength of the fiber as shown in Figure 5.6.

Broad classifications of optical fiber are based on the material composition used for the core are: step index fiber and graded index fiber. Both the types are further sub-classified into single mode and multi mode fiber as shown by the tree structure in Figure 5.7.

Step index fiber: In this fiber, refractive index of the core remains consistent throughout the length and varies abruptly only at the core cladding interface.

Graded Index fiber: in this fiber, the refractive index of the core varies radially from the center of the fiber.

Single mode fiber: This fiber supports only a single propagation mode. For such a fiber, core and cladding diameters vary around 10 μm and 120 μm respectively. Optical power can be launched into this fiber using laser source only.

Multi mode fiber: Multi mode fiber is mostly used for short distance communication, example inside a building or communication within a campus. Multi mode fiber support many propagation modes. Hence

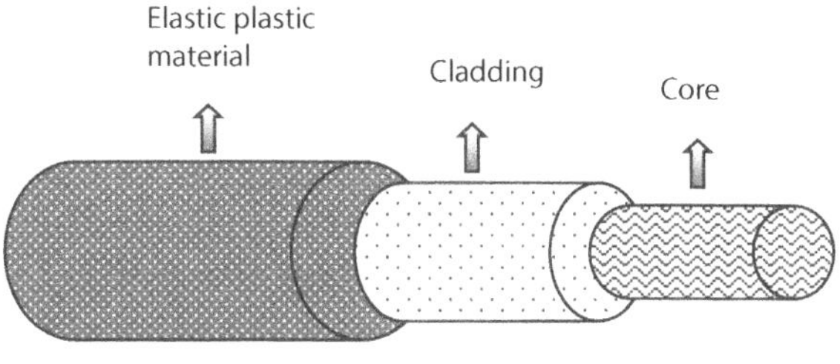

Figure 5.6 Schematic of a single optical fiber structure.

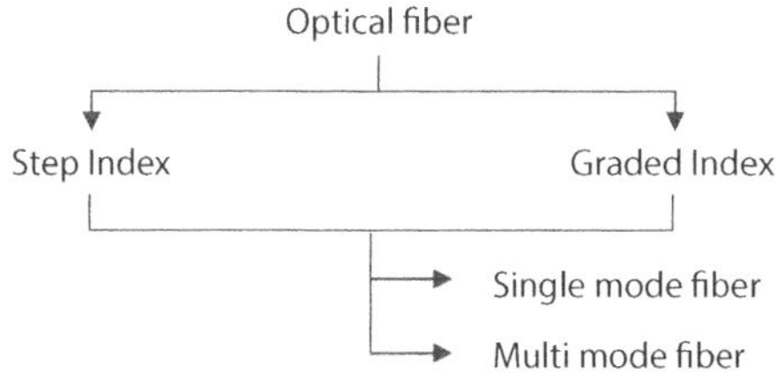

Figure 5.7 Types of optical fiber.

supports more data transmission at the given time. For such a fiber, core and cladding diameters varies around 100 μm and 250 μm respectively. The maximum transmission distance is around 550 m at the speed of 10 Gbps for multi mode fiber cables. Even LED source can be used to initiate the optical power into this fiber. When optical pulses are propelled into this fiber, different modes of light travel with different velocity along the fiber and arrives at the point of culmination at different times causing pulses to spread out in time. This results in intermodal dispersion in multimode fiber. This problem can be overcome using graded index profile for multi mode fiber. Multi mode fibers are classified into OM1, OM2, OM3, OM4 and OM5 fibers based on ISO 11801 standards.

OM1 fiber has a core size of 62.5 μm and uses LED light source. It supports up to 10 Gigabit Ethernet at 30 m length and is generally used for 100 Megabit Ethernet applications.

OM2 fiber has a core size of 50 μm and uses LED light source. It supports up to 10 Gigabit Ethernet at 80 m length and is generally used for 1 Gigabit Ethernet applications.

OM3 fiber has a core size of 50 μm and is optimized for equipment's based on laser. It supports up to 10 Gigabit Ethernet at 300 m, 100 Gigabit Ethernet at 100 m length and is generally used for 10 Gigabit Ethernet applications. It has an effective modal bandwidth of 2000 MHZ*km at 850 nm for 10 Gbps transmission.

OM4 fiber was specifically developed for transmission of VSCEL laser. It supports up to 10 Gigabit Ethernet at 550 m. It has an effective modal bandwidth of 4500 MHZ*km at 850 nm for 10 Gbps transmission for longer distance.

OM5 fiber is the most recent type of multimode fiber intended specifically for high bandwidth and short to medium range applications. It is first of its kind to get approved as Wide Band Multimode Fiber (WBMMF). It is designed to handle high speed data center applications by making use of two fibers to transmit 40 Gbps and 100 Gbps. It supports short wavelength division multiplexing (SWDM) technique. SWDM technology uses many VCSELs at different wavelengths more or less around 850 nm region to produce individual data stream that are all multiplexed onto a single fiber pair.

OM5 operates at 850 to 950 nm range and provides 100 GB data streams using just two fibers in parallel. Thus, the multiplexed design permits OM5 to significantly decrease parallel fiber counts. Reduction in fiber count reduces crosstalk and overall loss. OM5 is downward compatible with OM3 and OM4 systems.

5.8 Fiber Materials

Selection of materials to make fiber is based on various criteria's as listed below:

- The material must be flexible in making long and thin fiber.
- The material needs to be transparent at a precise wavelength for it to guide the light efficiently.
- Compatible materials to make core and cladding with different refractive indices.

Materials that meet these requirements are glass and plastic. Signal attenuation is higher in plastic fiber when compared to glass fibers. Hence, plastic fibers are less widely used. Since plastic fiber offer higher mechanical strength than glass fiber, they are preferred for short distance communication (range of few hundred meters).

Silicon dioxide or silica that has a refractive index of 1.5 at 850 nm is the most preferred material in making glass fiber. The varying core refractive index and cladding is obtained from doping SiO_2 with dopants like GeO_2 or P_2O_5 (the refractive index of SiO_2 increases) and with $B2O_3$ (the refractive index of SiO_2 decreases). The various combinations of core and cladding glass fiber are listed in Table 5.3 [1].

Polymethylmethacrylate or perflourinated polymer is used to make the core of plastic optical fiber. They are tough and reliable and have relatively more signal attenuation compared to glass fiber. The core diameter of plastic fiber is 10–20 times larger than glass fiber hence coupling from optical sources is easier.

Table 5.3 Core cladding combinations in glass fiber.

Core	Cladding
SiO_2 doped with GeO_2	SiO_2
SiO_2 doped with $P2O_5$	SiO_2
SiO_2	SiO_2 doped with $B2O_3$
SiO_2 doped with GeO_2 and $B2O_3$	SiO_2 doped with $B2O_3$

5.9 Benefits of Optical Fiber

The benefits of using optical fibers are listed below:

- Signal attenuation is lower.
- Provides higher bandwidth that enables more data transmission.
- Light weight compared to metallic cable.
- Stray interference problems associated with metallic cable is avoided.
- Power loss is less. Hence suitable for long distance communication.
- Provides higher security for data transmission.
- Optical fiber is immune to electromagnetic interference.
- Space occupied and weight of optical fiber is lower than metallic cable.
- Material used in manufacturing optical fiber cable costs lower than that for metallic cable.

5.10 Drawbacks of Optical Fiber

The drawbacks of optical fibers are listed below:

- Installation cost for optical fiber is higher.
- Number of repeaters required increase with distance.
- More mechanical strength is to be provided than metallic cable.

5.11 Recent Advancements in Fiber Technology

Recent inventions on optical fibers have shown usage of long tapered fibers have added advantage in suppressing the nonlinear effects of the fiber and helps maintaining superior beam quality compared to uniform fiber[2]. Tapering in fiber is achieved by either varying both cladding and core sizes at a constant ratio called core-cladding diameter ratio or by retaining cladding size uniform and tapering the core with varying core-cladding diameter ratio.

Another variation of optical fiber includes development of bending loss insensitive fibers using hole-assisted fiber, a type of holey fiber technology. The developed fiber provides higher resistance to bending loss compared to conventional optical fiber.

Bandwidth that forms the backbone for optical fiber communication systems is continuously being increased year by year. At present, transmission capacity of optical fiber is increased by using Wavelength Division Multiplexing (WDM) technique. In WDM, increase in the density of signal multiplexing increases the power density per cross section of the fiber. However, higher power density leads to nonlinear effects that degrade the signal characteristics. It also leads to thermal destruction of optical fiber by causing fiber fuse. By considering the power density threshold and optical amplifier bandwidth, the conventional single mode fiber has the maximum signal transmission capacity of around 100 Tbps. To overcome the capacity limitation, Space Division Multiplexing (SDM) is being developed as an alternative to wavelength division multiplexing. Multi Core Fiber (MCF) technology for transmission is widely studied as the simplest type of SDM. MCF forms the next-generation fiber technology to achieve higher speed and capacity.

Multi-Core Fiber (MCF): It contains many cores within a cladding. MCF can be classified as step index, graded index or trench (hole) assisted types. In step index MCF, all the cores have refractive index variation only at core cladding interface [6]. If at the least one core has varying refractive index profile throughout its length, the fiber is classified as graded index MCF. Trench assisted type of multi core fiber has a step wise refractive index variation in the cladding. This helps in reducing the mode coupling or inter-core crosstalk in-between the adjacent modes. MCF supports both single mode as well as multi mode transmission within the fiber. Three to seven cores can be made to fit inside the same cladding and seven cores is preferred as ideal.

In each core, only LP01 transmission mode is supported within a single mode MCF [3]. If several LP transmission modes are supported in a fiber, it is termed as multimode MCF. IF the MCF supports only first three or four LP modes (LP01, LP11, LP21, LP02), the fiber is called few mode MCF.

Based on the spatial homogeneity, MCF can be classified as homogeneous MCF where all the cores within a cladding have same refractive index profile and heterogeneous MCF where one core has different refractive index profile at the least.

The crosstalk level in the inter-core amongst the LP modes of the core is determined by the distance between core to core. In MCF, if the cores are very closely spaced, the overlap between the modes becomes significant and the light couples from one fiber into another. The modes become degenerated and support super modes. MCF is then called coupled core MCF. If the cores are fairly far apart, the overlapping between the modes

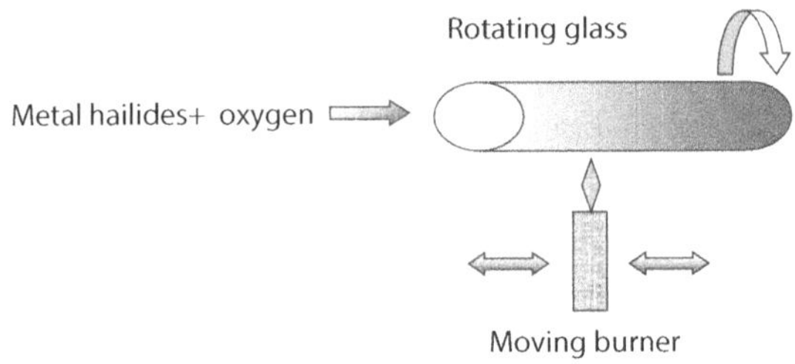

Figure 5.8 Modified chemical vapor deposition.

is negligible and each fiber can be considered as single mode fiber. MCF is then termed uncoupled core MCF.

MCF is all glass fiber containing air holes. Fabrication involves stack and draw technology that helps in making different fibers with same machinery. In this process, the stacking elements consist of tube core rods and adjusting rods. These rods are stacked inside the tube. First step of MCF fabrication involves development of all glass preform with multiple cores. Here bundle of rods are assembled to form multiple cores. Modified chemical vapor deposition process is used in perform fabrication as shown in the Figure 5.8.

The constituent metal halide gases and oxygen are made to flow through the rotating glass tube and the tube is heated at a high temperature (approximately 15000°C) by moving the burner back and forth along the length of the fiber. When the required thickness of the glass is deposited, the vapor flow is stopped and the tube is strongly heated to collapse it into a solid rod like structure called preform. The preform has the original silica tube as cladding and the core with vapor deposited material. The preform is then stretched such that core cladding ratio remains the same. The preform is then cut into seven performs. The performs are then stacked inside a tube to form a hexagonal structure. To maintain the shape, filling rods are used in between.

As the process is highly complex, it results in nonlinear effects on the process parameters. Few rods may be tough to be drawn into the required form resulting in higher levels of crosstalk.

5.12 Photodetector

Photodetector is the first element at the receiver part of an optical transmission link. It senses the light signals falling on it and transforms the variation of optical signal into the corresponding electric current. This is done by translating every photon in the received signal into a free electron to generate photo current which is linearly proportional to the optical signal power.

The signals emerging out from the optical fibers are generally very weak and distorted. The photodetector must be capable enough to detect such weak signals. In order to achieve higher signal to noise ratio (SNR), the photodetector should have higher quantum efficiency to produce large signal power. Photodetector must possess certain characteristics like high response in the emission spectrum of the optical source, quick response, sensitive to temperature variations, long operating life, and compatible to the size of the optical fiber.

Photodetector are of three different types:

- Photomultiplier.
- Pyroelectric detector
- Semiconductor based photoconductor, phototransistor and photodiode.

Photomultiplier consists of electron multiplier and a photocathode integrated inside a vacuum tube. Although it offers higher gain and lower noise, it is larger in size and requires higher voltage. Hence is not suitable for fiber optics.

A pyroelectric detector involves transformation of photons to heat. Temperature of the detector changes upon photon absorption and causes dielectric constant variation. This is usually observed as a change in capacitance. The response of pyroelectric detector is flat over a wide spectrum band. Pyroelectric detector finds application in sensing high speed laser pulses and is not well suitable for fiber optics.

Semiconductor photodiode is the principal form of photodetector used in optical communication system owing to its smaller size and faster detection with higher efficiency.

The photodiode consists of a PN junction like laser diode. The PN junction of a photodiode is reverse biased in contrast with a laser diode in order to allow a fraction of the reverse saturation current to pass through the diode without an input optical signal. Improvement in quantum efficiency of a photodiode is achieved by altering the device structure. For example, PIN photodiode and Avalanche photodiode.

PIN photodiode consists of an intrinsic layer (I-layer) between P and N types. On reverse biasing the device, the intrinsic region will be fully depleted of charge carriers. When the photon of energy greater than the band gap energy of the semiconductor material is incident on the PIN photodiode, energy is given out and an electron is excited from valence band to conduction band generating electron hole pair called photo carriers. The carriers get separated and accumulate across the depletion region due to the high electric field created across the depletion region. This result in the current flow called photo current.

Avalanche photodiodes achieves 5–10 dB improved sensitivity than PIN photodiodes, provided the gain-bandwidth product is sufficiently high and the multiplication noise is less. Avalanche photodiodes multiplies the photocurrent before it enters the following amplifier. This increases the receiver sensitivity. The photogenerated carriers must traverse through a high electric field region if carrier multiplication has to take place. In high electric field region, the generated electrons and holes gain enough energy and results in carrier multiplication. This is called avalanche effect. The most commonly used structure for achieving carrier multiplication is called Reach through Avalanche Photodiode (RAPD).

RAPD structure has a p-type material of high resistivity (π layer) on heavily doped p diffusion (p+). P-type diffusion is made on π layer followed by heavily doped n diffusion (n+) as shown in Figure 5.9. π layer is made of intrinsic material and acts like depletion region. It has some p doping due to imperfect purification.

RAPD operates in a fully depleted mode under normal usage. Light enters via the p+ region and gets absorbed by π layer which collects all photogenerated carriers. When photon is absorbed, it gives out energy creating electron hole pairs. The electric field in the π layer then divides the electron hole pair. The photogenerated electrons then drift into pn+ junction. This region has high electric field and aids in carrier multiplication [1].

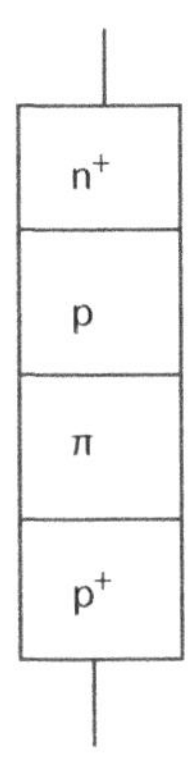

Figure 5.9 RAPD structure.

Multiplication (M) of the generated carriers in the photodiode is given by equation (i).

$$M = \frac{Im}{Ip} \tag{i}$$

Where Im indicates the average value of the multiplied current and Ip indicates the primary unmultipled photocurrent.

At present, high speed photodetectors that operate in telecommunication band from 1260 nm to 1625 nm are well studied. Recent innovations carried out on photonic systems have brought about new requirements on the photodetectors bandwidth. The cutoff wavelength must be extended from short wavelength infrared to long wavelength infrared. The performance of photodetectors at longer wavelength band is less studied and needs more innovations [4]. Efforts are also being made to improvise on prime figure of merit parameters of photodetectors like stability, sensitivity, response time, spectral selectivity using various nanostructures like inorganic semiconductors, optoelectronic diamond, hybrid halide perovskite and others.

5.13 Future of Optical Fiber Communication

At present, signal processing, switching and routing occurs in electrical domain. Hence, all optical signals are converted into electrical signal for processing that are later converted back to optical signal for long distance transmission. The process involves network overloading, delays and data rate limitation. This creates a demand for optical switches and routers. Future trend of optical fiber is "All Optical Network" (AON), where all the signal processing will happen in optical domain without the involvement of electrical or electronic processing. Only limited AON is implemented in commercial networks till now and needs lot of refinement in following areas:

- Developing an optical amplifier for only regeneration as against the electro-optical amplifier that performs regeneration, re-clocking and reshaping.
- Developing an optical coupler to split light from one fiber into more fibers or to recombine light from different fibers into a single fiber.
- Developing optical filters to filter a particular frequency from WDM signal.

- Developing optical switches to avoid optical to electrical and electrical to optical conversions.
- Developing other transmission technique like dense wavelength division multiplexing (DWDM). This is aimed at increasing the band (frequency interval) which further increases the quantity of data transferred though the connection in the specified time interval. Special attention could be given for optical fiber development in C band. In this band signal, loss is less, the fibers and the whole of the transmission system operates efficiently. Here many optical signals of different wavelength with frequency modulation are transmitted over a single optical fiber. Thus, a single optical fiber is transformed into many virtual channels.
- Radio over fiber (ROF) – Conventionally coaxial cables is used in RF transmission. The process is cumbersome, has limited capacity and is sensitive to electromagnetic interference. The drawbacks can be corrected using optical fibers to transmit radio frequency. Here the RF signals are modulated over the light in a single fiber. The drawback involved in ROF is the requirement for electrical to optical modulator at transmitter and optical to electrical converter at the receiver. ROF finds wide application in driverless cars.

5.14 Applications of Optical Fibers in the Industry

Optical fibers finds wide range of applications and few of them are listed below:

a. Medical Industry – A medical tool called gastroscope is used by doctors to detect illness inside the patient's stomach. The other uses are for X-ray, MRI, PET and CT scans. The images of arteries and organs which could not be accessed without surgery are now transmitted by the use of optical fibers.
b. Industry endoscopes – Fiberscope or boroscope used for examining the inaccessible pieces of machinery such as engines of an aeroplane.
c. Civil Engineering
    - Stress Sensors – To monitor the maximum stress in steel structures
    - Strain Sensors – To measure the relative displacement

- Crack Monitoring – To assess the concrete structures for detecting cracks beyond 0.15 mm.

d. Oil and Gas Industry – Optical fibers are used for sensing and communication.

5.15 Conclusion

Today's global businesses demands for more secured, rapid and high capacity communication systems for their network operations. Optic fiber technology plays a vital role in this growth. In industrial infrastructures, fiber optic cable finds its application from healthcare systems right up to marine environment. Optical fiber assemblies are also playing significant role in residential applications. High-speed internet access is expected by house owners as a part of their daily lives. Telecom industry is now trying to fulfill this requirement through fiber optic technology.

The future of optical fibers has scope for unlimited development in applicative contexts. This chapter provided a few such potentials of optical fiber communication to achieve impressive effects in near future.

References

1. Keiser, G., *Optical fiber communications*, 4th edition, McGraw-Hill Companies, Inc., New York, 2010.
2. Ortiz, A.M. and Saez, R.L., Multi-core optical fibers: Theory, applications and opportunities. Selected topics on optical fiber technologies and applications, Feb. 2018, doi: 10.5772/intechopen.72458.
3. Samir, A. and Batagelj, B., Stack-and-draw manufacture process of a seven-core optical fiber for fluorescence measurements, *Fiber Integr. Opt.*, December 2017.
4. Chen, B., Chen, Y., Deng, Z., Recent advances in high speed photodetectors for eSWIR/MWIR/LWIR applications. *Photonics*, 8, 1, 14, Jan. 2021, doi: 10.3390/photonics8010014.
5. Yu, H., Li, P., Zhang, L., Zhu, Y., Ahmad Al-Zahrani, F., Ahmed, K., Application of optical fiber nanotechnology in power communication transmission. *Alexandria Eng. J.*, 59, 6, 5019–5030, 2020.
6. Li, M.-J. and Hayashi, T., Advances in low-loss, large-area, and multicore fibers, *Optical Fiber Telecommunications VII*, 2020.
7. Hecht, J., High-power fiber lasers, *Optics & Photonics News (OPN)*, 29, 30–37, 2018.

8. Jiang, J.-M., Chen, X.-J., Gao, Y.-X., Liu, M., Bai, Z.-A., Luo, A.-P., Xu, W.-C., Luom, Z.-C., Bidirectional ultrahigh-repetition-rate ultrafast fiber laser. *Opt. Laser Technol.*, 142, 2021.
9. Singh, C., Sairam, K. V. S. S. S. S., Harish, M.B., Global fairness model estimation implementation in the logical layer by using optical network survivability techniques. *International Conference on Intelligent Data Communication Technologies and Internet of Things*, Springer, Cham, 2018.
10. Singh, C., Implementation of survivability by using tier modes approach. *Communication and Computing Systems: Proceedings of the 2nd International Conference on Communication and Computing Systems (ICCCS 2018)*, December 1-2, 2018, CRC Press, Gurgaon, India, 2019.

6

Evaluation of Lower Layer Parameters in Body Area Networks

Abhilash Hedge* **and Durga Prasad**

Department of ECE, NMAM Institute of Technology, Nitte, Karnataka, India

Abstract

The need for new low power, low cost and small device in the field of medical field made the researchers to go for Wireless Body Area Network (WBAN). Here, many sensors are appended on cloths or on the human being or it can be implants beneath the epidermis. These sensors of a WBAN determine physiological parameters such as heartbeat, body temperature etc., or trace a draw out Electro Cardio Gram (ECG). The project comprises of two parts namely software and hardware. In software part, the simulation is carried out in the tool Castalia, to study Physical (PHY) and Media Access Control (MAC) layers of WBAN in Baseline MAC. The study includes various interference parameters for throughput and the reasons for loss of packets. Also comparison of different radio models in terms of power is analyzed. While in hardware section, using Bluetooth Low Energy (BLE) device the sensor data is received in android application (app). Also demonstration of various protocols in the field of Internet of Things (IoT) is presented and finally, Message Queue Telemetry Transport (MQTT) protocol and Hyper Text Transfer Protocol Secure (HTTPS) protocol are implemented to read sensor data in the cloud. To make the utilization more reliable, an app is designed for this specific application of WBAN.

Keywords: Message Queue Telemetry Transport (MQTT) protocol and Hyper Text Transfer Protocol Secure (HTTPS), Bluetooth Low Energy (BLE), Internet of Things (IoT)

**Corresponding author*: abhilash.hegde1007@gmail.com

Chandra Singh, Rathishchandra R Gatti, K.V.S.S.S.S. Sairam and Ashish Singh (eds.) Modeling and Optimization of Optical Communication Networks, (99–124) © 2023 Scrivener Publishing LLC

6.1 Introduction

In the present globe inhabitants, 750 million inhabitants are grown old exceeding 60, it is also approximated that by the year 2050, this total will boost to 2 billion [1]. Health care of the elderly people this becomes the serious issue, this gives us a reason to look into several health care techniques that can provide quality of life to the aged. Recent advancements in electronics have found its era into the development of efficient medical sensors. These sensors are fit diminutive that can be also worn by a person or entrenched within the epidermis [2, 3]. A wireless interface will be an effective alternate for this use; these systems will be productive as well as facilitate the mobility of the patient. This will definitely prove to be an effective technique to push the diagnosis of patients and health care systems to a next level [4]. Several challenges have to be addressed when developing a Body Area Network (BAN). Power concern is a very important aspect in the developing technologies, when developing miniaturized devices that operate within or in the vicinity of the body, the form factor of the power sources that empower these systems is extremely small, yet still a very high life time of the device has to be planned [5].

Wireless Body Area Networks (WBAN) is the application for which the Castalia simulation tool has been designed and also for the Wireless Sensor Network (WSN) systems of small-power embedded strategies. Castalia tool is supported on the platform called OMNeT++ podium which is capable of utilizing by investigators and build upper layers who desires to investigate ones dispensed algorithms/codes and/or protocols in genuine non-wired guide with different radio representations and a genuine node performance principally communicating the admittance of the radio [1].

To obtain a very high data rate, requirement of bandwidth is necessary for WBAN. The frequency of operation of implanted chip is in the range of 402-405 MHz [5]. Medical implant communication services, determine this frequency range; it is seen that operation in this frequency is almost same as that of Medical telemetry services [6]. Even though high data rate is provided by Industrial Scientific and Medical (ISM) band, it suffers from interference issues of the devices operating with IEEE 802.11 and 802.15.4 standards [6].

In this innovative age of technology and communication, the hotheaded expansion of smart devices, mobiles and tabs which are conversed wired or non-wired has turn out to be the essential means of day by day living. This had led to the invention of linked things known as Internet of Things (IoT) [7]. Since the internet is widely used in the world today, with the use of smart phone and present day technology, the field of medical, military, sports etc., have developed to large extent. This has lead to the advancement in BAN technology providing better quality of life to the elderly people and patients suffering from chronic illness. But a low cost, low power device only motivates the people to use the technology in larger extent. Since, the requirement of low power and low cost has direct implication with lower layers of BAN, an effort has been made to design WBAN system analyzing the parameters of PHY and MAC layers of BAN.

6.2 Problem Definition

The application is real time and evaluation of network parameter is necessary for the efficient operation. The sensors have to transmit data to a server located locally and need to be monitored and accumulated. A physical link used in the wired network is too bulky which occupies some lofty money for employment and continuance. Conversely, usage of wireless medium facilitates an effortless function and is further cost efficient. The main issue addressed in the IEEE 802.15.4 is the constraint of Quality of Service (QoS) with power restraint in sensor nodes of WBAN. So, need for implementation of new technology is necessary such as Baseline MAC. And need for study in physical and Media Access Control (MAC) layer is necessary in more efficient working of this IEEE 802.15.6 standard. The scope of this project is in the field of medical applications and is not limited to this field only. Also the energy consumption, throughput, interference parameters need to be analyzed. The need for suitable radio for WBAN application is to be considered. Sensors are affixed to the epidermis. For people to wear these sensors, they must be of small size, low in consumption of power, efficient in terms of energy and very comfy to wear.

In this Software simulation of the WBAN scenario with interference parameters is carried out in Castalia to study throughput, latency and reasons for loss of packets. In the hardware part the sensors are connected to ESP 12-E board which is a Wi-Fi module that transmits sensed data to a website using Message Queue Telemetry Transport (MQTT) protocol designed in the website for special purpose applications. This store and monitor the sensor data in the internet cloud storage and also an android app built to monitor the data from the website. This enables the users to periodically monitor the situation and stay up to date. Also the users can monitor data at their ease whenever they require. This helps the users to grab the immediate Medicare during emergency condition which makes the life healthy and fit. It is necessary to study the PHY and MAC layer parameters to reduce energy consumption of the body sensor nodes. Further, it is necessary to implement the WBAN using IEEE 802.15.6 standards. So, considering mobility of nodes and interference parameters, study is on PHY and MAC layer of WBAN

Table 6.1 Modulation parameters for PLCP header and PSDU [14].

Frequency band	Packet component	Modulation	Symbol rate (Kbps)	Code rate BCH (n,k)	Information data rate (Kbps)
402 – 405 MHz	PLCP Header	π/2-DBPSK	187.5	(31,19)	57.5
	PSDU	π/2-DBPSK	187.5	(63,51)	75.9
	PSDU	π/4-DQPSK	187.5	(63,51)	303.6
420 – 450 MHz	PLCP Header	GMSK	187.5	(31,19)	57.5
	PSDU	GMSK	187.5	(63,51)	75.9
	PSDU	GMSK	187.5	(63,51)	151.8
863 – 870 MHz	PLCP Header	π/2-DBPSK	250	(31,19)	76.6
	PSDU	π/2-DBPSK	250	(63,51)	101.2
	PSDU	π/4-DQPSK	250	(63,51)	404.8
902 – 928 MHz	PLCP Header	π/2-DBPSK	300	(31,19)	91.9
	PSDU	π/2-DBPSK	300	(63,51)	121.4
	PSDU	π/4-DQPSK	300	(63,51)	485.7
950 – 956 MHz	PLCP Header	π/2-DBPSK	250	(31,19)	76.6
	PSDU	π/2-DBPSK	250	(63,51)	101.2
	PSDU	π/4-DQPSK	250	(63,51)	404.8
2360 – 2400 MHz 2400 – 2483.5 MHz	PLCP Header	π/2-DBPSK	600	(31,19)	91.9
	PSDU	π/2-DBPSK	600	(63,51)	121.4
	PSDU	π/2-DBPSK	600	(63,51)	485.7

system are carried out. Also, the requirement of low power and low cost has direct implication with protocol used in the network of BAN. An effort has been made to incorporate a MQTT protocol in order to bring out IoT based WBAN system is shown in Table 6.1.

6.3 Baseline MAC in IEEE 802.15.6

In IEEE 802 networking standards, MAC (Media Access Control) refers to the lower sublayer of the Data Link Layer. The MAC sublayer is responsible for controlling access to the physical medium and ensuring that multiple devices on the network can share the communication channel efficiently. There are several variations of MAC protocols defined in different IEEE 802 standards.

6.4 Ultra Wideband (UWB) PHY

The UWB PHY employs a variety of modulation techniques to transmit data, including pulse-based modulation and frequency-shift keying (FSK). UWB signals are typically transmitted as short-duration pulses with very low power levels. These pulses can be transmitted across a wide range of frequencies simultaneously, enabling high data rates and robust communication.

6.5 Castalia

6.5.1 Features

The simulation of Wireless Body Area Networks (WBAN) is well recognized with Castalia simulation tool, which has been designed for the systems of small-power embedded strategies. The Castalia tool is supported on the platform called OMNeT++ podium which is capable of being utilized by researchers to build various layers to investigate or experiment the

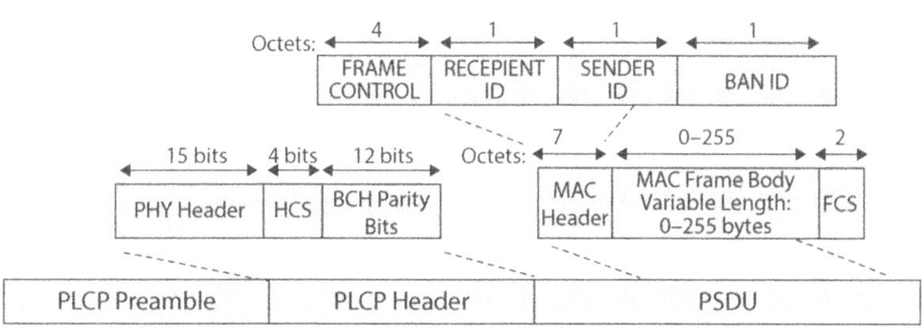

Figure 6.1 IEEE 802.15.6 NB PPDU structure [2].

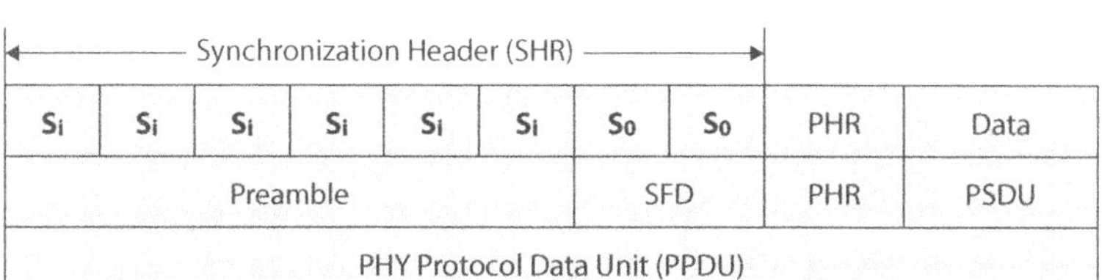

Figure 6.2 IEEE 802.15.6 UWB PPDU structure [2].

algorithms/codes and/or protocols in genuine non-wired guide with different radio representations and a genuine node performance principally communicating with radio. IEEE 802.15.6 NB PPDU structure is shown in Figures 6.1–6.3, superframe structure shown in Figure 6.4, CSMA/CA procedure in IEEE 802.15.6 shown in Figure 6.5, IEEE 802.15.6 standard shown in Figure 6.5, Connections of modules in Castalia shown in Figure 6.6. The main features of Castalia are [22] that is altered when messages are received and can send (or schedule) new messages. There are also composite modules.

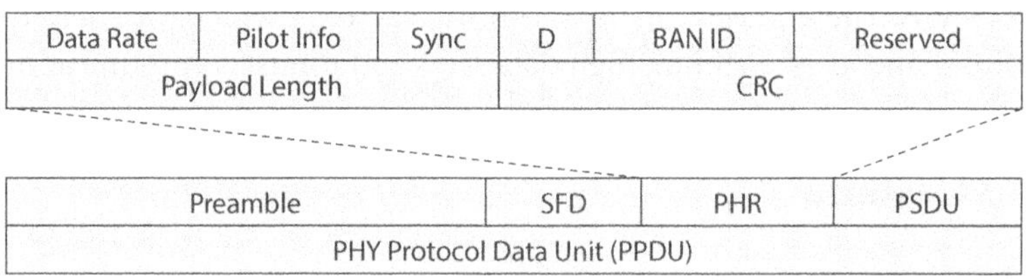

Figure 6.3 IEEE 802.15.6 EFC PPDU structure [2].

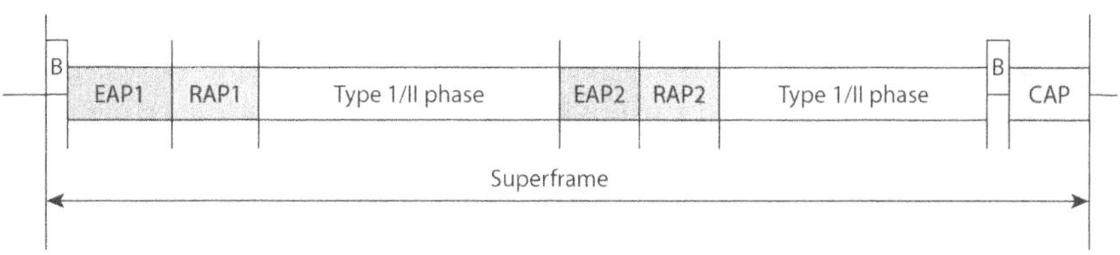

Figure 6.4 IEEE 802.15.6 superframe structure [2].

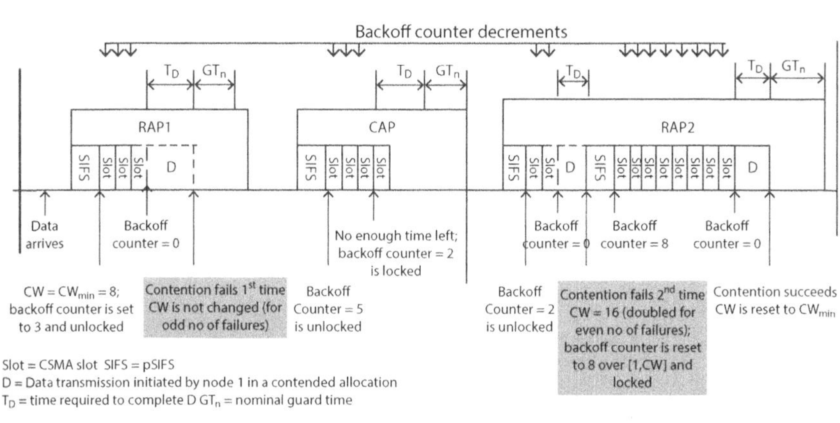

Figure 6.5 CSMA/CA procedure in IEEE 802.15.6 standard [2].

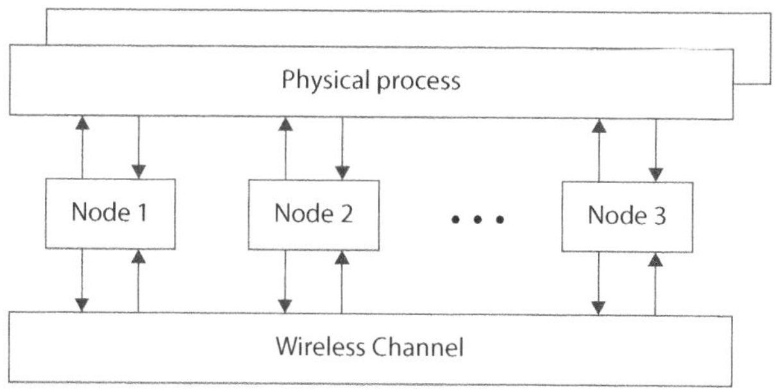

Figure 6.6 Connections of modules in Castalia [22].

6.6 Methodology

6.6.1 Simulation Method in Castalia

A real time scenario is created, considering five nodes with four nodes on the human body and one node as the access point as shown in Figure 6.8. Initially, the coordinate value of each node is identified. A room of 4mx4m size with an average height of human being to be 6 ft is considered. For node 0 mobility is set in the simulation while other nodes are kept stationary. Next, this scenario is simulated in Castalia using Interference Test1 with IEEE 802.15.6 (Baseline MAC) along with all node vary rate parameter. This parameter provides variable data rate for all the nodes [8, 9]. The parameters, useful for the present scenario are set in the file of Castalia.

6.6.2 Hardware Methodology

The block diagram for hardware setup is shown in Figure 6.9. It consists of a sensor, Wi-Fi/Bluetooth module, cloud storage facility and an app built on smart phone [10, 11]. The flow of data is considered from sensor to module via wired connection and module to cloud via wireless connectivity and from cloud to app again via wireless connectivity. Here, the temperature sensor and light level sensor are considered. ESP 12-E Wi-Fi module and HC-05 Bluetooth Low Energy (BLE) are used as collection and transmission centers the part of work for uploading sensor data to cloud. The sensed data can be stored in different serial monitors and cloud platforms, so that periodic monitoring of data is achieved. Processing the data has been achieved using Arduino IDE. Especially, MQTT protocol has been used for data transmission from ESP module to cloud [12, 13].

6.7 Results and Discussion

The simulation scenario created for WBAN environment designed especially considering more interference modules using 802.15.6 MAC [15, 16]. The performance is tested for various wireless channel conditions using PHY and MAC layer parameters of IEEE 802.15.6 standard the following are the commands used in the simulation to run the scenario of WBAN [17, 18].

```
abhilash@abhilash-Vostro-3550 ~/Castalia-3.2/Simulations/802.15.6
BAN $ ../../bin/Castalia -c
BaselineMAC,InterferenceTest1,allNodesVaryRate[GTSon,GTSoff]
[noTemporal,General] -r 5
Running configuration 1/4
Running configuration 2/4
Running configuration 3/4
Running configuration 4/4
```

```
abhilash@abhilash-Vostro-3550 ~/Castalia-3.2/Simulations/802.15.6
BAN $ ../../bin/CastaliaResults -i 171126-151229.txt

+--------------------+------------------------------------+----------
--+
|              Module |                             Output |
Dimensions |
+--------------------+------------------------------------+----------


--+
|         Application | Application level latency, in ms | 1x4(31)
|
|                     |         Packets received per node | 1x4
|
|   Communication.MAC |                Data pkt breakdown | 4x1(5)
|
|                     |         Mgmt & Ctrl pkt breakdown | 5x1(3)
|
|                     |            pkt TX state breakdown | 4x1
|
|                     |                         var stats | 5x1(2)
|
| Communication.Radio |                RX pkt breakdown | 5x1(5)
|
|                     |                         TXed pkts | 5x1
|
|     ResourceManager |                   Consumed Energy | 5x1
|
|      wirelessChannel |           Fade depth distribution | 1x1(14)
|
```

Further, comparing the results obtained for different radio system, it has been found that BAN radio is more suitable than CC1000 and CC2420 [18, 19]. This is evident from the energy consumption graph shown in Figure 6.10.

```
abhilash@abhilash-Vostro-3550 ~/Castalia-3.2/Simulations/802.15.6
BAN $ ../../bin/CastaliaResults -i 171126-151229.txt -s energy


ResourceManager:Consumed Energy BANRADIO
+-------------------+---------+---------+---------+---------+-------
--+---------+---------+---------+---------+
|                   | rate=14 | rate=16 | rate=18 | rate=20 |
rate=22 | rate=24 | rate=26 | rate=28 | rate=30 |
+-------------------+---------+---------+---------+---------+-------
--+---------+---------+---------+---------+
| GTSoff,General    | 0.313   | 0.313   | 0.313   | 0.313   | 0.313
| 0.313   | 0.313   | 0.313   | 0.313   |
| GTSoff,noTemporal | 0.313   | 0.313   | 0.313   | 0.313   | 0.313
| 0.313   | 0.313   | 0.313   | 0.313   |
| GTSon,General     | 0.313   | 0.313   | 0.313   | 0.313   | 0.313
| 0.313   | 0.313   | 0.313   | 0.313   |
| GTSon,noTemporal  | 0.313   | 0.313   | 0.313   | 0.313   | 0.313
| 0.313   | 0.313   | 0.313   | 0.313   |
+-------------------+---------+---------+---------+---------+-------
--+---------+---------+---------+---------+

ResourceManager:Consumed Energy CC1000
+-------------------+---------+---------+---------+---------+-------
--+---------+---------+---------+---------+
|                   | rate=14 | rate=16 | rate=18 | rate=20 |
rate=22 | rate=24 | rate=26 | rate=28 | rate=30 |
+-------------------+---------+---------+---------+---------+-------
--+---------+---------+---------+---------+
| GTSoff,General    | 2.242   | 2.242   | 2.242   | 2.242   | 2.242
| 2.242   | 2.242   | 2.242   | 2.242   |
| GTSoff,noTemporal | 2.242   | 2.242   | 2.242   | 2.242   | 2.242
| 2.242   | 2.242   | 2.242   | 2.242   |
| GTSon,General     | 2.242   | 2.242   | 2.242   | 2.242   | 2.242
| 2.242   | 2.242   | 2.242   | 2.242   |
| GTSon,noTemporal  | 2.242   | 2.242   | 2.242   | 2.242   | 2.242
| 2.242   | 2.242   | 2.242   | 2.242   |
+-------------------+---------+---------+---------+---------+-------
--+---------+---------+---------+---------+

ResourceManager:Consumed Energy CC2420

+-------------------+---------+---------+---------+---------+-------
--+---------+---------+---------+---------+
|                   | rate=14 | rate=16 | rate=18 | rate=20 |
rate=22 | rate=24 | rate=26 | rate=28 | rate=30 |
+-------------------+---------+---------+---------+---------+-------
--+---------+---------+---------+---------+
| GTSoff,General    | 6.193   | 6.193   | 6.19    | 6.193   | 6.19
| 6.189   | 6.191   | 6.191   | 6.19    |
| GTSoff,noTemporal | 6.183   | 6.183   | 6.186   | 6.183   | 6.178
| 6.18    | 6.186   | 6.181   | 6.179   |

| GTSon,General     | 6.193   | 6.193   | 6.19    | 6.193   | 6.19
| 6.189   | 6.191   | 6.191   | 6.19    |
| GTSon,noTemporal  | 6.183   | 6.183   | 6.186   | 6.183   | 6.178
| 6.18    | 6.186   | 6.181   | 6.179   |
+-------------------+---------+---------+---------+---------+-------
--+---------+---------+---------+---------+
```

The results of simulation with the details of the packets received per node for throughput test are shown in Figure 6.7, Figure 6.8, Figure 6.9, and Figures 6.10, 6.11, 6.12, 6.13, 6.14.

```
abhilash@abhilash-Vostro-3550 ~/Castalia-3.2/Simulations/802.15.6
BAN $ ../../bin/CastaliaResults -i 171126-151229.txt -s packets


Application:Packets received per node
+-------------------+--------+---------+---------+---------+-------
--+---------+--------+---------+---------+
|                   | rate=14 | rate=16 | rate=18 | rate=20 |
rate=22 | rate=24 | rate=26 | rate=28 | rate=30 |
+-------------------+--------+---------+---------+---------+-------
--+---------+--------+---------+---------+
| GTSoff,General    | 526.2   | 460.6   | 401.7   | 543.65  | 641.45
| 554.7   | 659.35  | 637.5   | 689.55  |
| GTSoff,noTemporal | 694.65  | 624.55  | 486.55  | 759.6   | 895.35
| 870.65  | 1222.75 | 934.85  | 1109.9  |
| GTSon,General     | 526.2   | 460.6   | 401.7   | 543.65  | 641.45
| 554.7   | 659.35  | 637.5   | 689.55  |
| GTSon,noTemporal  | 694.65  | 624.55  | 486.55  | 759.6   | 895.35
| 870.65  | 1222.75 | 934.85  | 1109.9  |
```

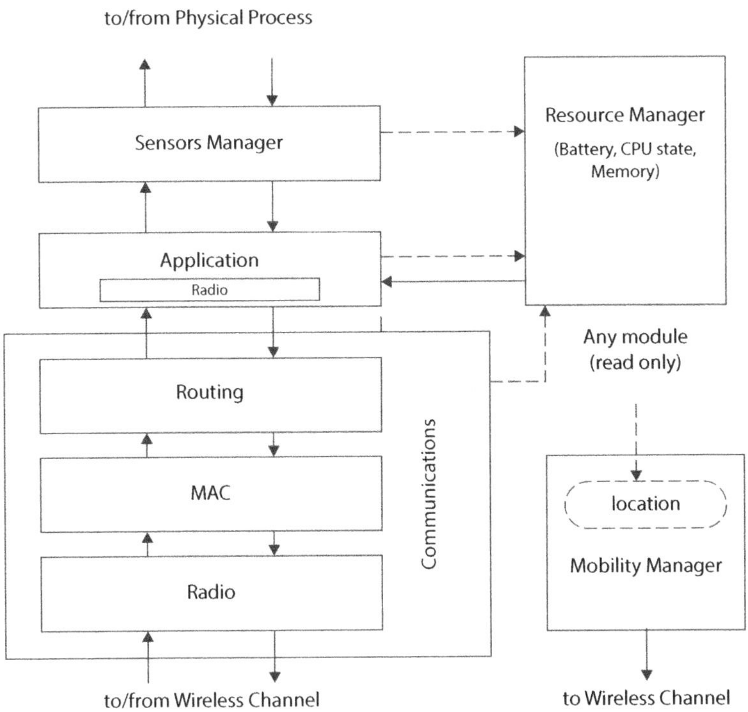

Figure 6.7 Block diagram.

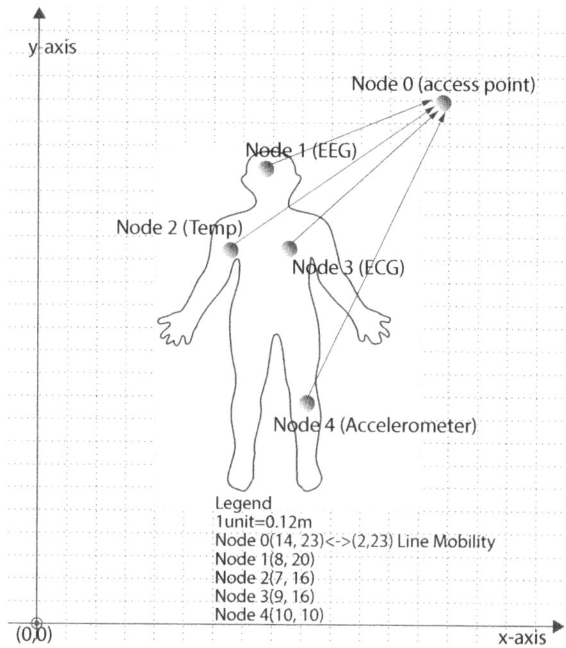

Figure 6.8 WBAN scenario.

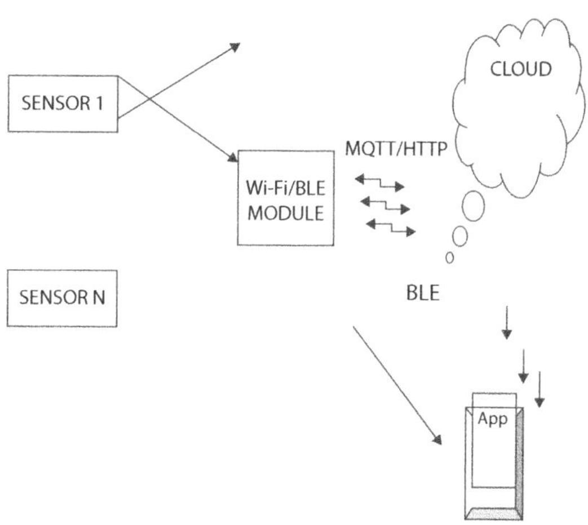

Figure 6.9 Hardware block diagram.

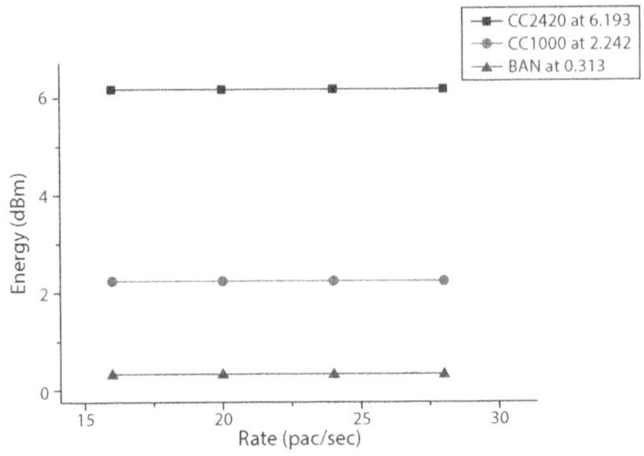

Figure 6.10 Radio comparison [20].

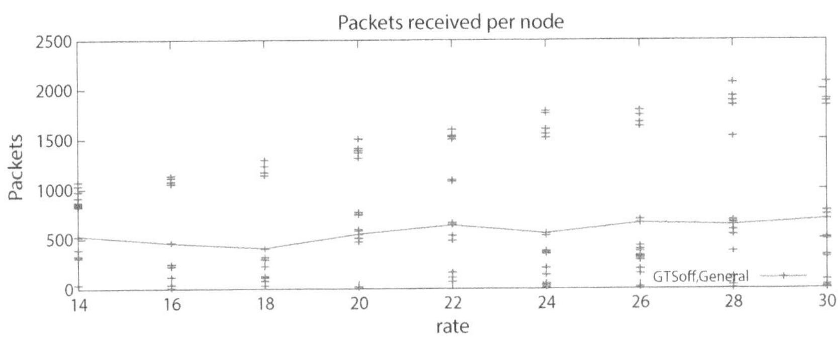

Figure 6.11 Packets received per node GTSoff, general.

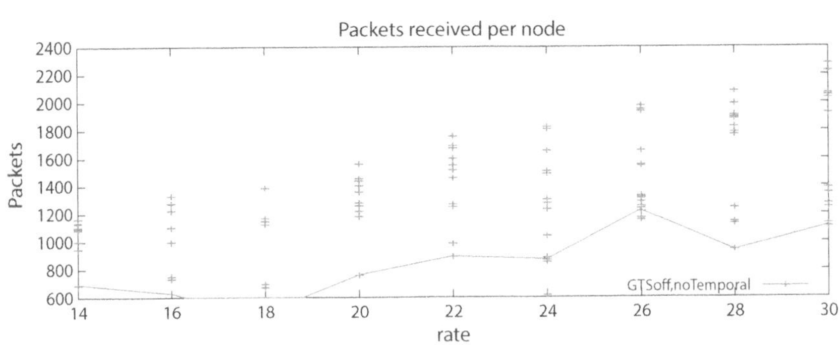

Figure 6.12 Packets received per node GTSoff, no Temporal [21].

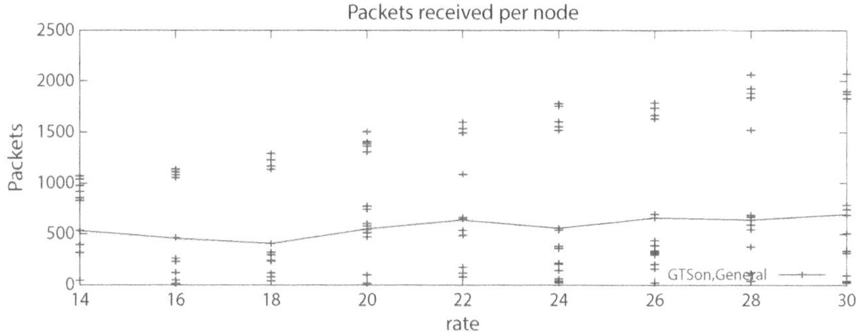

Figure 6.13 Packets received per node GTSon, general.

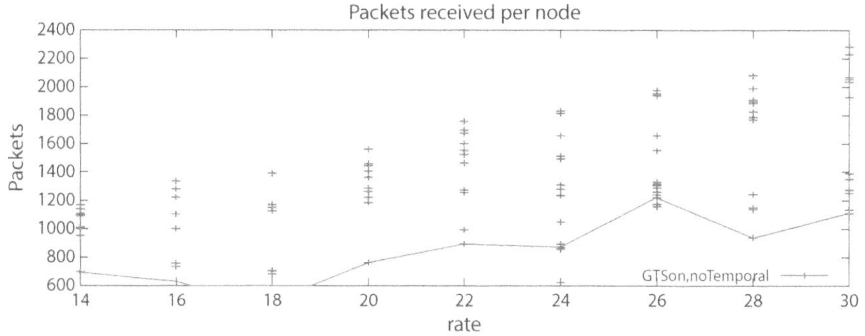

Figure 6.14 Packets received per node GTSon, no Temporal.

From the Figure 6.15 it is observed that for both GTSoff and GTSon, the performances are found to be similar. In Figure 6.15, axis y is the number of packets that is received per node (this means that node 0 which alone receives the number of packets, and it has been receives from the different nodes). The other axis x is the rate of transmission given for each number of node which is determined in terms of packets/sec. Here, the nodes are set to be transmitting the packets for only 100 secs. Therefore, considering the reception to be perfect, it is observed that to attain 1500 packets per each node, with 30 packets/sec/node. Observe if in favor of small traffic, GTSon no Temporal curves attains the utmost (e.g., 16 packets/sec/node giving rise to 800 number of packet per node has been received). Usually, it is seen that, protocol achieves superior whilst the Guaranteed Time Slot is rolled on. This result makes TDMA plots more capable than other plots when using a non-wired medium which results into reduced energy consumption by the nodes. The performance (packets received) is found to be

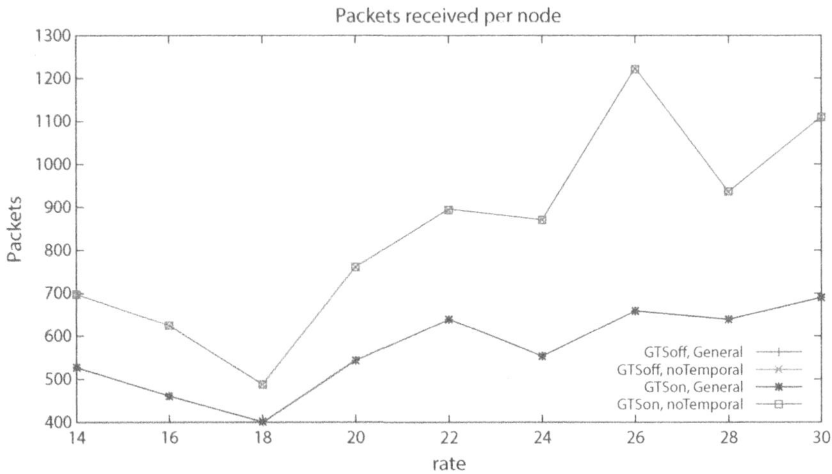

Figure 6.15 Packets received per node.

same in the simulation results acquired for no temporal variation channels. At last, when very high traffic is used, it has been observed that, there is an indication of saturation of packet flow. When observing at the histograms of latency, at the application-stage as shown in Figure 6.16, it is clear that there is a gradual decrease in receiving packets. It is because of the feature that most of the packets have been transmitted in the first frames created by MAC. It is also seen that, no Temporal and Temporal wireless models are achieving the same and no difference in performance is observed. Non-insignificant section of packets observed in Figure 6.16 shows that the after 600 packets, the saturation is observed. The number of packets have

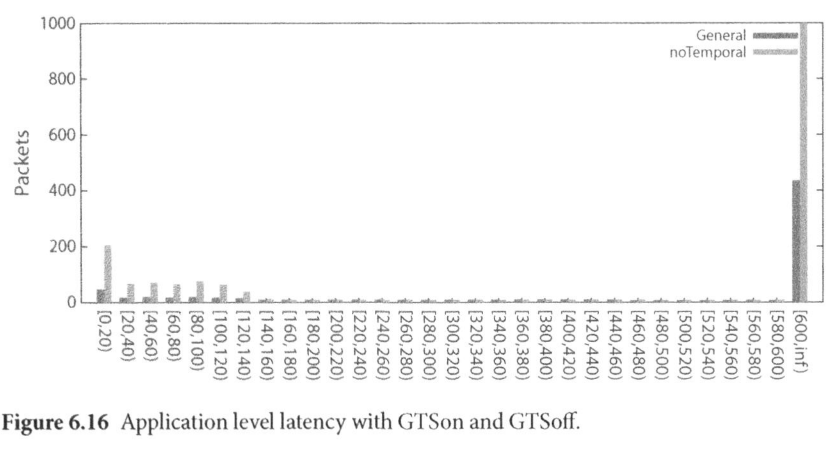

Figure 6.16 Application level latency with GTSon and GTSoff.

transmitted from nodes and received at the access point. Some of thepackets have been lost due to many reasons. Looking for the reasons, for loss of packets during transmission and reception, Castalia simulation gives the event to event information in the trace text and also in the created simulation file. Analyzing these files at the transmission and reception period, one can observe the reasons for loss of packets and learn from the behavior of the (node) transmitter and access point (receiver). Below is the result showing the reasons for loss of packets.

Figure 6.17 shows the MAC packet breakdown. This graph clearly shows the probability of transmission of packets and also distinguishes between transmitted packets and failed packets. Figure 6.18 shows receiver packets and also distinguishes between the packets received and packets lost in probability. Further, the analysis can be continued on this packet loss and can be avoided with particular measures. Here Y-axis is probability of packets received and x-axis is rate.

There is need to developing a smartphone app to view the sensor data from the website. Building an application have been carried out in website were in an individual can make his custom designed application based on the needs. It is not an easy task, since the design is done from the scratch. All the details should be user friendly, so that even a child or an old age person must feel comfortable to use the app with ease. The app has been designed in Thunkables. Thunkables is a platform especially developed for developing android/iOS application in website. Here, any person can

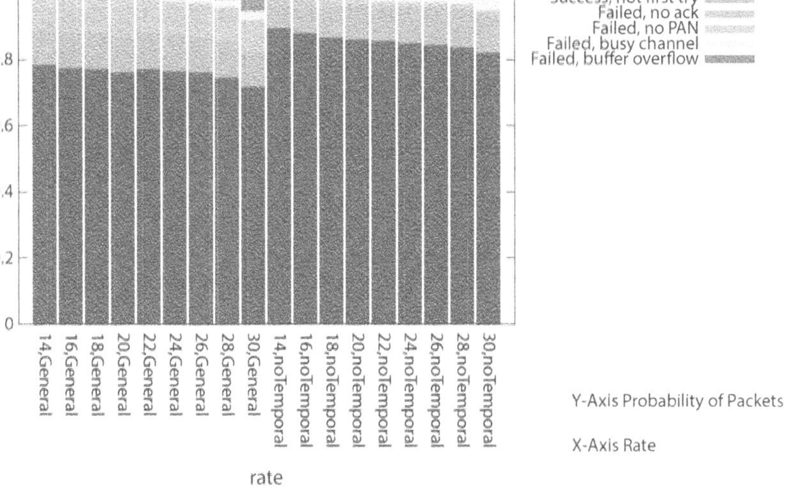

Figure 6.17 MAC packet breakdown.

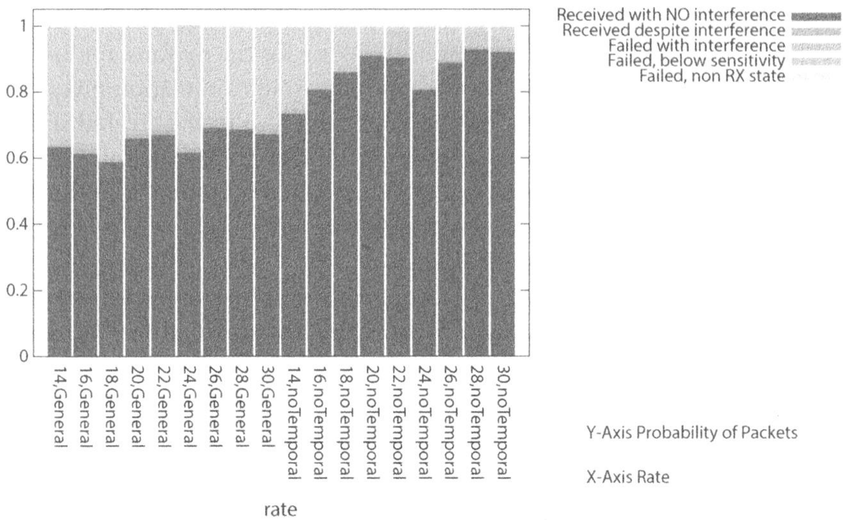

Figure 6.18 Packet received by the receiver.

design an app based on his requirement such as Internet of Things, image and voice Artificial Intelligence, modern user interfaces, etc. Before development one should have clear idea of about the objectives involved in the design of an app. The design procedure is explained below.

- Dragging and dropping the palette user interface such as box, button, notifier, text box, etc., into the development place.
- Placing those palettes on the place according to the design considerations.
- Editing and naming of these palettes is done as per the requirement.
- Each of these palettes should be functional and usable accordingly.
- The layout is done and palette should be adjusted.
- Also log in screen can be added if needed.
- Once the design is done it is next to move into the functional block where the functionality of the app is done.
- Drag and drop the blocks and giving the functionality to the app is explained in the figures.

Figure 6.19 shows the user interface screen which is designed as per the requirement. Here, the buttons GET Sensor1 and Sensor2 are clicked,

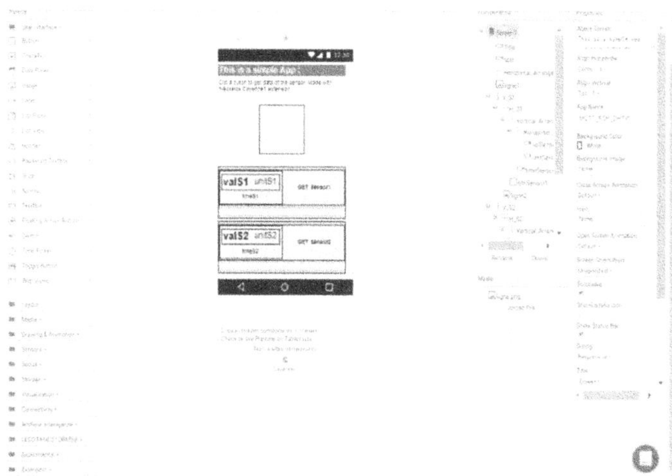

Figure 6.19 Design screen for app development in website.

which informs the website where the sensor data has been stored to send the values of the particular sensor to the subscriber app. These sensor data are then seen in the box S1 and S2 respectively, with its units and time of acquisition.

In Figure 6.20 date, time, month, day, device id and sensor id has been assigned to the app, whichbehaves as a subscriber ID to the app.

In Figure 6.21 the authentication to the app is given by assigning the email address and password.

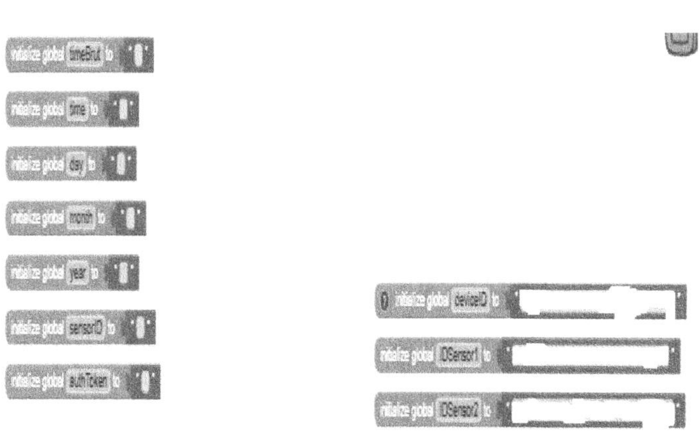

Figure 6.20 Block design for giving ID for app.

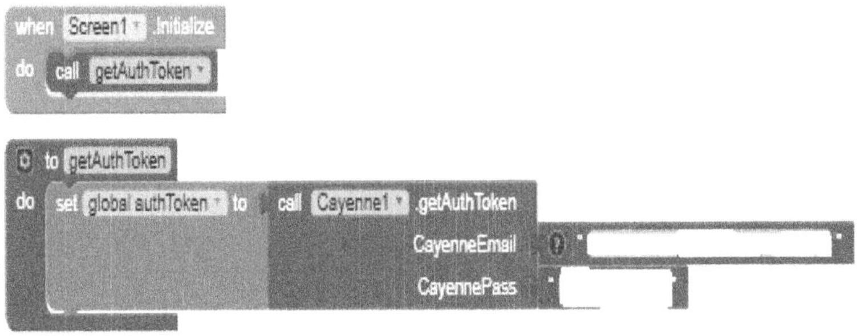

Figure 6.21 Block design for giving authentication and email pass for app.

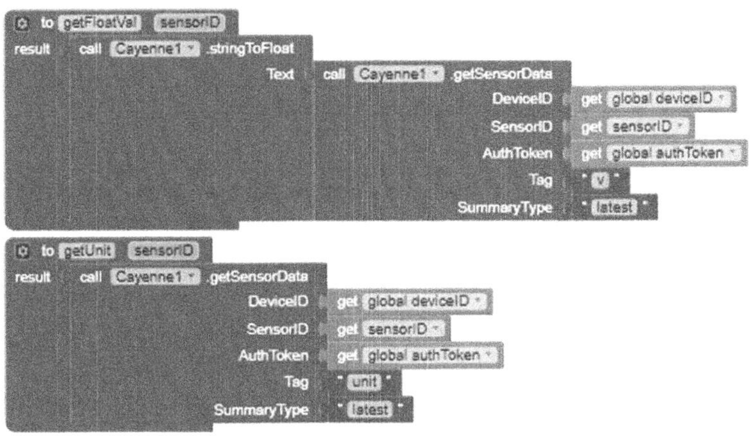

Figure 6.22 Block design for giving features for app.

Figure 6.22 explains the app features to get sensor data when the Get button is clicked. First, wait for Get button to be clicked upon, request it will look for which button is clicked, then the control goes to what sensor is attached to the particular button and it will look for its sensor id and requests the website to send the particular sensor data at that time.

Figure 6.23 explains the automatic update of values to the application by requesting the website to send data for every update of data in the website.

Figure 6.24 gives the functionality to the user interface which consists of buttons and boxes provided showing GetSensor1, GetSensor2, val1, val2, unitS1, unitS2, timeS1, and timeS2.

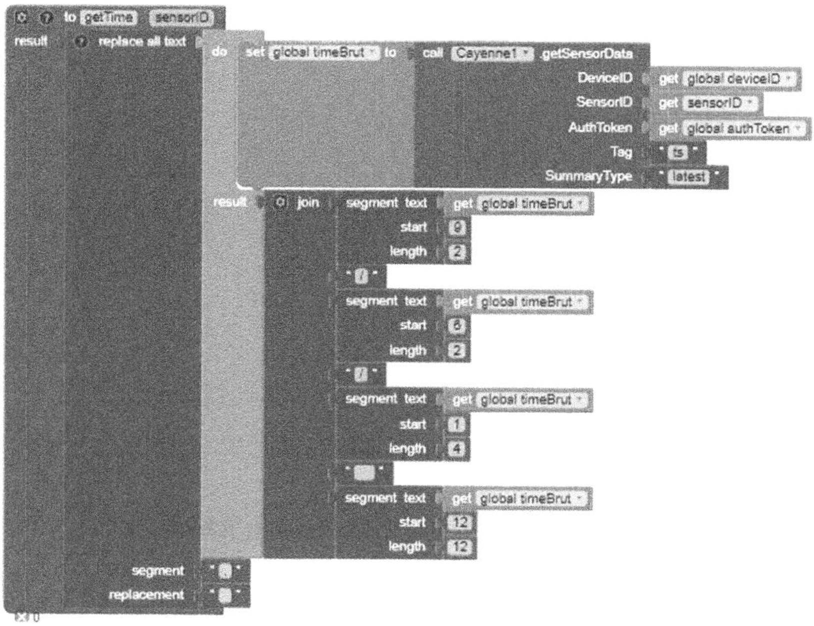

Figure 6.23 Block design for receiving text for app.

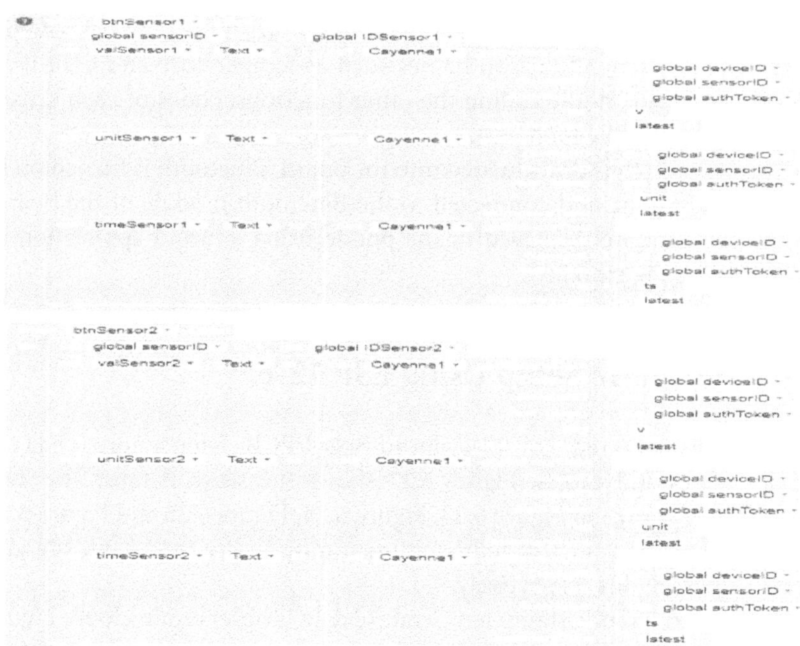

Figure 6.24 Block design for giving functionality for app.

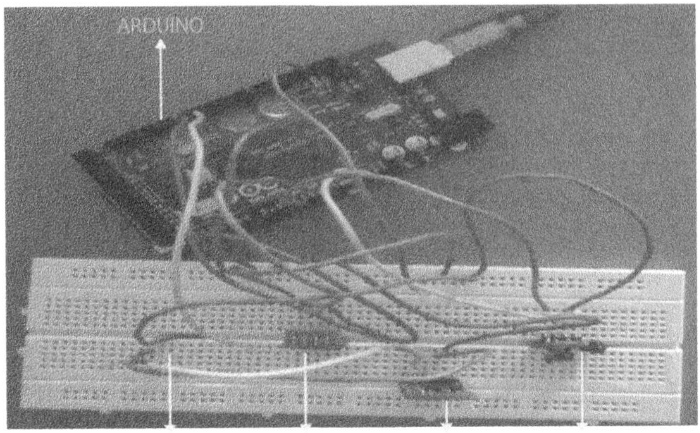

Figure 6.25 Hardware setup for Bluetooth module.

6.8 Hardware Setup Using Bluetooth Module

Components Required: Arduino Mega 2560, Bread Board, LDR Module (LM-393), Temperature (DHT11) Sensor, Bluetooth Module, wires and cables. The interface of the sensors to the Bluetooth module is achieved as shown below in Figure 6.25. Steps involved in interfacing: Setting the initialization of various pins, keeping the void set up. In the loop part of flow codes, the respective function names such as temperature and light level sensor are called. While calling the other functions, codes of each sensor are mentioned in the loop part.

Arduino sketch is downloaded into the board, Bluetooth is turned on in the android phone and connected to the Bluetooth module of the board, the sensor data are displayed in the phone using android application as shown in the Figure 6.26.

6.9 Hardware Setup Using ESP 12-E

Components Required: ESP 12-E, Bread Board/PCB, Temperature (DHT11) Sensor, wires and cables. Figure 6.27 shows the setup details. The data obtained are shown in Figure 6.23, Figure 6.24, Figure 6.25 and Figure 6.26 which shows the values of temperature and humidity in the website and also in smart phone respectively.

In Figure 6.28 time stamp temperature data is observed in cloud, Figure 6.29 humidity time stamp data is observed in cloud and in Figure 6.30 temperature data and humidity data is observed respectively, showing the

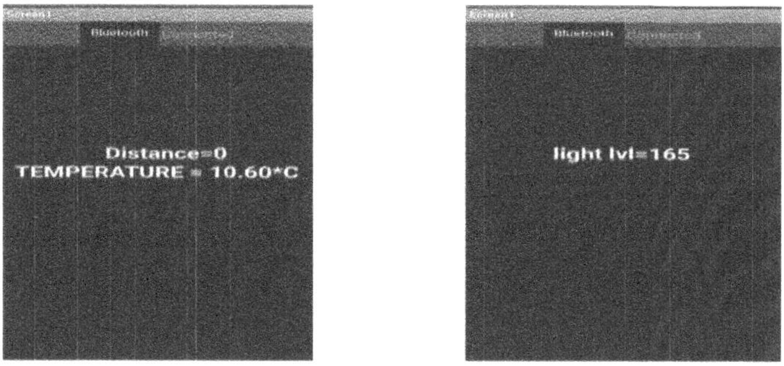

Figure 6.26 Sensor readings in Android.

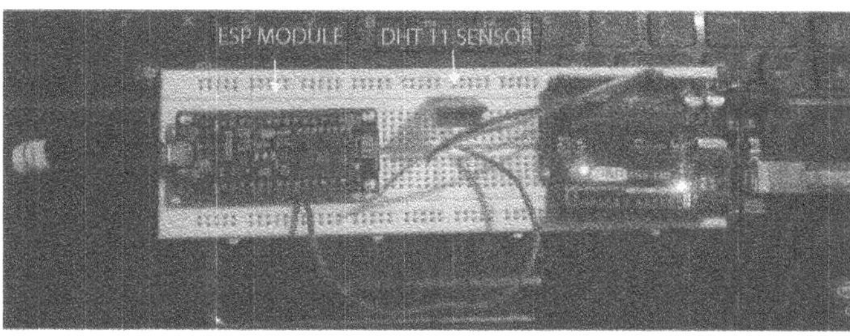

Figure 6.27 Hardware setup for ESP 12-E.

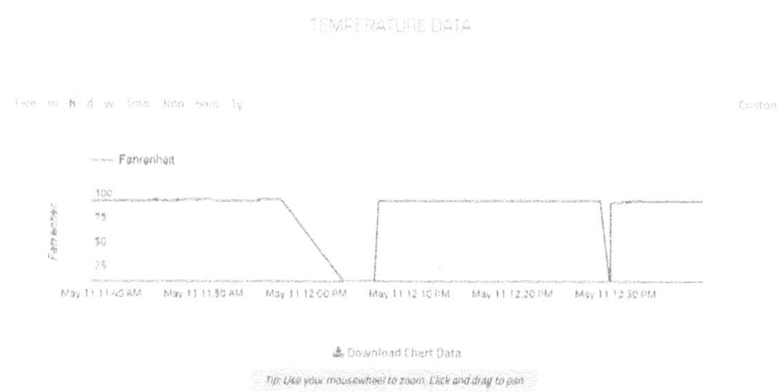

Figure 6.28 Temperature readings in website.

sensor data stored in the cloud. Figure 6.31 shows the sensor data obtained in designed app and works as planned. It is also possible to set threshold for temperature and humidity readings obtained. Therefore, alarm is turned on when the sensor readings crosses the threshold value. Hence, sensor data is monitored by the subscriber anytime and anywhere.

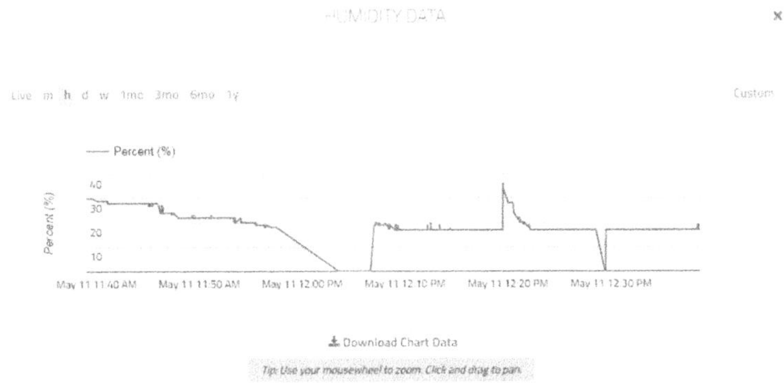

Figure 6.29 Humidity readings in website.

Figure 6.30 LED control/alarm, temperature and humidity readings in website.

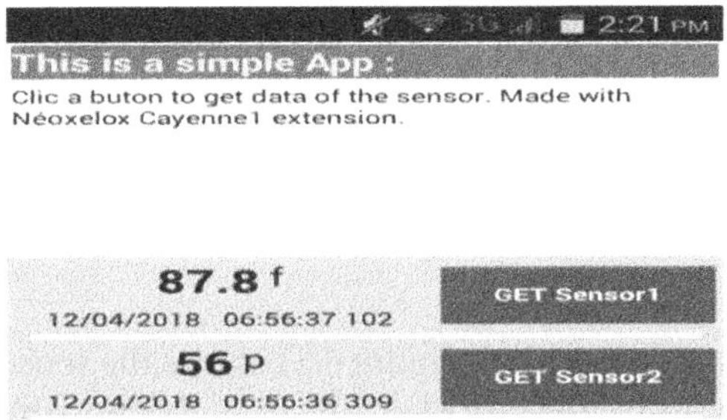

Figure 6.31 Temperature and humidity readings in smart phone.

In the algorithm shown in Figure 6.32 MQTT library has been included which is especially built for cloud platform cayenne to send sensor data. Other libraries such as cayenne, dht and esp header files have been added in the algorithm. Further, the connection name and password should be given and connection is established between ESP and router. If not the ESP tries to reconnect to the router. Once the connection is established the MQTT id, client id, server id should be entered as given by the platform. Now, the ESP is ready to publish the data into the subscriber through the MQTT broker provided by the platform for communication. Now, pins should be defined to send sensor data and loop is given to send the data continuously to the cloud without any delay [17].

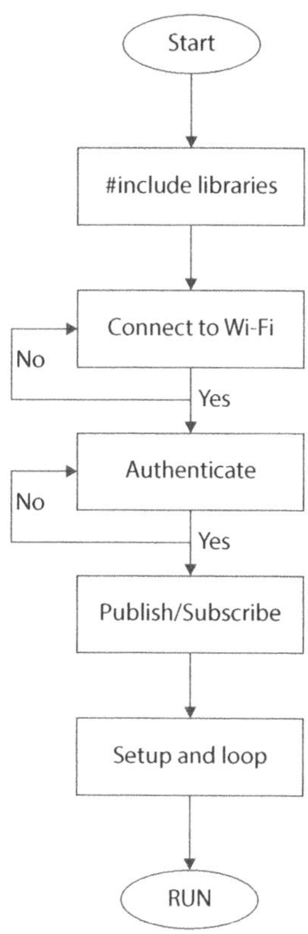

Figure 6.32 Algorithm incorporating MQTT.

6.10 Conclusions

A brief evaluation of parameters of the lower layers such as PHY and MAC layers in WBAN scenario using Baseline MAC of IEEE 802.15.6 standard are presented using Castalia simulation tool. The simulation results obtained evaluates BAN radio to be energy efficient when compared to CC1000 and CC2420. The reasons for loss of packets in the transmitter and receiver are observed and can be avoided by fine tuning of particular measures of lower layer parameters. The need for utilization of IoT and new technology will make a better living with the state of art of health management system. The design of this hardware implementation is one such example for IoT based technology. Requirement of low power and low cost has direct implication with protocols used in the network of BAN an effort has been made to design WBAN system using MQTT protocol. The cloud storage and android based app design facilitates a better health care management system with a provision of viewing measured data anytime, anywhere by the user.

References

1. Kurunathan, J.H., Study and overview on WBAN under IEEE 802.15.6. *Univ. Porto J. Eng.*, 1, 1, 11–21, 2015.
2. Kwak, K.S., Ullah, S., Ullah, N., An overview of IEEE 802.15.6 standard. *IEEE 23rd International Symposium on Personal Indoor and Mobile Radio Communications (PIMRC)*, pp. 402–751, 2012.
3. Ullah, S., Mohaisen, M., Alnuem, M.A., A review of IEEE 802.15.6 MAC, PHY, and security, *Int. J. Distrib. Sens. Netw.*, 2013, Article ID 950704, 12 pages, 2013. http://dx.doi.org/10.1155/2013/950704 12–24
4. Bradia, N., Belhaj, S., Chaari, L., Kamoun, L., Study of medium access mechanisms under IEEE 802.15.6 standard. *International Federation for Information Processing (IFIP) Wireless and Mobile Networking Conference (WMNC)*, pp. 1704–1714, 2011.
5. Ullah, S. and Kwak, K.S., Throughput and delay limits of IEEE 802.15.6. *IEEE Wireless Communication and Networking Conference (WCNC)*, pp. 25–31, 2011.
6. Gopalsami, N., Osorio, I., Kulikov, S., Buyko, S., Martynov, A., Raptis, A.C., SAW microsensor brain implant for prediction and monitoring of seizures. *IEEE Sens. J.*, 7, 977–982, 2007.
7. Nurmikko, A.V., Donoghue, J.P., Hochberg, L.R., Patterson, W.R., Song, Y.K., Bull, C.W., Borton, D.A., Laiwalla, F., Park, S., Ming, Y., Aceros, J., Listening

to brain microcircuits for interfacing with external world. *Proc. IEEE Prog. Wirel. Implantable Microelectron. Neuroeng. Devices*, *98*, 3, 375–388, 2010.
8. Jara, A.J., Zamora-Izquierdo, M.A., Skarmeta, A.F., Interconnection framework for mHealth and remote monitoring based on the Internet of Things. *IEEE J. Sel. Areas Commun.*, 31, 47–65, 2013.
9. Boyi, X., Li Da, X., Hongming, C., Cheng, X., Jingyuan, H., Fengli, B., Ubiquitous data accessing method in IoT-based information system for emergency medical services. *IEEE Trans. Industr. Inform.*, 10, 1578–1586, 2014.
10. Tabish, R., Ghaleb, A.M., Hussein, R., Touati, F., Ben Mnaouer, A., Khriji, L., A 3G/WiFi-enabled 6LoWPAN-based U-healthcare system for ubiquitous real-time monitoring and data logging, in: *Biomedical Engineering (MECBME), 2014 Middle East Conference on*, pp. 277–280, 2014.
11. Kulkarni, A. and Sathe, S., Healthcare applications of the Internet of Things: A review. *Int. J. Comput. Sci. Inf. Technol. (IJCSIT)*, *5*, 5, 16–22, 2014.
12. Elanthiraiyan, P. and Babu, S., Smart medicine and physical health system using IoT. *Int. J. Comput. Sci. Mob. Comput.*, 4, 3, 333–338, March-2015.
13. Ullah, K., Shah, M.A., Zhang, S., Effective ways to use Internet of Things in the field of medical and smart health care, IEEE, *2016 International Conference on Intelligent Systems Engineering (ICISE)*, pp. 15–19, 2016.
14. McCaffrey, C., Chevalerias, O., O'Mathuna, C., Twomey, K., Swallowable-capsule technology. *IEEE Pervasive Comput.*, 7, 1, 23–29, 2008.
15. Maduranga, K., Thotahewa, S., Redouté, J.-M., Yuce, M.R., Ultra wideband wireless body area networks, pp. 1–169, Springer Cham Heidelberg, 2014. https://link.springer.com/book/10.1007/978-3-319-05287-8
16. Wei, Z., Chaowei, W., Nakahira, Y., Medical application on Internet of Things, in: *Communication Technology and Application (ICCTA 2011), IET International Conference on*, pp. 660–665, 2011.
17. Talpur, M.S.H., The appliance pervasive of Internet of Things in healthcare systems. *IEEE Trans.*, 9, 468–486, 2013.
18. Dongxin, L. and Tao, L., The application of IoT in medical system, in: *IT in Medicine and Education (ITME), 2011 International Symposium on*, pp. 272–275, 2011.
19. Kortuem, G., Kawsar, F., Fitton, D., Sundramoorthy, V., Smart objects as building blocks for the Internet of Things. *IEEE Internet Comput.*, 14, 44–51, 2010.
20. Z. Jian, W. Zhanli, M. Zhuang, Temperature measurement system and method based on home gateway. Chinese Patent 102811 185 A, Dec. 5, 2012.
21. Sines, J.O., Permanent implants for heart rate and body temperature recording in the rat. *AMA Arch. Gen. Psychiatry*, 2, 2, 182–183, 1960.
22. Kaleem Ullah, MunamAli, Effective Ways to Use Internet of Things in the Field of Medical and Smart Health Care, 2015 International Conference on Identification, Information, and Knowledge in the Internet of Things, 978-1-4673-8753-8/16/$31.00 ©2016 IEEE.

7

Analyzing a Microstrip Antenna Sensor Design for Achieving Biocompatibity

Sonam Gour[1,2], Abha Sharma[1] and Amit Rathi[1]*

[1]*Department of EC Engineering, Manipal University Jaipur, Jaipur, Rajasthan, India*
[2]*Department of EC Engineering, Poornima College of Engineering, Jaipur, Rajasthan, India*

Abstract
Signal Transmission is the essential part of the wireless Communication. Biocompatibilty is very much required. For biomedical devices the Biotelemetry is used for transmitting the signal to a particular distance. Due to the planer profile and low cost Microstrip antenna is used for designing the plantable and wearable devices. The paper represents comparison of the different types of antenna used for analysis of different diseases in human beings. The comparison done for the optimization parameter like: Gain, SAR, Gain, Return Loss and Bandwidth. Different design used for gaining the accurate results. Paper discusses about the different design structure like Split Ring Resonator (SRR), Circular Ring Resonator (CRR) and Triangular Ring Resonator. The designed antenna performs accurately without harming any muscle tissue.

***Keywords*:** Biomedical, wireless, flexible structures, split ring resonator, bandwidth, return loss, SAR, circular resonator, antenna sensors, optimization techniques

7.1 Introduction

In the today's healthcare application the uses of Biocompatible devices increasing day by day. A Microstrip patch Antenna is used for designing any biocompatible design due to its reduce size, low power losses and planer surface. In the environment of wearable network an inside antenna is used for

**Corresponding author*: amitrathi1978@gmail.com

Chandra Singh, Rathishchandra R Gatti, K.V.S.S.S.S. Sairam and Ashish Singh (eds.) Modeling and Optimization of Optical Communication Networks, (125–136) © 2023 Scrivener Publishing LLC

transmission and reception of the data [1]. The EM waves' presence in the human body effect the working of the antenna. So, the efficiency is very much required for any plantable antenna [2]. So the updating in this type of devices required very much. A microstrip antenna is used in the Global tracking system and for the high 5G network. It contains a patch on upper side and ground plane structure with dielectric substrates. The requirement of the bandwidth is also increasing and the traffic near by the bandwidth is increasing day by day [3]. Microstrip Antenna has different type of advantage over the other antenna used for the same. The design of the Microstrip antenna is varied by the feeding technique of the antenna. The substrate of the patch is directly affects the performance of the Antenna. With the rapid development in the wireless technology in the last two decades the spectrum resources is reduced and very much traffic increases in band limit. The use of the microstrip antenna is increases due to the sustain bandwidth in the desired region [4, 5]. Biomedical devices can be used for the two band structure, the band structure is defined by the Federal Communications Commission (FCC). Two important band structure are as follows MICS (Medical implant communications) 402-405 MHz and ISM band (2400–2500 MHz). The feeding technique of the microstrip antenna is directly affected the characteristics of the Antenna. Microstrip Antenna can be designed by four different type of feeding techniques. Two different types of substrate are used in Aperture coupled and proximity type of feeding technique. These techniques increases gain and bandwidth of the antenna but the spurious radiation are present in the techniques. High impedance matching is required. In microstrip feed line technique strip of conducting material is connected with the microstrip patch [6–9]. It can also be designed for the different type of resonator circuit, some resonator is discussed under the paper the main designing of the resonator circuit as circular, triangle and circular.

7.2 Designing of Biomedical Antenna

By block the direct radiation of the Antenna the spurious radiation can be reduced [10]. The patch designed by the H type slot coupling provided the wider bandwidth and the spurious radiation pattern. This design also increases the bandwidth and reduces the radiation losses in the antenna. The effect of radiating element is also reduced [11]. The Aperture coupled antenna by using the U shaped and M shaped microstrip feed line provided the good impedance matching and provide stable performance [12]. The circularly polarized technique in the cross shaped microstrip antenna is proposed in three shapes which is circular, square and cross shaped.

Each design is analyzed by varying the rotation angle, arm width and the structure of the ground plane structure. The cross shaped structure desired larger perimeter to achieve cp radiation [13]. The flow diagram in Figure 7.1 defines or demonstrate antenna design for medical applications. The simulated results compared with the design and result will be stored for the designing purpose. The process will remain same for all the steps.

The proposed microstrip Antenna with multi band and low profile. ISM band used by the designer for simulated the antenna. The range of used band was 2.4 GHz and 5.8 GHz. The first design is suitable for WBAN medical applications; it was simulated in the ISM band. The second design is compatible with the 3.5GHz and 7.5GHz, in Ultra-Wide-Band (UWB) applications, it designed by using stacking the antenna. It is compatible for working in Size of the designed antenna was $11.15mm^3$ and $25.16mm^3$.

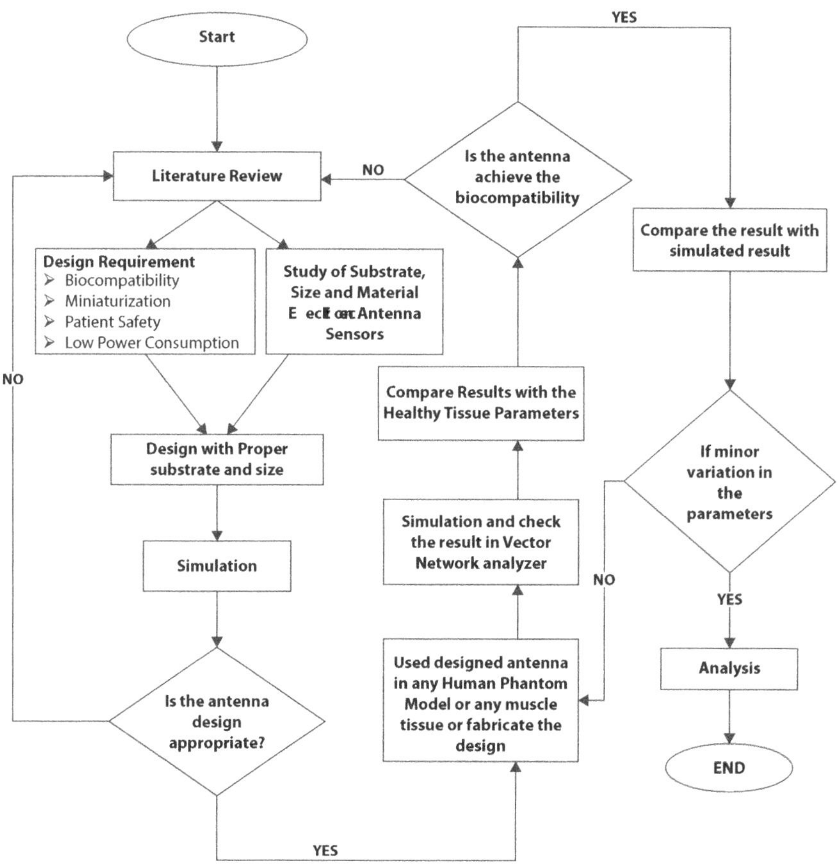

Figure 7.1 Block diagram for biomedical antenna sensors design.

A feeder design connected with 50 Ohm coaxial cable. Some miniaturization process designed in the second design. It is designed with 180° at the height of 0.5mm. One extra substrate height of 0.15mm added at the top for covered the structure. For the resonance frequency 2.46GHz the return loss was -13.3 dB and provides -10 dB bandwidth 150 MHz. The measured resonance frequency is 420 MHz with the return loss of -22.7dB. The return loss is -28.6dB for resonating frequency 2GHz and-24dB for 3.5GHz. In omnidirectional antenna the loss is -34.6 dB with resonating frequency of 7.5 GHz. The achieved SAR is calculated 0.05 W/Kg [14].

A Microstrip Patch antenna for dual band application is designed for Biomedical application in two band structure one is for ISM (2.4-2.5 GHz) and the other is MICS (402-405 MHz) band. The antenna size is 179.7 x 228.3 x 1.63 mm. Author reduced the size of antenna by using new serpentine shape and single feed point. The new antenna design is 31 x 25 x 1.63 mm. It reduced the size of Antenna by 52.81 times. The calculated bandwidth provides very narrow simulation results; for MICS band it is 0.56% and for the ISM band it is 0.2 %. The achieved resonance frequency is 2.47 GHz in ISM Band and 403 MHz for MICS Band. The calculated bandwidth is 0.2% for 2.45GHz for ISM band and 0.56% for the 403 MHz for MICS band. The bandwidth is improved after the measurement and but it provides the twice return loss. For checking the antenna compatibility, it is implanted over the muscle in the plastic pack body with the chicken skin. As per the addition of the plastic pack the frequencies shifted by some frequencies [15].

7.3 Sensing Device for Biomedical Application

The Biomedical sensing device can be used SRR (Split Ring Resonator). It Consist with two metal concentric metal ring separated by some gap, both the rings are use in the split structure at opposite side. In the ring oscillator magnetic resonance induced by split the ring between the inner and outer ring. It behaves like parallel Resonator circuit and two rings are coupled by gap between them. The capacitive and resonance properties of SRR make its favorable device for sensing applications [16]. Different type of SRR used by designer for reducing the size and optimizing the parameter. Some of them represented in Figure 7.2.

The CSRR (Complementary Split Ring Resonator) can also be used for designing the sensing device. CSRR rings are etched from the metallic surface. Its electric and magnetic properties can be changed with respect to SRR. CSRR design is designed to meet wearable antenna design specifications on 5G networks. The 5G network antenna required small size and

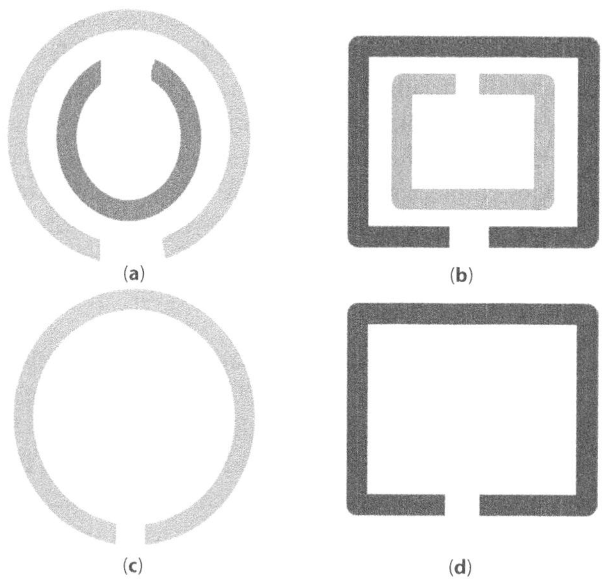

Figure 7.2 SRR Design. (a) Double circular. (b) Double rectangle. (c) Single circular. (d) Single rectangle [16].

compact structure for giving the best performance [17]. Use of metamaterial is increasing in the biomedical instrument due to its coefficient such as permittivity and permeability. It has the negative refractive index [18].

The next section discuses about the different design technique used by the researcher for designing the Microstrip Antenna for the Biomedical Applications. SRR element is used for transmitting the direct radiation in the EM wave propagation. The SRR is used for the different location in the loop [19].

Implantable Biomedical Antenna

The antenna was designed for operating at 915-MHz (ISM Band). The dimension of the Antenna is $\pi \times (4.7)2 \times 1.27 mm^3$. To increase certain current path many slots are inserted in the circular patch. For lowering the resonant frequency a shorting pin is inserted it would increase the circular polarization (CP) purity. Rogers has used for making the substrate and superstrate with $\varepsilon r = 10.2$, $\tan\delta = 0.0035$ and thickness is 0.635mm. By inserting a shorting pin, impedance bandwidth with $|S11|$ lower than -10 dB ranges, it is ranges from 866 MHz to 978 MHz. The achieved AR bandwidth 19.27%. Conventional design works on 3.6 GHz with two center-symmetric slots. The current path increased by inserting the rectangular slot in the design. were etched out from the patch to lengthening the current path. By this technique the achieved resonating frequency is 915MHz and impedance bandwidth ranges from 866 MHz to 978 MHz. Design influenced the circular polarization. First, a good AR performance can be achieved by changing the positions of the shorting and feeding. Second, with introducing the slots at each side of the patch brings the two orthogonal modes connectively realized CP radiation. When the antenna was implanted in the fat layer the resonant frequency shifted higher it is happened because the permittivity of the fat is much lower as compare to skin of humans. Value of the SAR is 778 W/kg [20].

Slot Dipole Rectangular Implantable Miniaturized Patch Antenna

Antenna was operated in for the ISM band. The design has tiny dimension and lower bendability. With these two features the designed antenna is capable for in-body applications. The antenna size is (8.20 x 12.5 x 0.045) mm as shown in Figure 7.3. The antenna designed in a silicon block structure and then it placed between the fat and muscle design. The return loss of the antenna is -38.42 dB. The bandwidth is found to be 127.9 MHz. For testing the bio-compatibility the bending test and SAR evaluation checked for the proposed design. The maximum SAR value is 0.8W/kg. The antenna was checked for the various substrates and patch materials for calculating efficiency and return loss. Antenna resonate at 2.45GHz, it is compatible to the ISM band [21].

Circular Polarized Implantable Antenna

A single feeding circularly polarized miniaturized antenna is defined in the paper. It is used in biomedical application for 2.4 GHz to 2.48 GHz band. For Miniaturization of Antenna Capacitive load and mender lines are used in the radiator. The size of antenna is (12*12*1.27) mm^3. Antenna has square slot in the center of the patch and meanders four lines. Patch in parallel connected with the four transmission lines. Four transmission

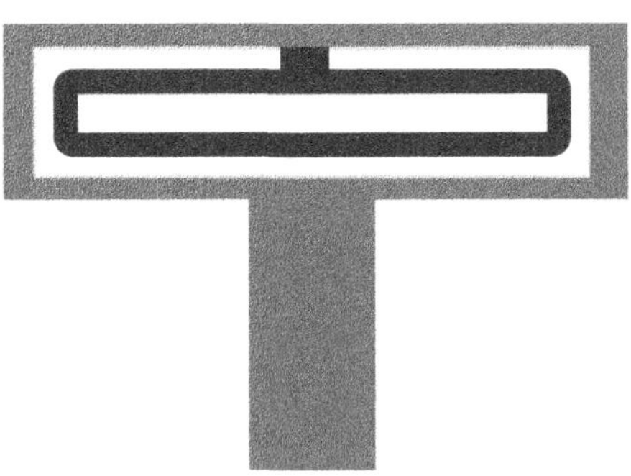

Figure 7.3 Shape of designed antenna [21].

Table 7.1 Performance analysis of different antenna.

Ref.	Size (mm^3)	Substrate	Frequency band	Return loss	Res. freq. (GHz)	AR<=3 dB	SAR	BW (-10dB)
[14]	Ist Design (11.15m m^3)	Alumina∈r = 9.4, tanδ = 0.006 h = 0.25 mm	ISM: 2.4 & 5.8 GHz	-13.3 dB -22.7 dB	2.46 GHz 5.77 GHz	-	0.05 W/Kg.	150 MHz 420 MHz
	IInd Design (25.16 m m^3)	Alumina∈r = 9.4, tanδ = 0.006h = 0.25 mm	ISM: 2.4 & 5.8 GHz	-28.6 dB -24 dB	2 GHz 3.5 GHz	-	0.05 W/Kg.	-
[15]	179.7 × 22.3 × 1.63 mm^3	∈r=4.3 h = 0.163 cm	ISM: 2.4-2.5 GHz MICS 402-405 MHz	-	2.47 GHz (ISM) 403 MHz (MICS)		ISM: (-17 dB) MICS: (-38 dB)	0.2% (ISM) 0.56% (MICS)
[20]	π × (4.7) 2 × 1.27 mm^3	Rogers 6010 εr = 10.2, tanδ = 0.0035h = 0.635 mm	915 MHz	-	0.915 GHz	19.7%	778 W/kg	17.5%

(Continued)

Table 7.1 Performance analysis of different antenna. (*Continued*)

Ref.	Size (mm^3)	Substrate	Frequency band	Return loss	Res. freq. (GHz)	AR<=3 dB	SAR	BW (-10dB)
[21]	8.20 × 12.5 × 0.045 mm^3	FR4 Polyamide Rogers	ISM: 2.4-2.5 GHz	-38.42 dB	2.4 GHz	-	0.8 W/kg	127.9 MHz
[22]	100 mm × 100 mm × (20 + d) mm^3	Alumina	ISM: 2.4-2.5 GHz	1.93 to 2.82GHz (B.W.)		2.38 to 2.51 GHz (B.W.)	217.849 w/kg (1- g avg) and 40.6027 (10-g avg)	-

lines are connected in parallel to the each sides of the patch. X axis is used for feeding the Antenna, the distance between the original and fed axis is 4.25mm. The proposed antenna was fed along the X-axis and the distance between the fed point and the original one is 4.25 mm. The superstrate and substrate are designed by Rogers 6010 (εr =10.2, $\tan\delta$ = 0.0035) height of 0.635 mm. The simulated S11 bandwidth is ranges from 2.31 GHz to 2.65 GHz and the 3dB axial ratio covers from 2.2 GHz to 2.5 GHz.

As per the designing of the antenna a wide bandwidth is achieved for the circular polarization can be achieved. For 1 W power the calculated SAR values is 503.2 W kg for the 2.4 GHz. So that the antenna can be operated for 3.18 mW [22]. The comparative analysis of complete papers is presented in Table 7.1.

7.4 Conclusion

This paper discussed about the different types and devices used for the Biomedical Applications. It represented a comparative analysis between different approaches in terms of bandwidth, frequency, gain and return loss. The biomedical devices have the issue of miniaturization. So the biocompatibility achievement is very big task for any biocompatible devices. The microstrip line based sensors also discussed in this paper. The performance of different sensors analyzed in the paper. The paper represents the detail analysis of Split Ring Resonator, Single circular SRR, Single Rectangle SRR, Double Circular SRR and Double Rectangle SRR. SRR devices consisted with two metal concentric metal ring separated by some gap and both the rings are use in the split structure. The biocompatibility of any implantable antenna is analyzed in terms of SAR, Radiation Loss and bendability. After reviewing the paper it was found that by adding some biocompatible substrate, design enhancement and miniaturization process the optimization in the design can be achieved.

References

1. Basir, and Yoo, H., A stable impedance-matched ultra wideband antenna system mitigating detuning effects for multiple biotelemetric applications. *IEEE Trans. Antennas Propag.*, 67, 3416–3421, 2019.
2. Greatbatch, W. and Homes, C.F., History of implantable devices. *IEEE Eng. Med. Biol. Mag.*, 10, 38–41, 1991.
3. Monà, D.F., Sakomura, E.S., Nascimento, D.C., Microstrip-to-probe fed microstrip antenna transition. *IEEE International Symposium on Antennas and Propagation*, pp. 1521–1522, 2018.

4. Kumar, G. and Ray, K.P., Broadband microstrip antennas, Artech house p. cm. — (Artech House antennas and propagation library, Boston, London 2002, ISBN 1-58053-244-6 (alk. paper).
5. Chen, Z., Dai, X., Luo, G., A new H-slot coupled microstrip filter-antenna for modern wireless communication systems. *IEEE International Workshop on Antenna Technology*, March 2018.
6. Sharma, A., Saini, Y., Singh, A.K., Rathi, A., Recent advancements and technological challenges in flexible electronics: MM wave wearable array for 5G networks. *AIP Conf. Proc.*, 2294, 1, 020007, 2020, AIP Publishing LLC.
7. Sharma, A., Suvalka, R., Singh, A.K., Agrahari, S., Rathi, A., A rectangular annular slotted frequency reconfigurable patch antenna, in: *International Conference on Communication, Devices and Networking*, Springer, Singapore, pp. 255–261, 2019.
8. Sharma, A. and Rathi, A., Analysing reconfigurable slot antenna for cognitive radio applications, in: *2020 8th International Conference on Reliability, InfocomTechnologies and Optimization (Trends and Future Directions) (ICRITO)*, IEEE, pp. 744–747, 2020.
9. Rathi, A. and Vijay, R., Expedite particle swarm optimization algorithm (EPSO) for optimization of MSA, in: *International Conference on Swarm, Evolutionary, and Memetic Computing*, Springer, Berlin, Heidelberg, pp. 163–170, 2010.
10. Noh, E. and Kim, K., Effects of MWT radiation on sidelobe levels in mm-wave microstrip array antenna. *IEEE International Symposium on Antennas and Propagation*, October 2019, pp. 873–874.
11. Chen, Z., Dai, X., Luo, G., A new H-slot coupled microstrip filter-antenna for modern wireless communication systems. *IEEE International Workshop on Antenna Technology*, March 2018.
12. Li, H., Kang, L., Wei, F., Mingcai, Y., Yin, Y.Z., A low-profile dual-polarized microstrip antenna array for dual-mode OAM applications. *IEEE Antennas Wirel. Propag.*, 16, 3022–3025, October 2017.
13. Naishadham, K., Bayat, A., Savi, P., Antenna integration with nanotechnology-based thick film circuit for sensor applications, IEEE, *International Conference on Electromagnetics in Advanced Applications*, pp. 1711–1714, 2017.
14. ElSalamouny, M.Y. and Shubair, R.M., Novel design of compact low-profile multi-band microstrip antennas for medical applications. *Loughborough Antennas & Propagation Conference*, 2015.
15. Grabssi, W., Izza, S., Azrar, A., Design and analysis of a microstrip patch antenna for medical applications. *5th International Conference on Electrical Engineering - Boumerdes (ICEE-B)*, 2017.
16. Nadin, A. and Hussein, M.I., Metamaterial-based sensor design using split ring resonator and Hilbert fractal for biomedical application. *Sens. Bio-Sens. Res.*, Published by Elsevier, 31, 2020.

17. Efri, S. and Baso, M., Complementary split ring resonator on the ground plane for wearable antenna. *International Conference on Radar, Antenna, Microwave, Electronics, and Telecommunications*, pp. 66–69, 2020.
18. Jahangiri, M. and Rajebi, S., Effects of split ring resonator (SRR) metamaterial on the radiation pattern and variation of the heating focus point of the microstrip patch antenna. *Adv. Sci. Technol. Eng. Syst. J.*, 5, 1, 307–316, 2020.
19. Liu, Z. and Guo, Y., Dual band low profile antenna for body centric communications with a split ring resonator. *IEEE MTT-S International Microwave Workshop Series on RF and Wireless Technologies for Biomedical and Healthcare Applications*, 2013.
20. Zhang, Y., Liu, C., Liu, X., Zhang, K., Yang, X., A wideband circularly polarized implantable antenna for 915 MHz ISM-band biotelemetry devices. *IEEE Antennas Wirel. Propag. Lett.*, 17, 8, 1473–1477, Aug. 2018.
21. AI Islam, N., Abrar, A.T., Arafat, U., Islam, A.J., Hoque, R., Design and performance measurement of an in-body implantable miniaturized slot dipole rectangular patch antenna for biomedical applications. *3rd International Conference on Advances in Electrical Engineering*, IEEE, 2015.
22. Yang, Z.-J. and Xiao, S.-Q., A single-fed miniaturized circularly polarized implantable antenna for ISM band biomedical application. *IEEE MTT-S International Microwave Workshop Series on Advanced Materials and Processes for RF and THz Applications (IMWS-AMP)*, 2016.

8

Photonic Crystal Based Routers for All Optical Communication Networks

T. Sridarshini[1]*, Shanmuga Sundar Dhanabalan[2†], V.R. Balaji[3], A. Manjula[4], S. Indira Gandhi[5] and A. Sivanantha Raja[6]

[1]*Department of Electronics and Communication Engineering, College of Engineering Guindy Campus, Anna University, Chennai, India*
[2]*Functional Materials and Microsystems Research Group, School of Engineering, RMIT University, Melbourne, Australia*
[3]*School of Electronics Engineering, Vellore Institute of technology, Chennai, India*
[4]*Department of Electrical and Electronics Engineering, Mohamed Sathak Engineering College, Kilakarai, India*
[5]*Department of Electronics Engineering, Madras Institute of Technology, Chennai, India*
[6]*Department of Electronics and Communication Engineering, Alagappa Chettiar Government College of Engineering & Technology, Karaikudi, India*

Abstract

All optical communication system aims to provide an efficient, secured, effective and consistent network, which is the critical need of the hour. Optical communication offers high speed and colossal bandwidth capacity and emerges as one of the technological successes of the 20th century. As an outcome of intensive research for more than three decades in the refined fabrication and technologies of optical communication & systems, scientist and researchers have reached a state of saturation in terms of the lowest possible attenuation and data rates that can be achieved, as opto-electronics devices and components imposes constraint on them. These scenarios have motivated optical engineers and researchers to move on in developing new optical medium with semiconductors such as silicon, germanium, gallium arsenide etc., called photonic crystals dates back to 18th century on which fully integrated optical circuits can be designed and developed. The fascinating photonic band gap property of photonic crystals has been employed to

**Corresponding author*: sridarshiniv@gmail.com
†*Corresponding author*: dshanmugasundar@gmail.com

Chandra Singh, Rathishchandra R Gatti, K.V.S.S.S.S. Sairam and Ashish Singh (eds.) Modeling and Optimization of Optical Communication Networks, (137–162) © 2023 Scrivener Publishing LLC

guide the flow of light inside the desired path of photonic crystals. In all-optical communication systems, the tactics required to have fully integrated optical networks is establishing complete links between the different nodes in the network. The overall performance of the network will be affected by the difficulties in establishing complete connectivity and complexity in the data transfer between the nodes. Photonic crystal based PIC serves a best method to overcome this difficulty. And also PIC designs with different topologies to obtain high throughput are providing solutions to chip multi-processors. Diverse components like splitters, multiplexers/de-multiplexers, couplers, circulators, modulators, routers, add-drop filters, pass band and stop band filters and modulators are to be unified into a single chip for PIC. Photonic crystal ring resonators forms the basic building blocks for almost every component mentioned. Routing of the optical signals between various components in the chip is the decisive necessity of PIC.

In this chapter, we focus our topic exploring the optical network component - router using photonic crystals, which will be a perfect candidate to be integrated in Photonic integrated circuits (PIC) for optical communication and networking systems. Different configurations of photonic crystal based routers have been detailed and reviewed based on the performance. We have also explained the basic concept and working of the photonic crystal based routers. All the reported routers are ultra-compact and it will be a suitable candidate which can be incorporated and integrated in future all optical components for enhanced optical communication systems and networks.

***Keywords*:** Photonic crystals, routers, ring resonators, photonic bandgap, gap map, photonic integrated circuits

8.1 Introduction

Routers basically are meant to provide or guide a suitable pathway for the transmission packets or signals from the source to the destination path. In case of a complex or integrated chip networks where an enormous operations are being done with various components, routing of signals from n transmitters to different receivers in the same chip is inevitable. Especially in case of optical integrated circuits and systems (all-optical integrated circuits or photonic integrated circuits, PIC), it needs the strategies for the pathway of optical signals between different nodes so as to establish a complete link in PIC. The efficiency, reliability, and complexity in providing the connectivity and data transfer in the integrated chip will decide the overall performance and efficacy of the entire system. All-optical integrated circuits designed with photonic crystals can be a breakthrough to overcome these factors and to provide a better routing in the integrated environment.

Photonic crystals! Why?

The technological growth in each and every sector of the globe, developing a consistent, highly reliable, efficient communication medium and the overall system is the pivotal need of the hour. All forms of communication intents to afford required key features such as high bandwidth, reliability, security, long epoch, cost effective etc., Amongst them, Optical fiber communication is envisaging a rapid and promising growth in its development for providing the above mentioned features and to replace the existing electronic integrated circuits. Communication by means of optical signals has substantiated itself over long distance with high data rate. This has also evidenced the effectiveness in both telecom and non-telecom applications, in medical field, defense sector, automobile industries and so on. Even though the state of fiber optic communication has reached a state of saturation in terms of least possible attenuation (~0.2db/km) and the fabrication of refined fiber optic cables to achieve the highest data rates, researches are in the process of realizing fully optical integrated circuits similar to electronic integrated circuits. Optical materials have been recommended to be used as a substrate for PIC, and this is been termed as Photonic Crystals (PhC), which dates back to 18th century. Moreover, various components and devices have been designed and simulated by optical researchers and scientist using photonic crystals and PhC based routers has also been validated to communicate with the components.

From When?

In 1887, Lord Rayleigh gave his impression on photonic crystals that, "In a Periodic Structure, the waves that would come along would experience the periodicity and would experience reflection and hence be totally reflected [1]". Followed by this, Wood presented the scientific article on the electromagnetic waves behavior on periodic structures [2]. After about 100 years, Yablonovitch and John independently validated about photonic crystals in their woks. Yablonovitch demonstrated 3D PhC for inhibiting spontaneous emission in LASERS with refractive index (RI) contrast of 1.9 [3]. Localization of photons in disordered periodic dielectric materials was verified by Sajeev John [4]. Following this, Photonic crystals attracted a lot of magazines and editors. Sir John Maddox, Editor-in-chief of NATURE journal in December 1990 has projected the qualities of PhC titled "Photonic band-gap bit the dust" [5]. SCIENCE magazine in December 1999 has predicted Photonic crystals as the future of optical communication in its article entitled "Breakthrough of the year".

"Seeing the Future in Photonic Crystals", The Industrial Physicist, an article by the American Institute of Physics, December 2001–January 2002 [6], highlights the conceptions of scientist and professors in Photonics, about the possible advancement and development which can be achieved by PhC in the field of Optical Communication systems and networking's.

What and How?
In simple terms we describe PhC as a periodic arrangement of different medium which vary in their refractive index (RI – Refractive index of a medium or material can be defined as the ratio of velocity of light in vacuum to velocity in the medium). And these medium or materials exhibits strong interaction with optical signal. This is being analogous to the semiconductor materials in electronics. Hence PhC is also called as "Semiconductor crystals for light". PhC exhibits periodicity due to the periodic arrangement of different RI values, will influence the propagation of optical waves inside them similar to the semiconductor materials affecting the flow of electronics in them. If we explore the properties and characteristics of PhC, we can note down various similar properties with respect to semiconductors in electronics starting with the periodicity, doping (defects as in case of PhC) ,control on the electrons and photons and energy bandgap. Photonic bandgap (PBG) is the term describing the range of frequencies or wavelength, where the light transmission becomes absolutely null. Defects are the process of breaking the periodicity in PhC to control and manipulate the propagation of photons inside the crystal as in the case of doping in semiconductors. These defects break or change the PBG structure of the design. These features of PhC enable itself to be a candidate to incorporate different designs of various optical devices and components for potential application in optical communication.

8.2 Photonic Crystals

Considering the arrangement of RI in the crystal, PhC is classified into three categories. That is based on the number of directions in which the periodicity of RI is observed, PhC are termed as one dimensional, two dimensional and three dimensional. Figure 8.1 shows the classification of PhC as 1D, 2D and 3D based on their periodicity exhibited in one, two and three directions. Different colors are used in the figure to describe the difference in their dielectric constant of the material.

8.2.1 1D Photonic Crystals

In 1D PhCs, the periodicity in the variation of dielectric constant or refractive index occurs in one direction, either X, Y or Z direction. Two or more

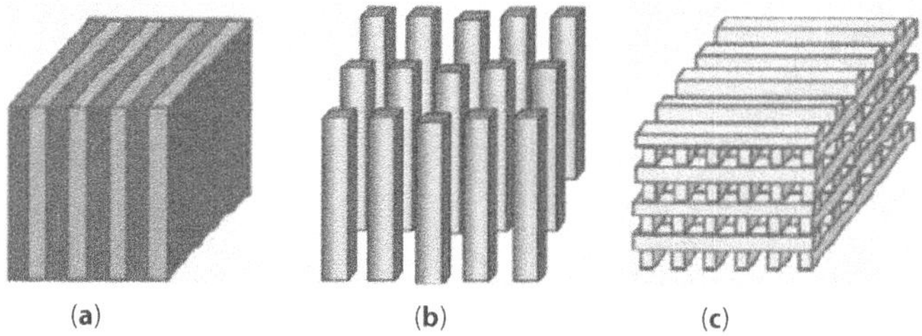

Figure 8.1 Classification of 1D, 2D, and 3D PhC [7].

materials which differ in their dielectric constant or refractive index values can be arranged to form a 1D PhC. Bragg mirror with alternating layers can be considered as a simplest example. In our daily life we can see these types of 1D arrangement such as staking up plates, arranging pile of files and so on. 1D PhC is used in optical filters [8–10], LASERS [11], waveguides [12] and in high efficiency mirrors [13, 14].

8.2.2 2D Photonic Crystals

In case of 2D PhC, different optical materials (obviously with different RI) are organized in a periodic fashion such that it reveals periodicity in two directions, as shown in Figure 8.1. The two materials used in 2D PhC are termed as background materials and rods. Usually the rods are placed in the background material to make 2D PhC. The background can be a low RI material and the rods can be high RI or vice versa. The arrangement depends on the application and the component to be designed and fabricated. 2DPhChavingdielectricrodsembedded in the background of air or air holes is a basic PhC design. Also, it is possible to have two or more than two optical materials to make a 2D structure. There are two main parameters that are associated with designing 2D PhC, they are lattice constant 'a' which is the distance between the radius of two adjacent rods and the thickness of the rods 'r' embedded. Depending on the arrangement of rods and background based on RI, the photonic bandgap of the structure will be for either TM or TE mode. High RI dielectric rods in air gives PBG for TM mode and PBG for TE mode can be obtained in the vice versa structure [15]. Various optical components such as filters [16–18], splitters, de-multipliers [19], logical gates [20], digital to analog converters, analog to digital converters, routers [21, 22], circulators [23, 24], etc. have been designed and simulated using 2D PhC.

Figure 8.2 3D photonic crystals in nature.

8.2.3 3D Photonic Crystals

A PhC structure with periodicity exhibited in all three directions is termed as 3D structure. 3D PhC structure possesses a PBG which behaves as an omni-directional reflector. A lot of 3D PhC is available in nature (Figure 8.2). The fluorescent colors in peacock feathers, vibrant colors in butterfly feathers are some of the examples of 3D PhC occurring in nature. Realization of a complete 3D PBG in laboratories are not done so far. This is because of the fact that fabrication of 3D PhC is very difficult and requires more complex geometry design. Moreover the state of art of laboratory instruments and technical facilities has far away beyond our reach. Hence designs of PC based structures are realized in 2D rather than in 3D PhC. But few smaller structures have been demonstrated and reported in 3D PC at telecommunication frequencies [25], whereas 3D PhC optical components remain challenging to realize.

8.2.4 Photonic Bandgap

Photonic bandgap in photonic crystals is an important characteristic that decides the operating wavelength of the designed structure. With the premeditated factors such as the number of materials to be utilized in the design, their RI values, lattice constant 'a' and radius of the rods 'r', the value of PBG remains unique for each design. PBG can be obtained by using gap map technique. Gap map can be defined as the plot of forbidden bandgap of the crystal arrangement with respect to one or more parameters [26]. The structure is designed such a way that the materials selected for the 2D photonic crystals and the arrangement of the crystal lattice are done as to have the photonic bandgap of the crystal structure to match with the operating wavelength of the application. Mathematically, Maxwell's equation for Electromagnetic wave propagation is used to obtain PBG. PhC dimensions being smaller than wavelength of light, wave optics is considered for mathematical calculations. The periodic arrangement of different dielectric media in PhC is considered to be non-conductive and non-magnetic [27]. Therefore relative permeability and current density in Maxwell's equation can be considered as zero. Starting with the initial conditions in Maxwell's equation and after performing basic mathematical and vector operations, combining s_0 and μ_0 to obtain the speed of light in vacuum C,

$$c = \frac{1}{\sqrt{\varepsilon_o \mu_o}} \tag{8.1}$$

and the overall master equation of photonic crystal is [26]

$$\nabla x\left(\frac{1}{\varepsilon(r)}\nabla x H(r)\right) = \left(\frac{\omega}{c}\right)^2 H(r) \tag{8.2}$$

where,

ε_0 = free space permittivity = 8.854 × 10^{-12} farads per meter,
μ_0 = free space permeability = 1.257 × 10^{-6} henry per meter,
c = light speed in vacuum = 3 × 10^8 m/s,
$\varepsilon(r)$ = dielectric function and
$H(r)$ = magnetic field vector.

This equation 8.2 represents the master equation of photonic crystals. It represents an Eigen value problem which can be solved for Eigen-frequencies. Similar to electrons in a periodic potential well, where the solution to Schrödinger wave equation is given by Bloch's theorem, the field inside the Photonic crystals will obey Bloch's theorem [28]. Bloch's theorem can also be used to analyze the same. It states that "The wave function of an electron within a perfectly periodic potential may be written as amplitude term multiplied by a plane wave" [29, 30] as given in equation 8.3.

$$\bar{F}(\bar{r}) = \bar{A}(\bar{r})e^{i\beta.\bar{r}} \tag{8.3}$$

where

$\bar{F}(\bar{r})$ = overall field,
$\bar{A}(\bar{r})$ = amplitude envelope with the same periodicity & symmetry of the structure,
$e^{i\beta.\bar{r}}$ = Plane wave.

The field inside the periodic structure takes up symmetry and periodicity of the structure. A portion of the electromagnetic waves traveling in the PhC will experience a delay after traveling through the dielectric object. The amplitude part of Bloch theorem that takes on the symmetry and periodicity of the structure [31]. Using these equations, the Eigen-value problem is solved in the frequency domain using Plane Wave Expansion (PWE) method [32, 33]. The solution of the Eigen-value problem is plotted as Band structure, also termed as Band diagram by PWE method. Band structure describes the electromagnetic waves propagating in the medium based on properties of the medium [26]. It also conveys the Electromagnetic properties of the periodic structure [27].

8.2.5 Applications

Photonic crystals find its application mainly in the areas of optical telecommunications. The PBG property is an important key factor in paving the way for PhC in every sector of the industry. PhC becomes a suitable candidate for designing both active and passive components for communication. This also possesses an advantage of designing all optical integrated circuits for modern world. In addition PhC finds its potential applications in (as in Figure 8.3).

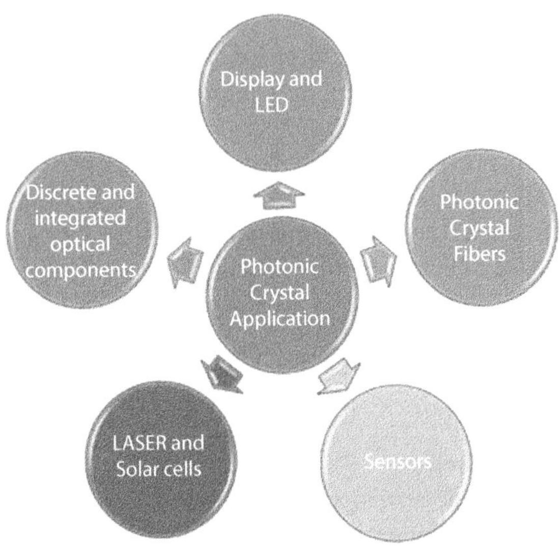

Figure 8.3 Photonic crystal applications [34].

- Displays and LED's
- Photonic Crystal Fibers
- Sensors
- LASER and Solar cells
- Discrete and Integrated optical components

8.3 Routers

Deciding over the pathway for propagation or transmission of signals between the different components and devices in integrated circuits is the crucial need of any circuit. Moreover this routing should be in such a way as to have an efficient, reliable, transmission of data without much loss and attenuation. This also requires having a shortest path that is possibly rapidest time for transmission, between any pair of sender and receiver. Usually router does this crucial [35–38] job and we do have routing techniques for the same. The characteristics of router also demands low insertion losses and crosstalk in case of multiprocessors chip [39]. Routing in PIC is a type of passive routing, where routing of the signals are based on its wavelength. In this chapter, the different designs of optical routers were discussed. Usually this passive routing is been achieved by the use of micro-ring resonators.

8.4 Micro Ring Resonators

Silicon photonics has been a promising suitable platform for photonic integrated circuits, and the main reason for this success is the ring resonators. They play an important role in all possible passive structures. Also in photonic crystals based optical components, most of the devised structures make use of photonic crystal ring resonators. The basic structure of any ring resonator has a waveguide which is looped back to itself and a waveguide. An optical ring resonator has an optical path or optical waveguide which is connected back to it, such that the resonance occurs when the optical path length is exactly or integral multiple of the wavelength. It also has an optical waveguide along with the ring resonator which acts as the input and output ports. There are also ring resonators with some complicated structures such as double ring resonator, circular ring, ultra-small race track, large folded- spiral ring etc., [40]. A coupling mechanism is involved to couple the signal to and from the loop.

There will be constructive interference of signal when the round trip phase shift is equals an integer times of 2π. This is termed as resonance in the cavity. The resonator shape can be circular (generally termed as ring resonators) or the shape can be slightly elongated along one direction as in case of ellipse, such resonators are called racetrack resonators. Also there are two basic configuration of ring resonators based on the structural arrangement. The simplest form of structure constructed with a single waveguide and a loop. This is all-pass ring resonator. In the other form there are two waveguides and a loop which acts as add-drop resonator. These structures are shown in Figures 8.4(a) and 8.4(b).

In case of all pass ring resonator when the wavelength of the input matches the resonant wavelength of the ring resonator then it is completely passed to the output port hence it is called as all-pass resonator. In the second type it is also possible to add or drop the signals based on the requirements. The mathematical equations of these structures have been well described in various articles [40].

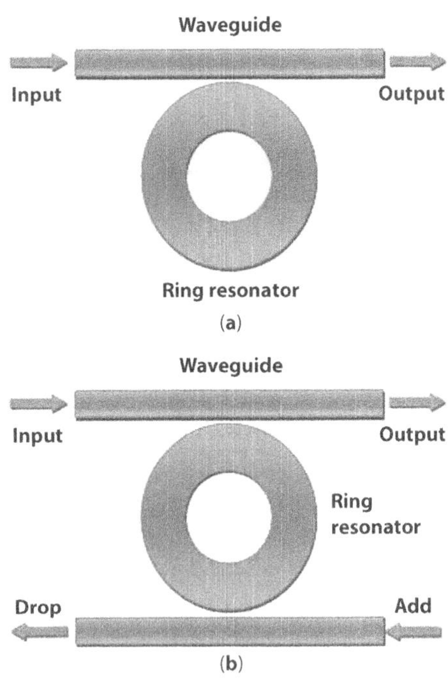

Figure 8.4 (a) All pass ring resonator. (b) Add-drop ring resonator.

8.5 Optical Routers

8.5.1 Routers Based on PCRR

Routers based on PCRR have been designed and demonstrated by different researchers. Starting with the basic 1 x 2 and 2 x 2 router configurations, the PCRR based routers have been explained for higher router configurations such as 4 x 4, and 6 x 6. The design of PCRR based router involves a series of steps starting with the application in which it's going to be employed and the wavelength spectrum of the design. The flow of the design procedure is given in Figure 8.5 below.

End Application of the Router
Initially the application for which the router is to be operated is defined by parameters such as the wavelength window or the range, size of the routing element required, size of the component and so on. Various research and data have been analyzed and collected to understand the basic application in which the router is to be employed. Considering the PC routers, the end application is in photonic integrated or all-optical integrated circuits where the size of the elements should be ultra-compact, in the order of nm. These factors form the problem statement of the design. Based on these parameters, router is designed and the requirement is achieved.

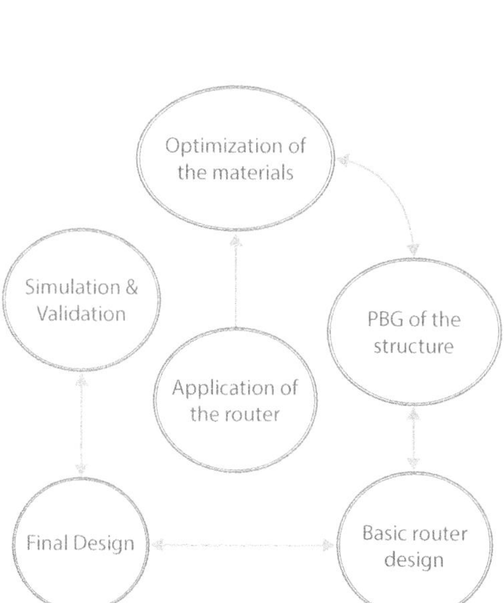

Figure 8.5 Flow of steps involved in router design.

Optimization

Based on the requisite conditions the initial material selection and lattice arrangement of the router is done. As 2D photonic crystal involves two dielectric materials, these are selected based on the refractive index difference. These materials can be arranged either in triangular lattice or square lattice. Two types of arrangement is also possible such as high RI material in the low RI background or vice-versa. These types are shown in the Figures 8.6(a) and 8.6(b). Square lattice gives an narrow PBG [41] of the structure and is used to have an symmetrical structure of the design.

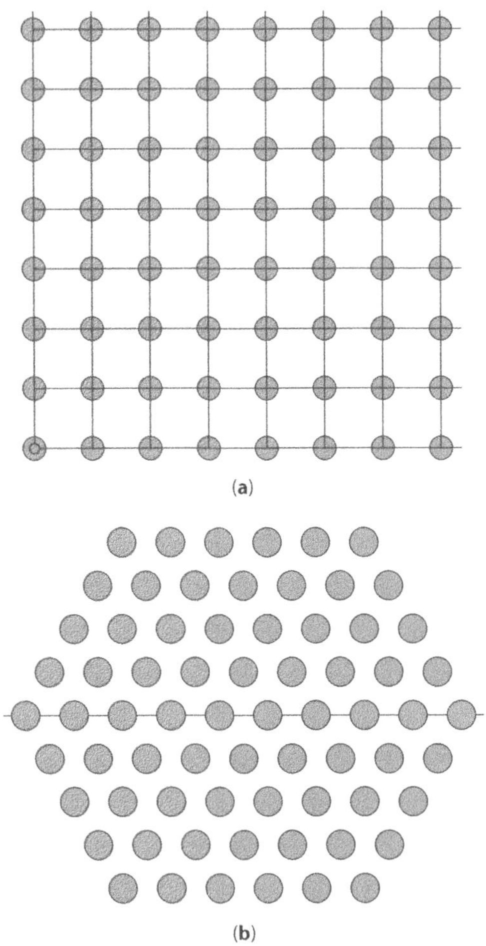

Figure 8.6 (a) Square lattice arrangement of PhC. (b) Triangular or hexagonal lattice arrangement of PhC.

The parameter of the PC such as the radius of the rods (r) to be placed in the background material and the lattice constant (a), the distance between the radius of two adjacent rods are also optimized to have the necessitate PBG region. This can be done by either trial and error method or can be obtained from the gap map of the structure from the simulation.

Photonic Bandgap or Forbidden Bandgap, Design, and Simulation
After this, the basic ideal layout formed from the optimized materials, lattice constant and radius of the rods is been simulated to check for the PBG range. Further process involves in designing the smallest basic router which forms the building blocks for the higher order routers. Along with it the number of transmitter and receiver ports was defined along with their positioning. Routing of each and every transmitter to the receiver with their pathways from input to output ports is also to be defined in the design procedure, such that the path between one pair doesn't affect the pathways of other links. The assigned wavelength between the input and output pairs and the ring resonator structure for those resonating wavelengths were also processed and validated separately before handling the final structure of the router. After this is been achieved the final simulation and validation of the parameters are carried out using FDTD software and the designs are accredited.

8.5.2 N x N Router Structures

In this section, PCRR based routers that have been designed and validated using FDTD software have been discussed. We will start with the basic 1 x 2 and 2 x 2 routers for better understanding. The schematic of both basic routers are shown in Figures 8.7(a) and 8.7(b). In 1 x 2 router there is a ring resonator which resonates at resonant wavelength λ_1. When input signal appear at input port the signal with the resonant wavelength will pass through the ring resonator and appears at the output port. Whereas the other signals will be given out at the straight output port. In 2 x 2 router, there will be two input and two intended output ports. It has two ring resonators resonating at different frequencies or wavelength. Based on the resonant condition the input wavelength satisfying the condition will be given at the intended output port. These two structures form the basis in designing N x N routers. The ring resonator is designed such that the resonating wavelength of the structure matches with the signal wavelength that has to pass through the resonator.

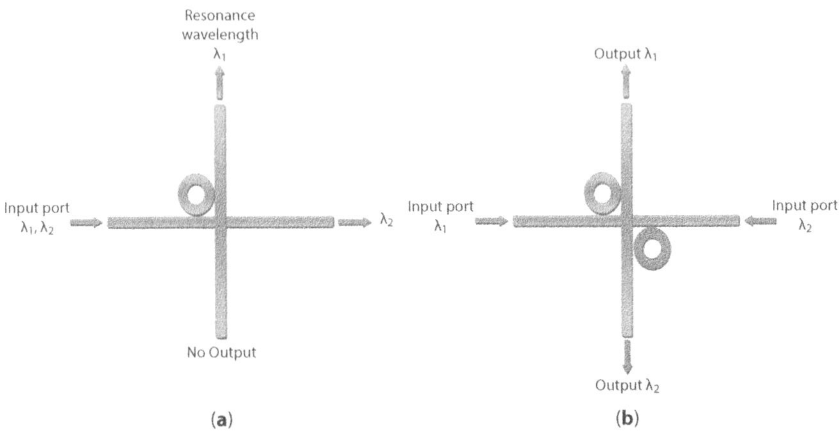

Figure 8.7 (a) 1 x 2 router schematic. (b) 2 x 2 router schematic.

8.5.2.1 3 x 3 Router

Two different structures of 3 x 3 have been reported using 1 x 2 routers. Both the structures use PCRR concept, aligning the resonant wavelength of the PCRR to the wavelength assigned to the link. 2D photonic crystal with silicon in air background is devised. The PBG of the photonic crystal is obtained in the optical III window wavelength of operation. The topology of router is shown in Figures 8.8(a) and 8.8(b). The input and output ports are indicted by the 'in' (black) and 'out' (blue) arrows. Difference in topology arises with respect to the number of RR used. In Figure 8.8(a) there are four RR with two different configurations. R1 and R3 was tuned to resonant for λ_1 and R2 and R4 for λ_2. Three wavelengths are utilized for the router where two wavelengths are designed and routed using the RR whereas the third link uses a wavelength from the PBG range. In the next topology there is a trade-off between the wavelength utilization and the complexity of the structure. Wavelength resource is reserved at the cost of increased RR and increased structure size and fabrication complexity. It uses two wavelength to route between 3 input and output ports with 6 RR. R1, R3 and R5 are tuned to one wavelength and R2, R4 and R6 for the second wavelength. The maximum crosstalk and insertion loss obtained for this structure in simulation is about −16.2973 dB and 1.8709 dB for router in Figure 8.8(a), and −13.5485dB and 2.5026dB for router in Figure 8.8(b). Both the routers possess propagation delay in terms of femtoseconds.

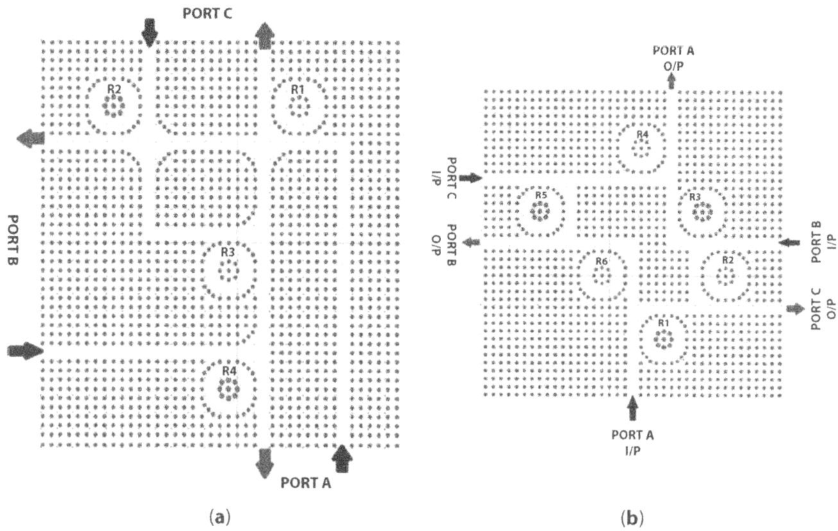

Figure 8.8 (a) 3 x 3 router with 4 PCRR [39]. (b) 3 x 3 router with 6 PCRR [39].

8.5.2.2 4 x 4 Router

Studies have been reported on various 4 x 4 routers with different designs. These routers differ in the basic elements which are used as the building block. All the routers have PC ring resonator for tuning the desired wavelength into the pathway for routing. Silicon rod in air background is applied for the devised structure. The lattice constant and radius of the rods are optimized to have the PBG of the structure around 1550 nm. This falls in the III window wavelength of operations. In 4 x 4 router reported by Calo *et al.* 2013, eight 1 x 2 routers were used which follows the broadband wavelength crossing. The broadband crossing is analyzed against the simple waveguide crossing and is found that these broadband crossing possess low crosstalk and hence makes its geometry suitable for practical use in photonic networks [36]. The Broadband crossing and simple waveguide crossing is shown in Figure 8.9. The structure of three different 4 x 4 router are shown in the Figures 8.10–8.12. The 4 x 4 wavelength router represented in Figure 8.10, the PCRR denoted by R1 and R2 have the resonant wavelength of λ_1 and λ_2. As shown in the Figure 8.9, the input on port W gets diverted to the pathway and to one of the output ports depending on the wavelength of the input signal. Signal with λ_1 will be resonated by the middle RR (R1) in the third row and will be given out at output port E. Wavelength λ_2 at W will be resonated by R2 and will be diverted to output

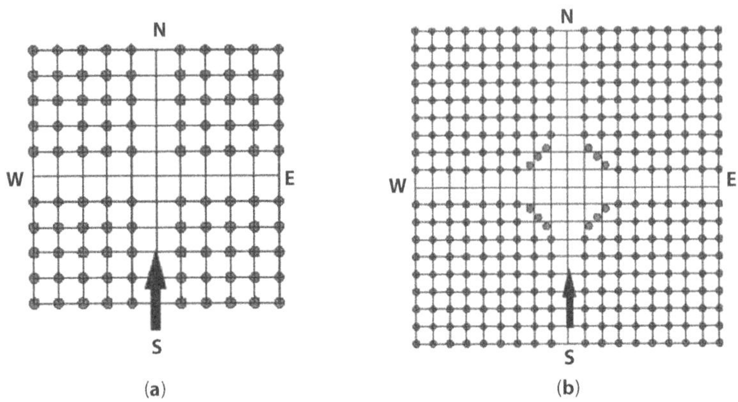

Figure 8.9 (a) Simple PhC waveguide crossing. (b) Broadband waveguide crossing.

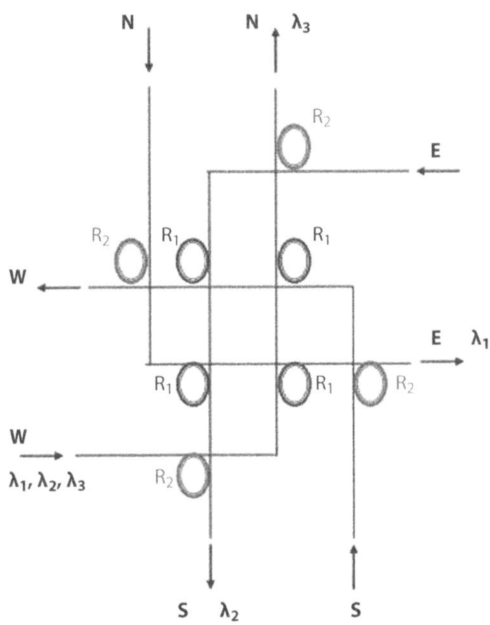

Figure 8.10 4 x 4 router [36].

port S. The third wavelength λ_3 will be propagated to the port N. This third wavelength does not require any RR for transmission.

In the other work reported by G. Calo [35], they have demonstrated the 4 x 4 wavelength router with 2 x 2 routers as basic building blocks of the structure (Figure 8.11). It consists of four 2 x 2 routers assembled near each

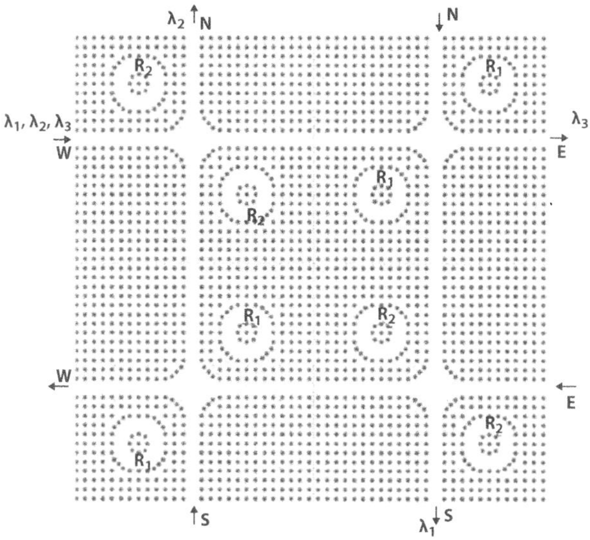

Figure 8.11 4 x 4 router [35].

input ports E, W, N and S. Input and output ports are described using the arrow headers. The wavelength assigned between each pair of the link can be obtained in detail from the research paper of Calo [35]. As mentioned before R1 resonators are tuned to a wavelength λ_1 and R2 to a resonant wavelength λ_2. Wavelength λ_3 does not require any RR and it can be any wavelength that falls in the PBG range of the 2D photonic crystal structure. Broadband waveguide crossing is structured and this router is symmetry about the middle horizontal and vertical axis. This design is also devised to operate around 1550 nm (III window wavelength). The propagation of input signals is same as that of the former router, which depends on the resonant wavelength of the RR which the signal encounters in its path.

Another group of researchers [37] have combined both 1 x 2 and 2 x 2 wavelength routers as the building blocks and have proposed a 4 x 4 wavelength router. All the problem definition such as the operating wavelength, rod material and the background air medium, the PBG range being the same, this team has designed a new structure. This structure is been simulated in FDTD simulation method and the parameters were highlighted. They have also analyzed the same structure for different optical materials such as germanium (Ge) of RI 4.0029, gallium arsenide of RI 3.3219 and indium phosphate (InP) of RI 3.457 for the router design. The results have shown that Ge pillared router has minimum crosstalk and that designed with InP has minimum insertion loss for the same structure. The router

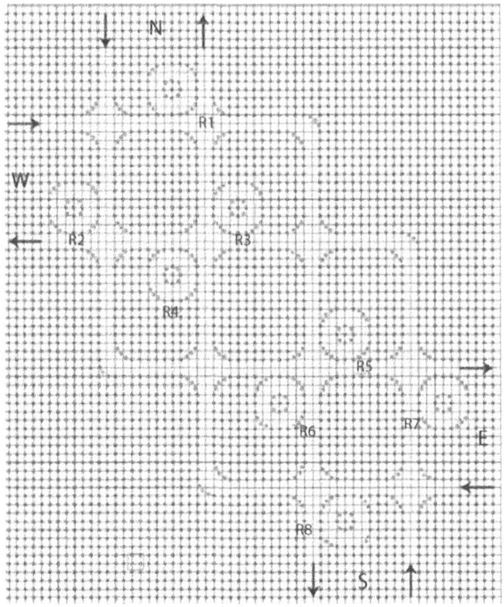

Figure 8.12 4 x 4 router [37].

is shown in Figure 8.12. It has four input and output ports as mentioned in the structure along with broadband crossing, four 1 x 2 routers and two 2 x 2 routers which contributes to a total of eight RR. The RRs are named from R1 to R8 and the resonant wavelength of RR are designed as per the requirement for transmission of the input signal. Performance characteristics of all the routers were tabulated in Table 8.1.

8.5.2.3 6 x 6 Router

A higher order 6 x 6 router is designed with 1 x 2 and 2 x 2 routers. 6 x 6 router topology is represented in Figure 8.13. It has two pairs of 1 x 2 routers and five pairs of 2 x 2 routers as basic blocks. Total of 24 RRs is utilized to route between six input output ports. Five wavelengths are used for routing out of which four wavelengths require RR structures to route it and one wavelength does not require RR. Different color of RR in the topology denotes that they are designed to have same resonant wavelength. Four groups of RR are grouped together and each has 4 RR. The structure is simulated using finite difference time domain software and the parameters were discussed. Although this structural design seems simple, the

Table 8.1 Photonic crystal based routers and their comparison of performance metrics.

Authors	N x N	Routing element	Resonated λ (μm)	Maximum insertion loss (dB)	Maximum crosstalk (dB)	Propagation delay (picosecond)	Area (μm^2)
Sridarshini *et al.* 2018	3 x 3 (topology 1)	Ring Resonators	1.50	0.9151	-16.2973	0.0021	524.88
			1.52	0.7058	-20.4096	0.0005	
			1.54	1.8709	-16.8883	0.00214	
	3 x 3 (topology 2)		1.50	2.5026	-13.5485	0.00101	428.49
			1.52	1.9540	-19.3254	0.00086	
G. Calo *et al.* 2013 (1 x 2 routers)	4 x 4	Ring Resonators	1.488	0.6	-15.4	0.2	900
			1.506	0.1	-13.9	0.2	
			1.540	0.3	-15.7	0.1	
G. Calo *et al.* 2014 (2 x 2 routers)	4 x 4	Ring Resonators	1.5000	0.160	-20.77	--	576
			1.5070	0.167	-22.37	--	
			1.5137	0.409	-20.13	--	
	4 x 4		1.540	0.36	-14.24	--	
			1.546	1.40	-10	--	
			1.550	0.47	-17.52	--	

(*Continued*)

Table 8.1 Photonic crystal based routers and their comparison of performance metrics. (*Continued*)

Authors	**N x N**	**Routing element**	**Resonated λ (μm)**	**Maximum insertion loss (dB)**	**Maximum crosstalk (dB)**	**Propagation delay (picosecond)**	**Area (μm²)**
Sathyadevaki *et al.* 2018	4×4 router	Ring Resonators	1518	0.98436	-20.8124	0.5	1216
			1550	0.35402	-22.1843	4.6	
	Increased radii 4×4 router		1529	2.41	-33.4683	4.6	1216
			1520	1.734322	-21.3637	2.02	
			1559	5.10545	-23.5218	0.6	
Jin Lu *et al.* 2015	4 x 4	Point defect micro-cavities	1.6408	0.087	-27.23	--	---
			1.6002	0.0098	-38.90	--	
			1.6280	1.085	-28.96	--	
Sridarshini *et al.* 2021	6 x 6	Ring Resonators	1.48	1.2569	-13.5044	415.0	4900
			1.49	0.6550	-12.3553	751.3	
			1.5	1.6115	-14.7712	641.1	
			1.51	1.2569	-13.2951	731.0	
			1.52	0.4915	-15.6644	810.0	

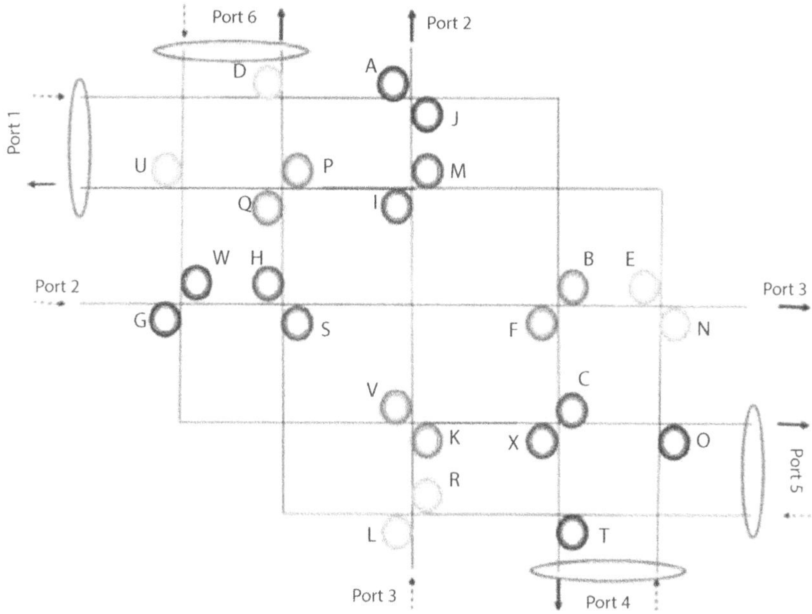

Figure 8.13 6 x 6 router [42].

complexity involved in real time designing and fabrication of 24 PC ring resonators in an ultra-compact size of 4900 square micro meter will be quite tedious and difficult process. The parameters of the design are compared with other order of N x N routers and are tabulated in Table 8.1.

8.5.3 Routers Based on PC Line Defect

Jin Lu in 2015 has demonstrated Photonic crystal based wavelength router using point defect micro-cavities instead of ring resonators. The 4 x 4 wavelength router has been designed and validated for its low cross-talk. This structure is designed with 2D PhC with silicon rods of dielectric constant 12.0409 in the background of air with dielectric constant 1. The parameters of the rods are selected as lattice constant as 540nm and radius of the rod as 94.5nm. Square lattice is used for the arrangement. With all these factors the PBG of the 2D PhC was in the normalized frequency range of $0.30178(c/a) < f < 0.44735(c/a)$ only for the transverse magnetic (TM) mode where, f is the normalized frequency, and c is the light velocity in free space. For the waveguide crossing, the rods are shifted 0.25a from their original lattice position along the horizontal and vertical directions [36]. The structure of the waveguide crossing and 1 x 2 router is shown in Figures 8.14(a) and 8.14(b). Two micro-cavities are designed in the PhC crossing. A channel drop micro-cavity is arranged with different values of the rod radius as shown in Figure 8.14(b). With this, channel add micro-cavity can be obtained by rotating the channel drop micro-cavity by 90° in an anticlockwise direction. These two micro cavities are connected with the 90° waveguide bend to obtain 1 x 2 router.

The routing mechanism in this case is the same as that of PhC channel drop filter. When the input wavelength matches the resonant wavelength of the micro-cavities, the signal (at input port say W here) is completed

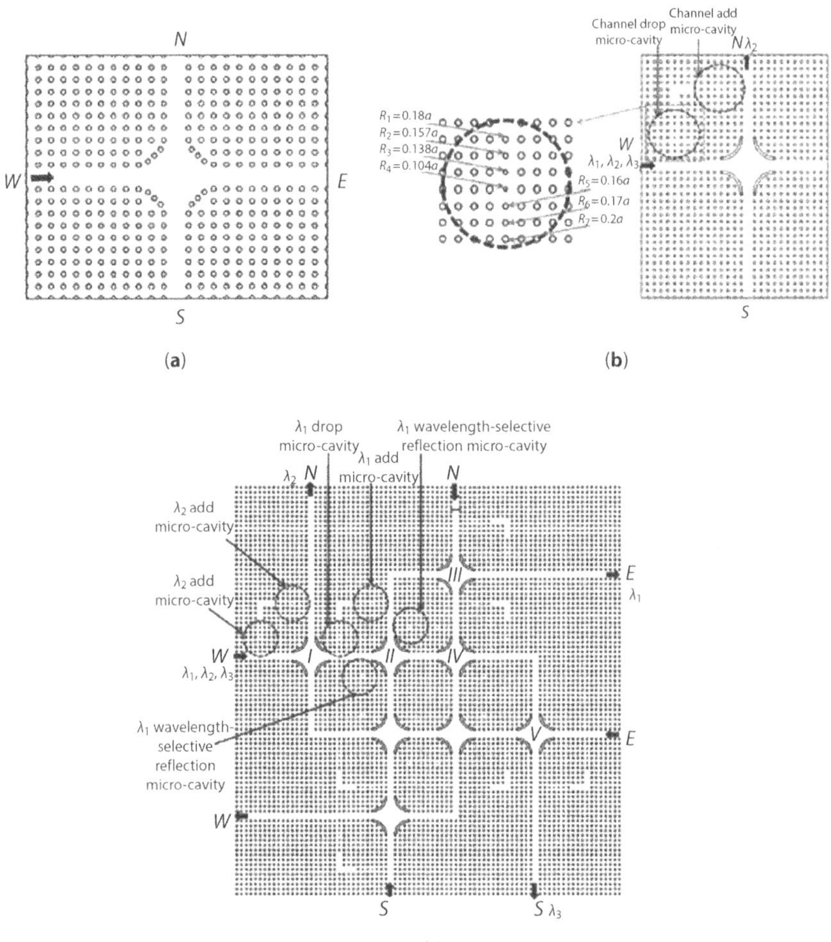

Figure 8.14 (a) Waveguide crossing [38]. (b) 1 x 2 router using micro-cavities [38]. (c) 4 x 4 wavelength router using photonic crystal point defect micro-cavities [38].

reflected by the crossing and is transferred to the 90° waveguide bend. This is done with the help of resonantly excited channel drop micro-cavity which is known as the drop process in the photonic crystal fiber.

At the end of the waveguide bend with the resonant excitation of the channel add micro-cavity, the light input signal is completely routed to the output port (port N). With this as the basic structure the authors have specified a 4 x 4 wavelength router which is displayed in Figure 8.14(c). In addition to add and drop channel micro-cavities, this structure possess two additional wavelength-selective reflection cavities which are side-coupled to the horizontal and vertical PhC waveguides which transmits the signal through the crossing with low crosstalk. Considering Figure 8.14(c) there are pairs of input and output ports in each direction with names E, W, N and S. 1 x 2 routers forms the basic building blocks of this structure. Low value of crosstalk has been reported than the PCRR based router by G. Calo [36].

8.6 Summary

This chapter has dealt with the various designs of photonic crystal based routers for photonic integrated circuits. These routers were either structured with ring resonators or point defect micro-cavities. Each design has its own merits and demerits with regards to the parameters. A comparison of the characteristics of the PC based routers is been tabulated for better understanding in Table 8.1. Routers, however designed with ring resonators or micro-cavities, involve realization complexity. Photonic crystal based components will be a lane to realize PIC. Where the ultimate aim of PIC is to design and develop each and every function as high functional miniaturized optical component on a single substrate. Even though a lot of research work and design of components is been done and simulated using simulations, the state-of-the-art of realizing these structures are still far behind its physical hardware realization of the complex structures. The other reliable factors after the fabrication of these structures, such as mechanical stability, stress, reliability, temperature dependence can be answered out only after realizing and testing the component in the real time environment. In spite of all these, the routers out of photonic crystals will be a suitable and desirable candidate for future all optical integrated circuits. The favorable characteristics such as loss crosstalk, insertion loss, propagation delay, effective utilization of wavelength, along with its ultra-compact size will make the routing in PIC to be in an efficient way.

References

1. Rayleigh, L., XVII. On the maintenance of vibrations by forces of double frequency, and on the propagation of waves through a medium endowed with a periodic structure. *Lond. Edinb. Dublin Philos. Mag. J. Sci.*, 24, 147, 145–159, 1887.
2. Wood, R., XLII. On a remarkable case of uneven distribution of light in a diffraction grating spectrum. *Lond. Edinb. Dublin Philos. Mag. J. Sci.*, 4, 21, 396–402, 1902.
3. Yablonovitch, E., Inhibited spontaneous emission in solid-state physics and electronics. *Phys. Rev. Lett.*, 58, 20, 2059–2062, 1987.
4. John, S., Strong localization of photons in certain disordered dielectric superlattices. *Phys. Rev. Lett.*, 58, 23, 2486–2489, 1987.
5. Maddox, J., Photonic band-gaps bite the dust. *Nature*, 348, 481, 6 December 1990.
6. Ouellette, J., Seeing the future in photonic crystals. *The Industrial Physicist*, pp. 14–17, January 2002.
7. Berman, O.L., Boyko, V., Ya. Kezerashvilli, R., Electromagnetic wave propagation in two-dimensional photonic crystals, in: *Wave Propagation*, Intech Open, United Kingdom, 2011.
8. Chen, M.-C., Luan, P.-G., Lee, C.-T., Novel design of organic one-dimensional photonic crystal filter. *Proceedings of CLEO/Pacific Rim 2003. The 5th Pacific Rim Conference on Lasers and Electro-Optics (IEEE Cat. No. 03TH8671)*, p. 605, 2003.
9. Nemec, H., Duvillaret, L., Garet, F., Kužel, P., Xavier, P., Richard, J., Rauly, D., Thermally tunable filter for terahertz range based on one-dimensional photonic crystal with a defect. *J. Appl. Phys.*, 96, 8, 4072–4075, 2004.
10. Lee, H.-Y., Cho, S.-J., Nam, G.-Y., Lee, W.-H., Baba, T., Makino, H., Cho, M.-W., Yao, T.F., Multiple-wavelength- transmission filters based on Si-SiO2 one-dimensional photonic crystals. *J. Appl. Phys.*, 97, 103111, 1–4, 2005.
11. Lu, T.-W., Chiu, L.-H., Lin, P.-T., Lee, P.-T., One-dimensional photonic crystal nano beam lasers on a flexible substrate. *J. Appl. Phys.*, 99, 071101, 1–3, 2011.
12. Taniyama, H., Waveguide structures using one-dimensional photonic crystal. *J. Appl. Phys.*, 91, 6, 3511–3515, 2002.
13. Bruyant, A., Lérondel., G., Reece, P.J., Gal., M., All-silicon omni directional mirrors based on one-dimensional photonic crystals. *Appl. Phys. Lett.*, 82, 19, 3227–3229, 2003.
14. Tecimer, M. and Holldack, K., Tunable terahertz-mirror and multi-channel terahertz-filter based on one-dimensional photonic crystals containing semiconductors. *J. Appl. Phys.*, 111, 073111, 1, 2012.
15. Passaro, V.M.N., La Notte, M., Troia, B., Passaquindici, L., De Leonardis, F., Giannoccaro, G., Photonic structures based on slot waveguides for nanosensors: State of the art and future developments. *IJRRAS*, 11, 3, 1–27, 2013.

16. Qiang, Z., Zhou, W., Soref, R.A., Optical add-drop filters based on photonic crystal ring resonators. *Opt. Express*, 15, 4, 1823–1831, 2007.
17. Sathyadevaki, R., Sivanantha Raja, A., Shanmuga Sundar, D., Photonic crystal-based optical filter: A brief investigation. *Photonic Network Commun.*, 33, 1, 77–84, 2018.
18. Sathyadevaki, R., Sundar, D.S., Raja, A.S., Design of dual ring wavelength filters for WDM applications. *Opt. Commun. Elsevier*, 380, 1, 409–418, 2016, https://doi.org/10.1016/j.optcom.2016.06.045.
19. Mehdizadeh, F. and Soroosh, M., A new proposal for eight channel optical demultiplexer based on photonic crystal resonant cavities. *Photonic Netw. Commun.*, 31, 65–70, 2016, https://doi.org/10.1007/s11107-015-0531-1.
20. Wu, C.J., Liu, C.P., Quyang, Z., Compact and low-power optical logic NOT gate based on photonic crystal waveguides without optical amplifiers and nonlinear materials. *Appl. Opt.*, 5, 5, 680–685, 2012.
21. Sherwood-Droz, N., Optical 4×4 hitless silicon router for optical networks-on-chip (NoC*). Opt. Express*, 16, 20, 15915, 2008.
22. Biberman, A., Lee, B.G., Sherwood-Droz, N., Lipson, M., Bergman, K., Broadband operation of nano photonic router for silicon photonic networks-on-chip. *IEEE Photon. Technol. Lett.*, 22, 12, 926–928, 2010, https://doi.org/10.1109/LPT.2010.2047850.
23. Dmitriev, V., Portela, G., Martins, L., Photonic crystal-based circulators with three and four ports for sub-terahertz region. *Photonic Netw. Commun.*, 33, 303–312, 2017, https://doi.org/10.1007/s11107-016-0641-4.
24. Umamaheswari, C., ShanmugaSundar, D., Sivanantha Raja, A., Exploration of photonic crystal circulator based on gyromagnetic properties and scaling of ferrite materials. *Opt. Commun.* Elsevier, 382, 1, 186–195, 2017, https://doi.org/10.1016/j.optcom.2016.07.065.
25. Jia, B., Buso, D., Li, J., Gu, M., Active three-dimensional photonic crystals with high third-order nonlinearity in telecommunication. *Proceedings of CLEO/Pacific Rim 2009*, Shangai, China, 2009.
26. Joannopoulos, J.D., Johnson, S.G., Winn, J.N., Meade, R.D., *Photonic crystals molding the flow of light*, Princeton and Oxford, New Jersey, 2008.
27. Sukhoivanov, I.A. and Guryev, I.V., Physics and practical modeling, in: *Springer Series in Optical Sciences*, Springer, London, 2009.
28. Yariv, A. and Yeh, P., *Optical waves in crystals*, A Wiley-Interscience Publication, New York, 2002.
29. Neamen, D.A., *Semiconductor physics and devices: Basic principles*, McGraw-Hill Companies, New York, 2002.
30. Ashcroft, N.W. and Mermin, D.N., *Solid state physics*, Harcourt College Publishers, United States of America, 1976.
31. Rumpf, R.C., Engineering the dispersion and anisotropy of periodic electromagnetic structures, in: *Solid State Physics*, vol. 66, pp. 213–300, Elsevier, University of Texas at El Paso, 2015.

32. Florescu, M., Torquato, S., Steinhardt, P.J., Complete bandgaps in two-dimensional photonic quasi crystals. *Phys. Rev. B*, 80, 155112, 1–7, 2009.
33. Gong, Q. and Hu, X., *Photonic crystals: Principles and applications*, Taylor & Francis, Florida, 2013.
34. Allied Market Research, Market report available from: "https"//www.allied marketresearch.com/press-release/photonic-crystaals-market.html" (February 2017), 2017.
35. Calo, G. and Petruzzelli, V., Compact design of photonic crystal ring resonator 2×2 routers as building blocks for photonic networks on chip. *J. Opt. Soc. Am. B*, 31, 3, 517–525, 2014.
36. Calo, G. and Petruzzelli, V., Wavelength routers for optical networks-on-chip using optimized photonic crystal ring resonators. *IEEE Photonics J.*, 5, 3, 1–12, 2013.
37. Sathyadevaki, R., Shanmuga Sundar, D., SivananthaRaja, A., Photonic crystal 4×4 dynamic hitless routers for integrated photonic NoCs. *Photonic Network Commun.*, 36, 1, 82–95, 2018.
38. Lu, J., Ren, H., Guo, S., Wu, Z., Qin, Y., Hu, W.S., Jiang, C., Wavelength routers with low crosstalk using photonic crystal point defect micro-cavities. *Optik*, 127, 6, 3235–3242, 2015.
39. Sridarshini, T. and Indira Gandhi, S., Compact 3 × 3 wavelength routing for photonic integrated circuits. *Photonic Network Commun.*, 36, 1, 68–81, 2018.
40. Bogaerts, W., De Heyn, P., Van Vaerenbergh, T., De Vos, K., Kumar Selvaraja, S., Claes, T., Dumon, P., Bienstman, P., Van Thourhout, D., Baets, R., Silicon microring resonators. *Laser Photonics Rev.*, 6, 1, 47–73, 2012, DOI 10.1002/lpor.201100017.
41. Lu, T.-W., Lin, P.-T., Sio, K.-U., Lee, P.T., Square lattice photonic crystal point-shifted nanocavity with lowest-order whispering gallery mode. *Opt. Express*, 18, 3, 2566–2572, 2010.
42. Sridarshini, T., Dhanabalan, S.S., Sannasi, I.G., Design and analysis of photonic crystal ring resonator based 6 x 6 wavelength router for photonic integrated circuits. *IET Optoelectron.*, 15, 40–47, 2021, https://doi.org/10.1049/ote2.12014.

9

Fiber Optic Communication: Evolution, Technology, Recent Developments, and Future Trends

Dankan G. Veeranna[1]*, M. Nagabushanam[2], Sridhara S. Boraiah[3], Ramesha Muniyappa[4] and Devananda S. Narayanappa[5]

[1]Department of Electronics and Communication Engineering, B.M.S Institute of Technology and Management, Bengaluru, Karnataka, India
[2]Department of Electronics and Communication Engineering, M.S Ramaiah Institute of Technology, Bengaluru, India
[3]Department of Computer science and Engineering, APS College of Engineering, Somanahalli, Kanakapura Road, Bengaluru, India
[4]Department of Electronics and Communication Engineering, GITAM School of Technology, GITAM University, Bengaluru, Karnataka, India
[5]Department of Electronics and Communication Engineering, PES Institute of Technology and Management, Sagar Road, Shimogga, Karnataka, India

Abstract
It is a communication method that, light pulses are used to transmit information from one location to another via an optical fiber, which is also known as fiber optic transmission. Thousands of these optical fibers are arranged in optical cable bundles and used to relay long distance light signals. Wide-scale signal transmission with low delay is the main necessity in today's world. With the use of optical fiber today, high-speed transmission in telecommunications networks is carried out via optical fibers, which have low latency and are the preferred medium. The new fiber that permits 21 times more bandwidth in communications networks is currently available. A major pre-requisite in current applications is wide data transmission (bandwidth) of signal with low delays. Optical fibers, which have modest expectations in terms of huge and incomparable transmission speeds, are the transmission medium of choice in media transmission systems because of their long separation and high information transfer rates. As a starting point, this

**Corresponding author*: dankan.v@bmsit.in

Chandra Singh, Rathishchandra R Gatti, K.V.S.S.S.S. Sairam and Ashish Singh (eds.) Modeling and Optimization of Optical Communication Networks, (163–178) © 2023 Scrivener Publishing LLC

chapter discusses fiber-optic communication systems and their fundamental technologies. It also discusses current developments as well as technological trends for the foreseeable future.

Keywords: Fiber optics, bandwidth, data transfer capacity, telecommunication, transmission systems

9.1 Introduction

When you think about it, it's kind of funny that communication is generally described as the transmission of information from one place to another. As the information is to be sent across long distances, it is generally necessary that a communication system be in place. A common method for transferring information inside a communication system is to modulate or superimpose an electromagnetic wave on top of the information signal to serve as a means of conveying information. Oscillating electromagnetic fields in the radio, microwave, and millimeter-wave ranges have been created for this procedure, utilizing both transmission methods (electromagnetic carrier waves) and reception techniques (electromagnetic waves oscillating at a certain frequency), while communication is often accomplished using electromagnetic carriers which fall inside the optical range of frequencies [1].

Visible optical carrier waves or light have been widely used for communication for decades. For long-distance communication, basic methods, such as signal fires, reflecting mirrors, and newer advances, such as lanterns, have shown to be effective, albeit restricted. More than four years after the creation of the telephone, Bell introduced the first image phone with a rotating picture frame. It was seen as cutting-edge technology. An optical diaphragm was used to modulate sunlight and transmit voice across a distance of 200 m. Even yet, research on optical communication did not cease throughout the first two-thirds of the twentieth century [2, 3]. It was, however, applied mostly to shorter, low-capacity communication connections on mobile platforms. As light sources were few, and light transmission was poor in the atmosphere, these conditions contributed to the issue. Some electromagnetic carriers, such as infrared light, can travel over long distances, but because of their frequencies, they can only communicate a limited amount of information. In principle, as the carrier frequency increases, the transmission bandwidth expands, increasing the system's ability to transport information. In the introduction of higher frequency radio communication, VHF and UHF (i.e. the VHF and UHF radio spectrum) were created, which led to the adoption of millimeter wave and subsequently microwave transmission. See the electromagnetic spectrum shown in Figure 9.1, With regard to this context, it is also noteworthy that optical-frequency communication yields an increase in useable bandwidth by approximately 104 compared to microwave-frequency transmission [4–6]. Additionally, greater carrier frequencies provide superior system performance by enabling more power to be concentrated in the transmitted electromagnetic wave, resulting in a more powerful system.

With the invention of the laser in the early 1960s, the area of optical communication saw a renaissance of sorts. The gadget, which produced a strong coherent light, also included the feature of high-frequency modulation. To make it possible to transmit free space optical signals, the laser's low beam divergence also increased that potential. Due to light transmission limitations in the environment, these devices were often confined to short-distance applications. Though there have been some difficulties with the implementation of certain small free space optical communication connections, they have also been used to construct a television camera connecting to a base truck, and to link buildings several hundred meters apart. Optical communication has also gained attention Using similar methods, spacecraft in outer space have been able to link up.

In spite of the laser being relatively restricted in its free space optical communication use, the laser's development began a huge effort in optical component research to develop dependable optical components capable of transmitting information using a light wave carrier [7, 8]. The fiber attenuation was extremely high when it was first developed (i.e. 1000 dB km−1) and as a result the optical fibers could not be used in lieu of the coaxial cables that they intended to replace (i.e. cables with attenuation of about 5 to 10 dB km−1). It was also necessary to properly junction the fiber cables in order to minimize loss, while simultaneously ensuring the procedure could be done in the field simply and frequently. Regardless, optical fiber

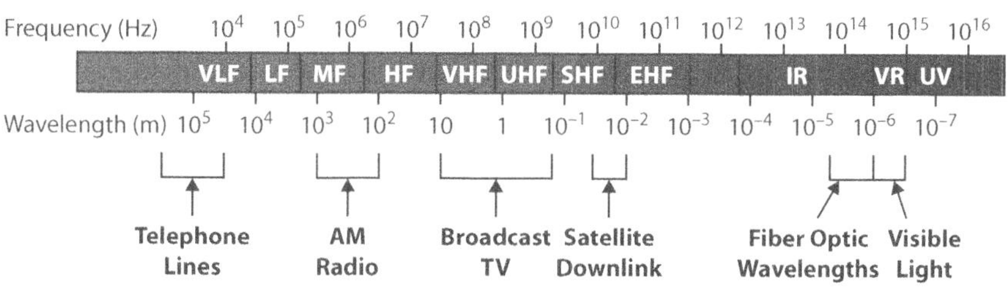

Figure 9.1 The portion of the electromagnetic spectrum that is used for optical fiber communications is shown in this diagram.

losses were decreased to below 5 dB km−1 within a ten-year time period and robust low-loss jointing methods were discovered.

Additionally, attention was given to the optical components that would comprise the optical fiber communication system as it was being developed. All these optical components have to be developed using a new technique because optical frequencies are accompanied by very tiny wavelengths. In order to succeed with the installation of the optical fiber system, injection lasers and light-emitting diodes (LEDs) as well as photodiodes and phototransistors, which were developed and manufactured to fit inside of optical fibers, were used. Initially, semiconductor lasers had limited lives that ranged from few hours to hours, but technological advancements in the laser design allowed lasers with lifetimes of over 1,000 hours by 1973 and over 7,000 hours by 1977. They initially came from gallium arsenide alloy sources (AlGaAs). Since the wavelength range between 1.1 and 1.6 μm had been expanded, the following semiconductor alloys were used to take advantage of the optical fiber's improved performance characteristics across this range [9, 10].

With telecommunication applications in mind, the material and manufacturing technology has been built to handle the expansion of that industry. Even though laser devices are useful for telecommunications, it is beneficial to include a thermoelectric cooler in addition to a monitoring photodiode to simplify current and temperature management. In the telecommunications industry, it should also be noted that this fiber type soon emerged as the primary option in telephone applications after the first field trial demonstration in 1982 [10]. However, at 1.57 μm (the other longer wavelength window), the lowest silica glass fiber losses recorded so far are 0.1484 dB km^{-1}. The higher chromatic dispersion, however, limits the maximum bandwidth that can be realized with traditional single-mode fiber. Notwithstanding the tremendous advancement in fiber technology over the last several years, certain

The areas of interest previously preferred, such as extended-infrared (10 to 12 μm) and mid-infrared (2 to 5 μm) operations using silica-based fibers, have seen a significant drop in support due to the unexpectedly low fiber losses coupled with the development of optical amplifiers capable of utilizing silica-based fibers. Though it has been long established that optical fibers may cause the phenomena of photonic bandgaps, this has only been proven with the finding of photonic bandgap formations, which include optical fibers. One specific kind of photonic crystal fiber has a microstructural array of air-filled cavities that runs the length of the fiber. Conventional fiber only transmits one mode of light, but there is a single-mode fiber variant, known as a "holey" fiber, that can transport more optical power. An additional type of photonic bandgap fiber is one in which the light is directed via a huge hollow core. Long-distance fiber connections with dispersion compensation or high-resolution imaging need the use of dispersion correction, adjustable imaging spectrum filters, photonic bandgap devices may use such air guiding or hollow-core optical fibers.

Telecommunication networks and other localized communication application sectors all now have dependable, high-performance optical fiber communication systems installed.

9.2 Basic Principles

When information is transferred from one location to another over an optical fiber, it is referred to as fiber optic communication. Telephony, cable television, and computer systems are all capable of producing digital information, which is then distributed primarily through the telephone network. Light is transmitted via an optical fiber, which is a cylindrical dielectric waveguide constructed of low-loss materials such as silicon dioxide and utilized for this purpose. When light pulses go through a waveguide with a refractive index that is slightly higher than that of the cladding, they are completely internalized by the waveguide core. This allows the light pulses to be guided along the fiber. Fiber optic communication systems transform electrical impulses into optical signals using optical transmitters, which are then sent via optical fibers, with fiber optic amplifiers to increase the strength of the signals, and optical receivers to turn the signals back into electrical signals. A communications system using an optical fiber basically functions in the same way. To successfully transfer the information, a communication system needs a transmitter, modulator, and source of information; a method of transmission (such as physical cables, radio waves, or data lines); and a receiver, demodulator, and destination. An electrical signal, such as sound, is sent to a transmitter that consists of electrical and electronic components. This converted signal is propagated across the transmission medium. Modulation is possible in the carrier of an electromagnetic wave, as previously established. When a signal is transmitted, In order to be transmitted to a receiver, it must first be transformed into an electrical information signal (demodulated), and then converted back into the original information signal (modulated). A pair of wires, a coaxial cable, or even empty space may serve as the transmission medium. However, while it is true that signals can be attenuated, lost, or degraded as a result.

It is also true that signals can be attenuated, lost, or degraded irrespective of the transmission medium, all communication signals are subject to attenuation, distortion, and noise. This means that as distance between transmitter and receiver increases, so does the communication system's ability to transmit and receive information. Repeaters or line amplifiers must be added at regular intervals in order to eliminate signal distortion and increase signal strength before the transmission can be resumed successfully.

A semiconductor laser or light-emitting diode may be used to perform the electrical-optical conversion (LED). An optical fiber connection connects the transmitter to the reception. It also comprises an optical detector that drives an additional electrical stage, which in turn performs the function of optical demodulation in the receiver. P–n, p–i–n, and avalanche photodiodes (also known as avalanche diodes) are semiconductor devices that are used for light detection and to aid in the electrical-to-optical conversion of light. Nowadays, the signal processing is typically done by electrical means since the interface from one end of the optical link to the other must include electrical connections.

Fiber optic data transmission systems use a basic but sound set of concepts. Figure 9.2 shows the three core elements found in all fiber optic transmission systems.

In fact, this basic diagram is no longer relevant for contemporary fiber optic transmission systems, as illustrated in Figure 9.2.

Next, various wavelengths of light were introduced to fiber optic transmission systems. At the receiver end, the different optical filters enable us to break out each wavelength of light into its individual "colors." the WDM method is known as (Wavelength Division Multiplexing). The example shown in Figure 9.3 illustrates a simple WDM system that utilizes two wavelengths that are sent in the same direction on the cable.

WDM systems transitioned from systems that could handle one wavelength to systems that could handle several wavelengths (colors). The DWDM (Dense Wavelength Division Multiplexing) systems were referred

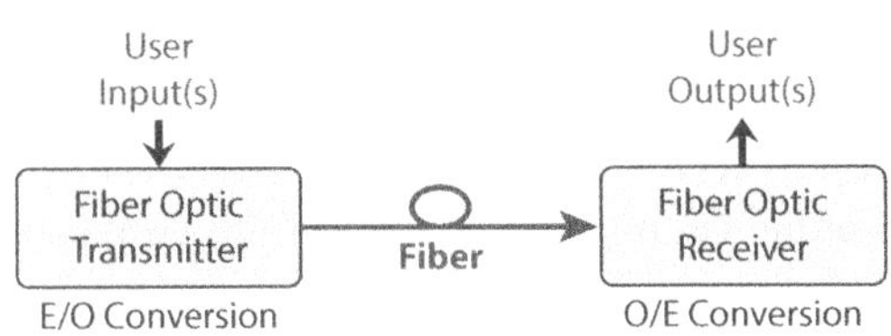

Figure 9.2 The fundamental components of a fiber optic transmission system.

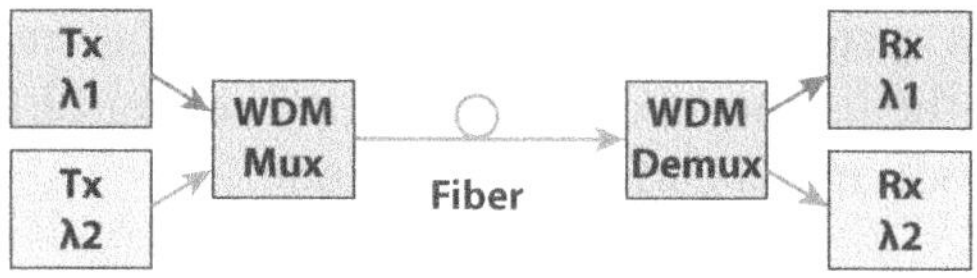

Figure 9.3 WDM Fiber Optic Transmission System at its most basic level.

to as these. EDFA invention had a critical role in making DWDM systems viable (Erbium-Doped Fiber Amplifier). The EDFA has made it possible for the many wavelengths on the fiber to be enhanced at the same time, resulting in the systems being very cheap. Four to eight wavelengths was common in early systems, and then sixteen wavelengths came along and revolutionized the whole industry. To accommodate this increasing amount of Internet traffic, the increase in capacity per fiber was substantial, but was still inadequate. Figure 9.4 illustrates a typical DWDM system setup.

Digital modulation requires a much lower signal-to-noise ratio at the receiver, while analogue modulation using an optical fiber communication system requires a much greater signal-to-noise ratio. Semiconductor optical sources may not always offer the necessary linearity for analogue modulation; this is especially true at high modulation frequencies. In most cases, analogue optical fiber communication connections are limited to shorter distances and lesser bandwidth. Operations, while digital links are capable of longer-distance and higher-bandwidth options.

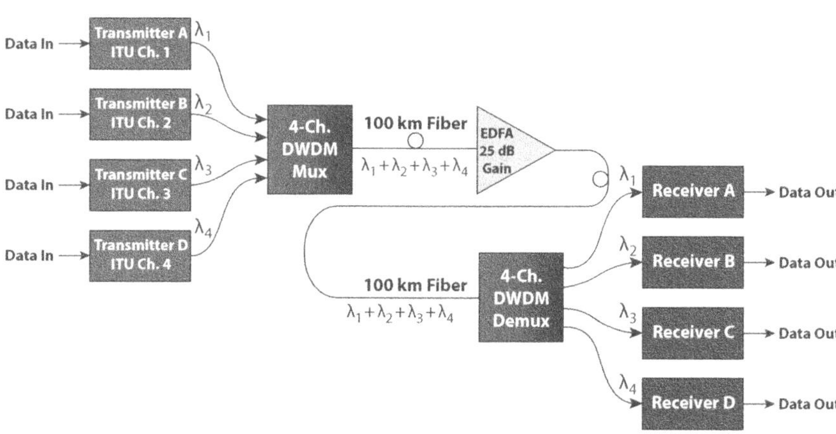

Figure 9.4 DWDM fiber optic transmission system with 4 channels at a low cost.

With DWDM systems, channel count and distance may be extremely high because of the use of advanced features. The typical DWDM components seen in modern DWDM systems include:

System element	Benefit/usage
Soliton Transmitter	The pulse-shape of solitons is distinctive, and it enables them to go much further than other media.
FEC Coding	Forward Error Correction is called FEC. In addition to the main data load, you transmit extra information so that any mistakes that arise during transmission may be rectified.
LiNbO3 Modulators	Clean signals at extremely high data speeds are generated using Lithium Niobate Modulators. No longer feasible to directly tune the laser at these prices.
Pre-Chirp DCF	This is a fiber with a certain length that generates an opposite kind of distortion in relation to long transmission fiber. By implementing this feature, we are able to significantly extend the maximum distance.
Raman Laser Pump	A Raman optical amplifier is also known as a Raman Amplifier. This new development lets the fiber transmit its own signal amplification, which means there is more signal travel and better coverage. This invention enhances the length of the fiber, allowing it to extend further.
VOA	Variable Optical Attenuator.
Dispersion Compensation	Fibers of certain length or a grating that eliminates transmission distortion.
Gain Equalization	This filter maintains the amplitude of all wavelengths at exactly the same level.
PMD Compensation	Switch to Polarization Mode prefetching PMD is a difficulty to overcome at greater data rates on lengthy lengths of fiber (OC-192 and beyond). The pulse-shape of solitons is distinctive, and it enables them to go much further than other media.

9.3 Future Trends in Fiber Optics Communication

The utilization of fiber optics for data transport is unquestionably the wave of the future. The advancement of technology, as well as an increase in demand for fiber optic communication, has driven the expansion of fiber optic communication systems. New and more advanced communication technologies will continue in the foreseeable future. Some of the most promising advances in fiber optic communication may be seen in the diagram below.

All Optical Communication Networks Building an all-optical communication network is the ultimate aim in the long-term. The term "optical networks" refers to the fact that these networks process all signals visually, without the need for any electrical manipulation, thus the name. However, the conversion of optical signals to electrical signals, as well as signal routing to their destinations, must be accomplished first before any processing or switching activities can take place in the electrical domain. The post-processing and routing transforms the initial analogue impulses into optical signals, which are subsequently sent across long distances to their destination, as shown in the diagram. This electrical to optical conversion, as well as the inverse, causes a rise in network latency, which makes it impossible to transfer data at very high rates, as well as the reverse. However, in order for this to become a reality, these barriers must first be cleared away. As a result of these difficulties, research is now being conducted in order to find an effective solution.

Multi - Terabit Optical Networks DWDM is a technology that paves the way for multi-gigabit communications. In response to the worldwide need for increased bandwidth availability, researchers are investigating and developing optical networks that are capable of sustaining multi-terabit/s rates. Currently, four 40 GB/s data rate terabit networks, each with 100 DWDM channels, are being utilized in conjunction with each other. Researchers are attempting to achieve bandwidths of up to 100 gigabits per second using their technology. Fiber optic components are becoming more affordable, making it possible to offer much more bandwidth choices in the future.

Intelligent Optical Transmission Network Traditional optical networks are now unable to keep up with the constantly growing demand for online data services, mostly due to the continuous and unpredictable allocation of bandwidth throughout the network. The traditional method of configuring optical networks is by hand, which results in a lengthy period of time to adjust to changing network requirements. Intelligent optical networks

(IONs) are a future growth trend in optical network development that will have a variety of applications in many fields.

Ultra – Long Haul Optical Transmission It is necessary to investigate the problems that may arise as a result of flaws in the transmission medium. It has prompted researchers to investigate the potential benefits of soliton propagation, which can be achieved via the elimination of the dispersion effect.

Improvements in Laser Technology The increased usage of current semiconductor lasers is expected to be a future trend. It is now possible to purchase spectrally structured laser sources that include chirp management. In order to reduce chromatic dispersion, it is necessary to manage the laser's wavelength changes such that a sudden rise in wavelength happens when the pulse is being generated. To properly characterize lasers of this kind, it is essential to develop instruments that can do so. Furthermore, single mode tunable lasers are essential for the development of future coherent optical systems. In this laser, there is a single longitudinal mode that may be modified to produce frequencies in the range of several gigahertzes to more than ten gigahertz.

Laser Neural Network Nodes The laser neural network is a good option for building optical network nodes because of its high efficiency. The adoption of specialized hardware configurations working in the optical domain, as well as the use of ultra-fast photonic sections, are expected to result in further increases in the capacity and speed of communications networks. The use of optical laser neural nodes in future optical networks may prove to be a useful strategy.

Polymer Optic Fibers Comparing polymer optical fibers to other data transmission alternatives such as copper cables or wireless communication systems or glass fiber, there are many benefits to using polymer optical fibers. The polymer optical fibers used in plug connections are more flexible [and allow for more efficient and less expensive processing of optical data]. Because of the broad range of uses for polymer optical fibers, many research and development teams are now investigating the benefits of utilizing polymer optical fibers in aeroplanes. According to the results of its testing, the German Aerospace Center thinks that "using Polymer Optical Fibers multimedia fibers for future aircraft applications is possible."

High – Altitude Platforms Currently, both inter-satellite communications and ground-to-orbit linkages are available, but the latter cannot be utilized due to poor weather conditions. It is now being studied if optical communication from and to high-altitude platforms is possible. The platforms, which are situated between 16 and 25 kilometers above sea level, have lesser atmospheric effects on laser beams than the ground.

Improvements in Optical Transmitter/Receiver Technology As a result, high-quality transmission must be generated throughout the transmission process, even when the waveform is distorted and the signal strength is low. Improved chromatic dispersion and optical signal to noise ratio (OSNR) tolerance are being delivered as a result of the efforts of the optical transceiver community as part of the ongoing research effort to develop new and improved optical transceivers. New and improved modulation technology is being used as part of the ongoing research effort to develop new and improved optical transceivers. It is also anticipated that improved error-correcting codes will be available in the near future, and that they will be more effective.

Improvement in Optical Amplification Technology In optical fiber communication systems technology, Erbium Doped Fiber Amplifiers (EDFAs) are a key component of the technology (EDFA). According to current predictions, the development of improved EDFA technologies will take place in the near future. In order to improve the EDFA bandwidth, higher-accuracy optical amplification technology will be developed in the near future. A strong pumping laser with excellent optical amplification characteristics and an output of at least +20dBm is currently being developed, which will result in a lower noise figure due to the decreased noise figure.

Advancement in Network Configuration of Optical Submarine Systems The development and testing of a technology for creating mesh networks in optical underwater communication systems is anticipated to take place in the near future as a first step toward improving the flexibility of network design in optical underwater communication systems. Unlike a mesh network, a ring network links stations that are not directly linked to one another on the same ring. A ring network connects stations that are physically connected to one another on the same ring. For the time being, the ring-configuration is the most often used arrangement for large-scale optical underwater systems. Configuration of the MESH network.

Improvement in WDM Technology A wave division multiplexing research project is now being carried out in order to broaden the wavelength range across which WDM systems may operate. For the time being, the C-band wavelength range is between 1.53 and 1.57 microns. The wavelength range of 1.30-1.65 microns is extended by using a low-loss fiber with slow attenuation.

Improvements in Glass Fiber Design and Component Miniaturization It is possible to alter the light transmission characteristics of glass fibers by introducing or removing certain impurities within the fiber. This enables the development of customized glass fibers to meet specific traffic engineering requirements depending on the transit speed of light through the fiber. In order to improve the uniformity and efficacy of glass fibers, it is anticipated that this long-term trend will continue. It is also anticipated that the trend toward smaller optical fiber communication components would continue.

9.4 Advantages

When optical carrier wave communication was first envisioned, many advantageous characteristics became evident. Thus, it can be said that the current advancement in technology is already above even the most hopeful forecasts, and has a number of added benefits. Conventional electrical communications sometimes fall short of optical fiber communications because of its general drawbacks, so you should analyze the advantages and specialized characteristics provided by optical fiber communications when deciding which communications technology to employ. When introducing a concept in this context, we begin with the benefits that were initially envisioned and then identify additional characteristics that became evident as the technology was developed.

Enormous potential bandwidth. Compared to metallic cable or even millimeter wave radio systems, the near-infrared optical carrier frequency offers significantly wider potential transmission bandwidth. As of 2000, the average optical fiber connection bandwidth multiplied by link length was approximately 5000 gigahertz kilometers, compared to the standard coaxial cable bandwidth and link length product of approximately 100 megahertz kilometers. While this provided an increased capacity over a longer distance, the improved bandwidth provided by optical fiber was 50,000 times that of coaxial cable, while also exhibiting better data transfer over longer distances. Therefore, the bandwidth utilization of an optical fiber may be significantly increased when it is used to simultaneously carry multiple optical signals, each at a distinct center wavelength. Using the well-defined optical wavelengths (or, to be more precise, narrow frequency spacing) in a WDM system provides the possibility for much increased fiber information-carrying capacity compared to copper cables or a wideband radio system.

Small size and weight. In the majority of cases, optical fibers have a diameter that is less than the width of a human hair. Because they are so much smaller and lighter, these fiber optic cables are wrapped in protective coatings and are known as mini- and micro-fibers. This is a significant advantage toward decreasing duct congestion in cities, as well as enabling mobile devices like as aeroplanes, satellites, and ships to expand their signal transmission.

Electrical isolation. Unlike metallic wires, glass fibers do not suffer earth loop and interface issues. In addition, the above characteristic is especially beneficial in situations where electrical shock is a threat since the optical fibers do not spark or arc at abrasions or short circuits.

Immunity to interference and crosstalk. For this reason, optical fibers are excellent candidates for use as waveguides, as they do not suffer from electromagnetic interference (EMI). As a consequence, electronic noise in the environment has no effect on the operation of an optical fiber communication system, and the fiber cable does not need shielding against electromagnetic radiation. Electromagnetic interference is a term used to describe the phenomenon of electromagnetic interference. In order to be lightning-proof, the fiber cable must be run above ground rather than underground. Also available is a straightforward way of guaranteeing that there is no optical interference, which guarantees that crosstalk is kept to a minimum even when a large number of fibers are cabled together in a single configuration.

Signal security. The fact that optical fibers produce relatively little light allows them to provide very secure transmissions to their users. It is not feasible to collect an optical signal sent via a fiber without causing damage to the fiber. As a result, any efforts to intercept optical signals should be noticeable, at least theoretically. This capability is obviously beneficial in a variety of applications such as the military, banking, and data transmission via a computer network, among others (i.e. the internet).

Low transmission loss. In recent years, technological advancements in optical fiber production have resulted in optical fiber cables with lower transmission or attenuation losses than copper conductors. Loss levels as low as 0.15 dB km-1 have been achieved in optical fiber communications fibers, which is a very appealing feature of optical fiber communications technology. By establishing communication links with very wide optical repeater or amplifier spacing, it is possible to reduce both the cost and complexity of the system as well. Because modulations have shown higher bandwidth capacity when dealing with optical fiber cables, there is a compelling argument for utilizing optical fiber communications in a majority of long-haul telecommunications applications rather than copper cables or even satellite communications.

System reliability and ease of maintenance. Because optical fiber cables are capable of carrying signals with fewer losses, this feature is mostly derived from the fact that they are less expensive to purchase. This results in the need for fewer repeaters or line amplifiers to enhance the strength of the broadcast signal being transmitted. As a consequence, when optical repeaters or amplifiers are not required, the overall system reliability is improved. Considering that most components have useful lifetimes of about twenty to thirty years, the problem of component dependability has been addressed. Furthermore, both of these qualities may be beneficial in terms of lowering maintenance costs and saving time.

Potential low cost. However, rather of being made of a valuable natural resource, the glass medium through which optical fibers are most commonly carried is composed of sand, not glass. This has the potential to significantly reduce the costs of long-distance communications by a considerable margin. Although optical fiber cable is much more cost-effective than twisted-pair cable when purchased in large quantities, this cost-effectiveness has not yet been reached in all other areas that are critical to optical fiber communications. Many things in this industry, such as the price, continue to be excessively expensive, despite recent improvements (demountable connectors, couplers, etc.).

However, despite the fact that these techniques are feasible in terms of frequency bandwidth, they are not practical since "line of sight" transmission needs costly aerial structures to be spaced at intervals no more than a few tens of kilometers away from one another. Thus, optical fiber has exceeded all other transmission mediums in the main industrialized nations, with the exception of communications access networks, in part due to the high cost of installation. However, optical fiber has outperformed all other transmission mediums in the developing world. A lightwave carrier, when used in combination with an optical fiber transmission medium, has a number of advantageous properties to it. The actual application of these fundamental principles, as well as the conceptual underpinnings that serve as their basis, are discussed in detail in the following chapters. Understanding the fundamentals of light, on the other hand, is expected.

9.5 Conclusion

This last decade has seen significant development in the fiber optics communications sector. While challenges exist in improving and expediting network operations, as well as in meeting the needs for higher bandwidth rate increases, novel having switched strategies, and thus smarter infrastructure components that could dynamically reconfigure their hierarchy based on the traffic patterns while strongly impacted, but there is much more work to be done. This prediction is supported by future advancements in the lab that will be applied to industry for widespread use, giving rise to a new age of fiber optics communications.

References

1. Saleh, A.A., Mustafa, A.B., Osman, A.A., Proposal laying fiber optic for cables along railways tracks in Sudan. *IOSR J. Comput. Eng.*, 5, 2, 90–94, 2015.
2. Matthews, V.O., Uzairue, S.I., Noma-Osaghae, E., Analysis and implementation of fiber to the home network using peace estate Lagos as a case study. *Int. J. Sci. Technol. Manage. Res.*, 3, 8, 1–4, 2018.
3. Kiminza, M.M. and Were, S., Factors affecting successful implementation of fiber optic cable projects in Kenya: A case of Nairobi City County. *Int. Inst. Sci. Technol. Educ. (IISTE)*, 6, 10, 52–59, 2016.
4. Tang, B.H. and Zhou, Z.X., The design of communication network optical fiber cable condition monitoring system based on distributed optical fiber sensor. Paper Presented at: *2018 International Conference on Electronics Technology (ICET)*, May 2018, IEEE, pp. 97–101.
5. Sastrawidjaja, L. and Suryanegara, M., Regulation challenges of 5G spectrum deployment at 3.5 GHz: The framework for Indonesia. Paper Presented at: *2018 Electrical Power, Electronics, Communications, Controls and Informatics Seminar (EECCIS)*, October 2018, IEEE, pp. 213–217.
6. Silva, B.N., Khan, M., Han, K., Towards sustainable smart cities: A review of trends, architectures, components, and open challenges in smart cities. *Sustain. Cities Soc.*, 38, 697–713, 2018.
7. Nyarko-Boateng, O. and Adekoya, A.F., Evaluation and analysis of key performance indicators which affect QoS of mobile call traffic. *Int. J. Comput. Netw.*, 9, 1, 770–777, 2019.
8. Florkowska, B., Jackowicz-Korczynski, A., Timler, M., Analysis of electric field distribution around the high-voltage overhead transmission lines with an ADSS fiber-optic cable. *IEEE Trans. Power Deliv.*, 19, 3, 1183–1189, 2004.
9. Henriques, H.O., Barbero, A.P.L., Ribeiro, R.M. *et al.*, Proposal of fault prediction system for underground installations. *Measurement*, 81, 232–240, 2016.
10. Zhou, X., Liu, H., Urata, R., Zebian, S., Scaling large data center interconnects: Challenges and solutions. *Opt. Fiber Technol.*, 44, 61–68, 2018.

10

Difficulties of Fiber Optic Setup and Maintenance in a Developing Nation

Dankan G. Veeranna[1]*, M. Nagabushanam[2], Sridhara S. Boraiah[3], Ramesha Muniyappa[4] and Devananda S. Narayanappa[5]

[1]*Department of Electronics and Communication Engineering, B.M.S Institute of Technology and Management, Bengaluru, Karnataka, India*
[2]*Department of Electronics and Communication Engineering, M.S Ramaiah Institute of Technology, Bengaluru, India*
[3]*Department of Computer Science and Engineering, APS College of Engineering, Somanahalli, Bengaluru, India*
[4]*Department of Electronics and Communication Engineering, GITAM School of Technology, GITAM University, Bengaluru, Karnataka, India*
[5]*Department of Electronics and Communication Engineering, PES Institute of Technology and Management, Shimogga, Karnataka, India*

Abstract
In India, the telecommunications industry has undergone many transition phases. In recent decades, the industry has expanded exponentially. Effective infrastructure implementation and management are the key to sustaining this development. Fiber optics has become the principal infrastructure and the means of connectivity for the present and evolving technologies, offering greater bandwidth capacities at higher speeds. As the need for new technologies and services increases, fiber optics provides the promise of a network platform that is flexible, scalable, comprehensive in service, and has the potential for limitless capacity. Mobile network providers have made significant and strategic investments in fiber optic infrastructure; nevertheless, the frequency of network disruptions caused by recurrent fiber optic network failures, such as fiber cable breakage, has risen as a result. In India, cable cuts have been the single most common cause of telecommunications transmission failure or interruption, which has a significant impact on the subscriber's overall experience with the service provider. This chapter focuses on the difficulties associated with fiber cable deployment in India, with a particular emphasis on

**Corresponding author*: dankan.v@bmsit.in

Chandra Singh, Rathishchandra R Gatti, K.V.S.S.S.S. Sairam and Ashish Singh (eds.) Modeling and Optimization of Optical Communication Networks, (179–192) © 2023 Scrivener Publishing LLC

the economic, regulatory, and managerial difficulties. It is possible that external causes, such as dig-ups during road building, are a result of the problems associated with frequent fiber cutting. A lack of fiber deployment and management regulatory guidelines and policies poses a significant challenge to fiber management in the region.

Keywords: Fiber optics, bandwidth, telecommunication, scalability, transformation

10.1 Introduction

Over the years, India's telecommunications sector has experienced many phases of change. In the late 1990s, the industry's use of mobile communication was initially implemented in India. Though the industry participants came to understand the limitations of a microwave transmission system in a long-haul infrastructure, they decided to use a different method of transmission. The weakening of the radio signal over long distances and environmental influences are a few of the disadvantages. As a result, the fiber optic cable transmission system was implemented in the latter half of the 2000s. This recent advancement in optical network deployment has led to an exponential increase in telecommunications service delivery in the past few years. Until recently, fiber optics was the dominant means for internet connection, whether it was in urban or rural regions. While organizations, businesses, and households may benefit from the use of internet services, they rely on internet services to conduct their day-to-day operations, improve productivity, and achieve established goals in a specific area [1, 2]. It has become a fundamental requirement for individuals in the modern world, as well as in future communication systems in all forms, to have fast internet connection. To provide good service, we have to install and maintain our infrastructure efficiently. The competitive advantage gained by mobile network operators (MNOs) because of their capacity to strategically build and maintain their network infrastructure is considerable. The network that supplies more bandwidth capacity for cable and wireless broadband services is made up of fiber optics. Due to the use of high capacity fiber, industries and other companies have taken use of the ability to transmit voice, video, and data traffic. Fiber optics offers the promise of a flexible, scalable, convenient, and high-capacity network [3, 4]. On the other hand, much like any other technology, fiber optic cable deployment adds complexity and difficulties into the network architecture. In India, it has been the factors of both technological and economic interest that have pushed the deployment of fiber optics. Various MNO approaches and tactics have been used for installing fiber optics cable, with the primary goal of gaining competitive advantage in a digital market that is developing. The MNOs perform an effective and dependable network that offers cutting-edge services while minimizing disruption. Installation difficulties, early installation expense, fiber cable cuts, and trouble in locating defects plague the optical network and transmission system [5, 6]. At this time, there are no clear regulations in place for optical cable deployment.

In the last 50 years, only optical fiber has provided dependable backhaul for contemporary backbone network systems. Due to the excellent properties of fiber optics cable, you can be confident that you will always have a constant speed of transmission of services via fiber optics cable. The quality of the cable, as well as the throughput of the light signal, is mainly dependent on the quality of the optical cable's handling during installation. Favorable legislation, operators' investment capacity, market requirements, and the presence of appropriate urban infrastructures are all required in order for an optical cable rollout to succeed [7, 8]. This study is investigating fiber optic cable deployment, with a focus on its technical, regulatory, and management issues. This paper has two purposes: 1) identifying and analyzing issues related to fiber optic cable deployment, and 2) offering possible solutions to those issues. Furthermore, the article also aims to examine the regulatory framework of fiber cable deployment, as well as the shortcomings of the Indian mobile operators in managing fiber optic infrastructure. It will also offer methods for deployment and management of efficient techniques in fiber optic infrastructure.

10.2 Related Works

Connectivity at a speed high enough to provide high-end services, such streaming video, requires the presence of smart technology, IoT, smart city, and big data. The issue at hand revolves on whether the existing legacy network is capable of handling the significant volume of data required. Service providers have upgraded their network infrastructure in order to meet the rising demand. Based on the findings in [9], marketers deal with different marketing difficulties based on the market's stage of development. [10] Fiber optic cable deployment is expensive and time-consuming to install. To be adaptable, you need governance that is both strong and flexible. An improved communication infrastructure is required to ensure that fiber-optic cable does not develop breakage or other associated problems. This research on resilient network infrastructure was authored by authors in Reference [9]. By strictly adhering to this technological method, the whole network and the communication system may avoid delays, jitters, and conceptions. The method of fiber optic cable transmission is important according to Reference [10]. Overhead, subterranean, and submarine transmission systems have been used to deploy end-to-end optical infrastructure.

[8] More studies [9] on optimizing end-to-end fiber optic cable network architecture for efficient communications have been published. Low-loss fiber-optic submarine cable transmission is required to transmit signals over large distances, such as 34,000 km across the Atlantic Ocean. The repeaters are put at 8-km intervals along the undersea cable in order to compensate for the losses.

10.3 Fiber Optic Cable

Fiber networks have transformed the telecommunications industry during the last three decades. Due to new advancements coming rapidly, the technologies were well ahead of real demand by the end of 2000. In April 1977, the first commercial fiber optic link in the United States was put into operation in Long Beach, California, which was capable of transmitting data at 6 Mbps. Speed has been increased over a single fiber to about 3.2 Terabit/s, and now lets you send a million times the amount of data. An optical cable is used to transmit data by transmitting light. The optical detector receives this light and translates it into an electrical pulse by means of the associated detectors. It's commercially available lasers that have been able to achieve data transfer rates of up to 10 gigabytes. The latest commercial networking technologies and current study on the subject have shown that 400 Gb per rack using a single fiber-optic cable color is feasible. On the other hand, most difficulties to achieving these high speeds are based on the light to electrical pulse conversion detector. It is also feasible to use wavelength division multiplexing to mix many colors onto a single fiber. To put things in perspective, it is possible to purchase systems that enable the use of up to 160 colors on a single fiber, resulting in a maximum total of 150 Terabit/s for that one fiber line. Most labs have achieved speeds of up to 25 Tbit/s. Multimode fiber optic cables are divided into two categories:

10.3.1 Single-Mode Cable

Single-mode cable is a long layer of silica material that includes a thin core optical material for electrical signals, with a core diameter ranging from 8.3 to 10 microns. Single-mode cable is used for signal transmission in telecommunications networks. The fiber core has a diameter of 1310 and 1550 nm, with single-mode fiber, which helps it carry signals in the 1310 and 1550 nm wavelength ranges. Fiber-optic cables that have just one mode of transmission are better able to preserve their respective signal's spatial and spectral integrity across long distances, enabling for data to be sent at a faster pace. The bandwidth of this fiber type is greater than multi-mode fiber. Fiber optic cable is most often utilized for longer-distance and higher capacity applications.

10.3.2 Multimode Cable

Multi-mode has a diameter of 50–100 μm, and 62.5 μm is the typical light-carrying component size. High-speed bandwidth transfers across a short distance using multimode fiber. Light flows along several pathways in multimode fiber cable, usually with a wavelength between 810 and 1300 nm.

10.3.2.1 Step-Index Multimode Fiber

A huge core of up to 100 μm in diameter is present in multimode step-index fiber. This results in a phenomenon known as direct or normal bounce, in which some of the light rays in the digital pulse take a straight route, while others take a zigzag course as it bounces off the cladding. The various groups of light beams, known as modes, arrive to the receiving site by means of unusual routes. The light pulse has many modes, which cause the light pulse to scatter and dissipate, so it loses its distinct form. To avoid overlapping limits, it is essential to allow some space between pulses. This kind of fiber optic cable is most appropriate for short- distance transmission.

10.3.2.2 Graded-Index Multimode Fiber

In graded-index multimode fiber, the refractive index becomes progressively lower toward the core from the central axis. Since the refractive index is greater on the down-the-axis rays, the incoming light is travelling slower compared to that closer to the cladding. The light does not wander off the cladding since it is travelling a longer distance because the index is lower. As the light in the periphery is reached more quickly and the distance to the receiver is reduced, the peripheral light arrives at the receiver at the same time as the center core axis. These findings show that a digital pulse that is not as widely dispersed is created.

10.3.3 Deployed Fiber Optics Cable

For long-haul transmission, single-mode fiber optics cables are used, which conforms to the worldwide standard. Choosing the incorrect settings or picking unsuitable cable may cause technical difficulties. In a few cases, naked fiber optics cables were installed in subterranean trenches, according to this paper's research. Whenever there's road work on that section, there will be cable cuts.

10.4 Fiber Optics Cable Deployment Strategies

Fiber optic cable may be deployed in a number of ways. Selecting the installation type relies mainly on the surrounding environment, company growth, and population density. In order to help guarantee effective deployment and low maintenance costs once the cable is installed, several critical elements are included. These results reveal that, in the case of a fiber-optic network, between 60% and 80% of the total capital expenditures are attributable to civil construction, ducts, and cables. Fiber optic cables may be installed using three techniques: aerial, subterranean, and subsea. Aerial installation is done by attaching wires to utility poles or pylons, whereas subterranean installation is done at a certain depth below the surface of the ground, and undersea installation is performed underwater.

10.4.1 Aerial Installation

To experience stress, as well as additional strain produced by temperature fluctuations, wind, and in certain frigid areas, the weight of ice, wires in the air must be under tension. Only some cables intended for aerial installation have sufficient strength to enable direct attachment. For this job, a standard fiber cable would be best. However, in certain instances, a metallic stranded cable would be needed to sustain the wire. In order for the supporting cable to adequately sustain the fiber optic cable across the gap between support structures, it must have sufficient strength. sixteen, seventeen, and twenty-seven When dealing with length fluctuations, such as those resulting from straining in the wind or with temperature changes, more caution is needed when installing fiber optic cables with a cable assembly that supports the cable. Because fiber optic cables are not intended to stretch, slack is required at the supports to allow for movement in the messenger position without stretching the cable.

Additionally, there is an ADSS cable category that has a thicker jacket for better strength and rigidity in order to resist long-term high stress loading. A kind of aerial cable that goes through the building and is located on the outside is the optical power ground wire (OPGW). This is a hermetically sealed tube containing a high-voltage conductor with optical fiber cable in the middle. The cable is often utilized in a number of places for communications and energy. Connections are terminated and coiled on the tower like high-voltage wire, then transported to the ground.

10.4.2 Underground Installation

Fiber optic cables may be buried or placed within a duct buried beneath. Direct burying of cables is often seen in situations like cross-country installation. Steel armor is often used for cables, which are then placed into trenches that have been excavated to a specific depth. The steel armor is constructed to shield the fiber optic cable from the environment's harmful elements and burrowing animals.

10.4.2.1 Direct-Buried

The wire that is buried directly into the ground is designed to particular specifications to keep it from becoming too hot, too wet, conductive, or acidic. It is less expensive and simpler to install than other types of cable, such as buried cable, which must be protected from the ground. However, buried cable may also be damaged while excavating or when there is other excavation work being done. Since this cable is buried directly instead of wrapped around, it is located on side streets instead of the major streets. Direct-buried cable has the benefit of having minimal setup requirements since it does not need the installation of pipes or creating other provisions to lay the cable. So as far as voice and data transfers are concerned, a cable that is buried immediately below ground level and direct-buried is ready to use. Direct bury cables also provide the advantages of fewer recurrent replacements and a higher possibility of retaining integrity even in the case of a natural catastrophe, because the fiber cable is self-contained.

10.4.2.2 Installation in Duct

Common subterranean applications include the use of ductwork installation in metropolitan regions where it is difficult to excavate the ground. Since it is possible to utilize fiber optic cables that are not coated in steel armor, it is simpler to put fiber optic cables in established ducts. The duct protects the optical fiber cables from the most difficult operating conditions. Adding new fiber cable is as simple as digging a few shallow holes.

Underground fiber optic installation's beneficial features include being protected from the bad weather conditions, and that will aid in the widespread use of underground fiber optic installation. With these kinds of cables, moisture and mechanical stress are of significant concern. Damages may be expensive in terms of service interruption and the expense of new parts. This also applies to lengthy lengths of continuous cabling that need a protected and leak-free environment.

To minimize the possibility of being inadvertently dug up, the wires are typically buried at a depth of 3–4 feet (1–1.2 m). Typically, trenching is done first to bury the conduit, which is typically 4-inch (102 mm) plastic pipe, which may also have an inner duct added for pulling tape purposes. If there are many cables already in the conduit, pulling one more wire may increase pulling strain and possibly damage cables. If the total cable fill and pulling strain are not excessive, then several cables may be pulled at simultaneously.

10.5 Deployment of Fiber Optics Throughout the World

There has been extensive deployment, extension, and relocation of long-distance fiber-optic cable throughout the globe. These actions have helped increase the reach of the internet, allowing users to communicate over the phone, and they've enhanced the flexibility of telecoms interconnection. Most of the world's telecommunications providers use submarine and subterranean fiber-optic cable networks to link their networks.

Verizon, a US fixed-line provider, built an extensive fiber optic network underneath the FiOS brand, which contributes to a boost in ARPU. Although mobility provider Mobility, with more than 100,000 subscribers, does not have sufficient copper infrastructure focused on fiber optic development, it has grown its customer base by a hundred thousand in 2010. Due to the growing need for high-quality service delivery, the MNO has recently extended its optical network infrastructure. As a result, the group's aim of hitting over 100 Mbps by the end of 2011 was realized by using optical network infrastructure. Deutsche, Germany's largest MNO, has expanded its area of service offering by investing significantly in fiber optic infrastructure development. As a result of using fiber optics for communication and internet service delivery across many nations, international MNOs (e.g. China, France, Dubai, and Australia) have achieved a major advancement in the use of smart systems. A massive fiber optic cable rollout in the UK will reach two-thirds of UK households by the end of 2015. Singapore has spent millions of dollars on a FTTB/FTTH (fiber to the home/fiber to the curb) network, which covers 95% of the population by 2012. Eleven and eighteen Finland was the first nation to make broadband infrastructure available to everyone in the country via fiber optic cable in 2015. The more sophisticated countries are capable of putting up a sound regulatory framework for providing, providing access to, and providing access to fiber optic cable. Effective measures have been implemented to control the growth of optical networks in the industrialized nations.

In underdeveloped nations, there is a lack of motivation to promote and support legislation that governs the use of optical networks. In the developing and low-income nations, optical infrastructures have been deployed haphazardly. This study has attempted to uncover various issues impeding optical network deployment in low-income nations.

10.5.1 Fiber Optics Deployment in India

An increased number of private companies invested, purchased, and operated telecommunications infrastructure to provide communication services to consumers after the liberalization of the telecommunications industry in India. In 2000, optical infrastructure (i.e., fiber optic cable and satellite terminals) was commercially deployed in India. Most of the previous radio wave transmission infrastructure was replaced by fiber deployment. It's now the year 2015, and commercial and governmental organizations, as well as people, may enjoy the benefits of a fully dispersed fiber optic infrastructure.

10.5.2 Submarine Fiber Optic in India

There are presently four undersea cables in India with various capacities. The SAT-3 cable is the first cable in the world to touch the Indian coast with a total distance of 28,800 km. The cable is able to transmit 340 gigabits per second from Portugal all the way to South Africa and across the Indian Ocean. This new cable, which is capable of transmitting 5.12 Terabits per second, is commercially active in West Africa today. Both internet and phone traffic is sent via these cables from India to other areas of the globe. There are about 7,000 kilometers of cable stretching along the

West African coast from India, Nigeria, and Portugal, with the main cable having 7000 km.

The GLO-1 cable, which stretches from the UK to Spain and Portugal and connects 14 African nations with the US, has a capacity of up to 2.5 gigabytes per second. There are 15 terminal stations for WACS cable, throughout the length. This new undersea cable system utilizes the latest technology, and it will link India to Europe via South Africa. WACS uses next-generation transmission technology to deliver a high-capacity optical signal with a data rate of 5.12 Terabits per second.

10.5.3 Installation of Fiber Optic Cable in the Inland

As previously discussed, all five MNOs have used fiber-optic cable to connect their networks with the backbone and the metro fiber networks in India to meet customer demand. 34 Volta River Authority's Vodacom subsidiary built the country's first backbone fiber optic network. In a way, Vodacom used pylons that were already in place and is now in the process of deploying an aerial fiber optic cable. Most of the country's electricity system is covered with optical fiber, which was placed in both the south and middle regions.

To boost data and voice traffic, the countrywide mobile telecommunication network, MTN, has built an optical network utilizing an underground deployment method to link all of India's 16 regions. There are about 38 underground fiber metro networks (FTTx) connected to MTN and Vodafone India. Several telecom companies, including Airtel India, Tigo, and Glo, have laid fiber-optic cable underground in order to offer sufficient capacity on the backbone route as well as making up for any failures on microwave connections.

10.6 Fiber Deployment Challenges

The introduction of fiber deployment is accompanied by processes and procedures that, unless rigorously adhered to, may pose a danger to the whole operation. Multifunctional units are a need for this project, and therefore there has to be a more thorough consultation. Technical and management problems may be divided into project-related and external difficulties.

10.6.1 Deploying Fiber has a Number of Technical Difficulties

In a developing nation such as India, there are a number of difficulties when it comes to deploying underground fiber. Mostly fiber optic cable is placed along major highways and city streets underground. Nevertheless, in an unauthorized structure, route planning and following planned routes are more challenging. Fiber as first envisioned becomes tricky. If several MNOs are using an existing ROW, there are logistical issues due to frequent fiber cable cuts.

10.6.2 Right of Way

A very costly and time-consuming ROW acquisition is required in order to install optical cables on the roadway. Laying cable is a complicated operation, which may delay project deadlines for MNO. An important factor in the development of optical infrastructure is ROW management. A number of mobile network operators have claimed that excessive demands for information hampered the rollout of new networks, the process of applying for a permit, exorbitant fees to utilize ROWs, and extensive cleanup and maintenance requirements.

Many others have pointed out that installing infrastructure beyond municipal borders is both expensive and time-consuming due to the complicated patchwork of processes across municipalities. However, the situation required some flexibility when it came to the regulation of the use of public rights-of-way to handle and track fault in underground fiber optic cable in order to guarantee the safety of the streets in the communities. City officials note a shortage of staff as a fundamental challenge to issuing licenses.

10.6.3 Administrative Challenges

This present system of permits and licenses makes implementing a fiber project very difficult. Lack of procedural rules to govern the actions of local authorities in addition to bureaucratic procedures at the local level impede the advancement of optical infrastructure deployment. Many municipal governments have realized the need of harmonizing the ROW processing period, which includes the time from when a permit application is first filed to when the facilities are completed. This particular implementation in the U.S. reduces installation time by limiting registration approval/denial dates to just before the installation deadline. Some other localities have enacted blanket permissions, no longer requiring individual facility registration. Three other states, including Kansas, Indiana, and Ohio, imposed 30-day permit processing deadlines while Michigan and Virginia set longer timeframes at 45 days.

Additionally, administrative difficulties include costs connected with the permit procedure and the excessive delays when it comes to managing permissions. In the opinion of this article, pricing disparities across administrative regions should be avoided since they impede optical infrastructure development procedures. 34 Inadequate fee structure and best practices lead to fee charges that have not been authorized and lead to unnecessary delays, which extends the whole project cycle. When booking travel, several nations including the US, Europe, Australia, and a significant portion of Asia have various price structures in place for ROW use and other amenities. As government organizations are required to utilize well- known fees, most applicants should have already chosen their preferred pricing model.

In Africa, including Egypt, South Africa, the procedure of acquiring permission hasn't been implemented across the whole country. It is not uncommon for the federal government to handle all processing costs. The project frameworks needed to execute optical initiatives are lacking. To note, optical initiatives have damaged several road networks in India and Nigeria. Road users' safety is at risk due to road damage. Poor administrative measures cause inadequate project execution in portion of Africa, contributing to a significant amount of traffic accidents.

Other significant administrative issues which impact the optical deployment process include inadequate information management. Certain current utility infrastructure has unnecessary information, which brings with it some problems. 34 Knowing what is presently accessible underground helps in anticipating what should be included and how it should be implemented and where it should be done. As a consequence of not having GIS data on the current infrastructure, utility disruptions continue because of trenching.

10.6.4 Post-Fiber Deployment Management

A methodology for preservation, operation, maintenance, repair, and renewal of ageing fiber infrastructure that relies on engineering concepts. Fiber optic communication systems are excellent for transmitting voice, data, and video across bandwidth-intensive networks. Because of this, MNOs and the end-user require efficient control of this infrastructure. Telecommunication services were hampered in India when they discovered that large numbers of underground cable cuts had been made. The MNOs in India deliberately destroyed their fiber optic connection to cause outages for tens of thousands of customers. Protection against deliberate or accidental damage is a need for fiber optic networks.

10.6.5 Fiber Optic Cable Deployment and Management Standards and Best Practices

Until the end of 1998, an optical cable installation standard did not exist. While fiber optic cable deployment did provide a number of difficulties, a standard measuring length was still required. This included an ANSI program and the National Electrical Installation Standards' guidelines. The Fiber Optic Industry Association (FIA), a body that defined the installation, operating, and maintenance standards for optical fiber cabling in the UK, established standards for fiber optic cable deployment.

Technical Specifications for fiber Optic Cabling are available in India, contains specifications and methods for setting up for all aspects of the cabling installation, including planning, installation, surface and underground surveying, and clearing; underground work like excavation and trenchless construction; trenching; reinforced concrete or Hume pipe installation; marking; backfilling; underground cable installation and Deployment, along with similar issues with internet connections, are all focused on network performance, accessibility, affordability, and dependability.

10.7 Conclusion

Based on the results of this study, it can be deduced that the majority of construction project failures are caused by extraneous variables such as digs, activity of excavators, and work on public roads. A lack of procedures and standards for installing fiber cable was shown to be a significant underlying cause of frequent fiber failure. Due to the absence of any regulations or rules regarding fiber deployment and administration, India is at risk of losing control of its fiber optics management. A last shortcoming of this idea is that there are currently no fiber engineers with the necessary technological know-how. Exposure of optical fiber cables to chemicals has been shown to reduce the quality of the work produced. assault and damage, this has a detrimental effect on the performance of the organization.

References

1. Saleh, A.A., Mustafa, A.B., Osman, A.A., Proposal laying fiber optic for cables along railways tracks in Sudan. *IOSR J. Comput. Eng.*, 5, 2, 90–94, 2015.

2. Matthews, V.O., Uzairue, S.I., Noma-Osaghae, E., Analysis and implementation of fiber to the home network using peace estate Lagos as a case study. *Int. J. Sci. Technol. Manage. Res.*, 3, 8, 1–4, 2018.
3. Kiminza, M.M. and Were, S., Factors affecting successful implementation of fiber optic cable projects in Kenya: A case of Nairobi City County. *Int. Inst. Sci. Technol. Educ. (IISTE)*, 6, 10, 52–59, 2016.
4. Tang, B.H. and Zhou, Z.X., The design of communication network optical fiber cable condition monitoring system based on distributed optical fiber sensor. *Paper Presented at: 2018 International Conference on Electronics Technology (ICET)*, May 2018, IEEE, pp. 97–101.
5. Sastrawidjaja, L. and Suryanegara, M., Regulation challenges of 5G spectrum deployment at 3.5 GHz: The framework for Indonesia. *Paper Presented at: 2018 Electrical Power, Electronics, Communications, Controls and Informatics Seminar (EECCIS)*, October 2018, IEEE, pp. 213–217.
6. Silva, B.N., Khan, M., Han, K., Towards sustainable smart cities: A review of trends, architectures, components, and open challenges in smart cities. *Sustain. Cities Soc.*, 38, 697–713, 2018.
7. Nyarko-Boateng, O. and Adekoya, A.F., Evaluation and analysis of key performance indicators which affect QoS of mobile call traffic. *Int. J. Comput. Netw.*, 9, 1, 770–777, 2019.
8. Florkowska, B., Jackowicz-Korczynski, A., Timler, M., Analysis of electric field distribution around the high-voltage overhead transmission lines with an ADSS fiber-optic cable. *IEEE Trans. Power Deliv.*, 19, 3, 1183–1189, 2004.
9. Henriques, H.O., Barbero, A.P.L., Ribeiro, R.M. *et al.*, Proposal of fault prediction system for underground installations. *Measurement*, 81, 232–240, 2016.
10. Zhou, X., Liu, H., Urata, R., Zebian, S., Scaling large data center interconnects: Challenges and solutions. *Opt. Fiber Technol.*, 44, 61–68, 2018.

11

Machine Learning-Enabled Flexible Optical Transport Networks

Sridhar Iyer[1]*, Rahul Jashvantbhai Pandya[2], N. Jeyakkannan[3] and C. Karthik[4]

[1]Department of Artificial Intelligence, KLE Technological University, Belagavi, Karnataka, India
[2]Department of Electrical Engineering, Indian Institute of Technology, Dharwad, WALMI Campus, Karnataka, India
[3]Department of Bio Medical Engineering, Rathinam Technical Campus, Coimbatore, Tamil Nadu, India
[4]Department of Mechatronics, Jyothi Engineering College, Cheruthuruthy, Thrissur, Kerala, India

Abstract
This chapter overviews the various existing solutions for optimizing the Space Division Multiplexed-Elastic Optical Networks (SDM-EONs). Firstly, in view of enabling the realization of SDM-EONs enabled by the development of appropriate fiber solution to ensure long haul signal transmission, the chapter will review various physical layer models which have been proposed to maintain the desired quality of transmission in an SDM-EON under consideration of the fiber solution which has been adopted. Secondly, the chapter will survey new resource assignment strategies and the algorithms used to explore added freedom degrees in multiplexing of the signals within the SDM-EON. Such added freedom degrees are presented by the latest advances in the various fiber solution systems, aiming to exploit the advancing networking profits arising from the aggregate spatial and spectral domains. Lastly, the chapter will detail various open research avenues within the SDM-EONs which require timely solutions.

***Keywords*:** EON, SDM, OTN, spectrum, spatial multiplexing

**Corresponding author*: sridhariyer1983@klescet.ac.in

Chandra Singh, Rathishchandra R Gatti, K.V.S.S.S.S. Sairam and Ashish Singh (eds.) Modeling and Optimization of Optical Communication Networks, (193–216) © 2023 Scrivener Publishing LLC

11.1 Introduction

With the passing of every year, exponential growth in the number of telecommunication network users and connected devices is occurring. With the users mainly interested in bandwidth-intensive services such as, Big Data, real-time gaming, High Definition (HD), audio-video streaming, content distribution, cloud computing, etc., the consequence is a major growth in the overall network traffic volume. CISCO predicts that global IP traffic will increase three times over the next five years [1]. Simultaneously, it can also be noticed that communication networks have become an indispensable part of society as they support various human activities, including entertainment, social life, finance, medical care, and business. Hence, for their ubiquity and relevance, it is essential to provide an undisrupted functioning of these networks. Also, to follow the network trends and the increasing users' requirements, it is crucial to implement remarkable innovations in the currently applied network solutions. Accordingly, it is required to ensure efficient approaches in terms of both the physical network architectures and the algorithmic solutions for the network design and optimization.

Currently, the deployed optical transport networks (OTNs) rely mainly on the wavelength division multiplexing (WDM) technology, as shown in Figure 11.1, where the transmission multiplexing is realized in the frequency domain. In WDM, each transmission occupies a set of frequencies (i.e., a channel) whose total size is fixed to 50 GHz for each channel [2]. Working in conjunction with the C-band Erbium-Doped Fiber Amplifiers (EDFAs), the main aim of the WDM-OTNs is to provision the channels (or lightpaths) for the transmitted signals. However, the currently deployed single-mode fiber-enabled fixed-grid WDM-OTNs does not provide any resource assignment flexibility as they transmit simultaneous lightpaths via non-overlapping and fixed-spaced wavelength channels. Hence, with increasing traffic demands, the infrastructure of such 10Gbps WDM-OTNs needs enhancement by increasing the capacity and adapting the advanced modulation formats, having increased spectral efficiency at a higher data rate [3]. In this regard, a mixed line rate (MLR) network, shown in Figure 11.2, supports

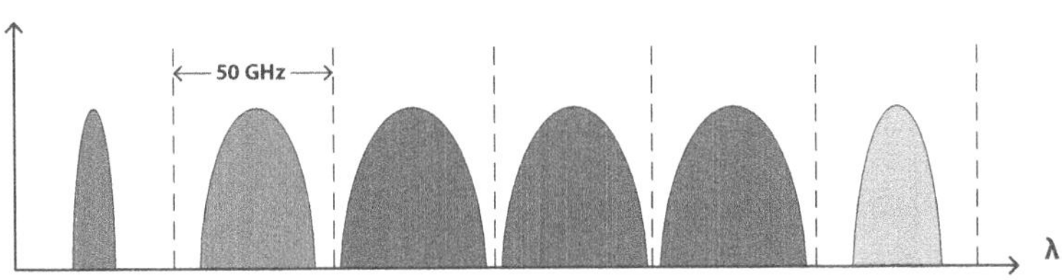

Figure 11.1 The fixed-grid WDM OTN with every color representing a different connection.

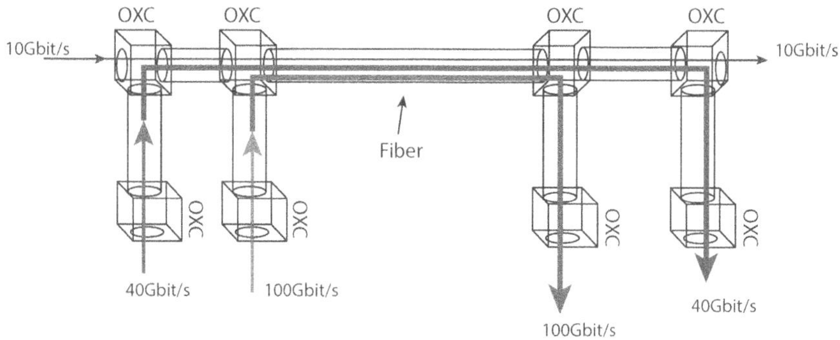

Figure 11.2 The multi-line rate optical network.

10/40/100 Gbps on different wavelengths over the same fiber thereby, avoiding the provisioning of low-bandwidth connections over high-capacity lightpaths, and using an optimal combination (number/rate) of the wavelengths on each link which addresses both, the traffic and the network asymmetry [4, 5]. However, the fixed frequency grid technique (i.e., allocation of 50 GHz spectrum for all transmissions regardless of their volume) provides low spectrum efficiency, and hence, the WDM/MLR technology is not expected to be efficient for the increasing traffic volumes [6].

To provision flexibility in the spectrum usage through the frequency slots (FSs), the research community has proposed the adoption of spectrum flexibility through the deployment of the Elastic Optical Networks (EONs), which resort to the implementation of spectrally efficient modulation formats using the super-channel switching technique to enhance the network performance. Hence, the EONs, in comparison to the WDM technology, provide superior spectrum efficiency owing to the new spectrum division/provisioning manner and the application of advanced modulation and transmission techniques [6]. Specifically, in the EONs, the entire spectrum width is divided into narrow and equal size segments (6.25 GHz or 12.5 GHz [7]) called frequency slices [8], as shown in Figure 11.3.

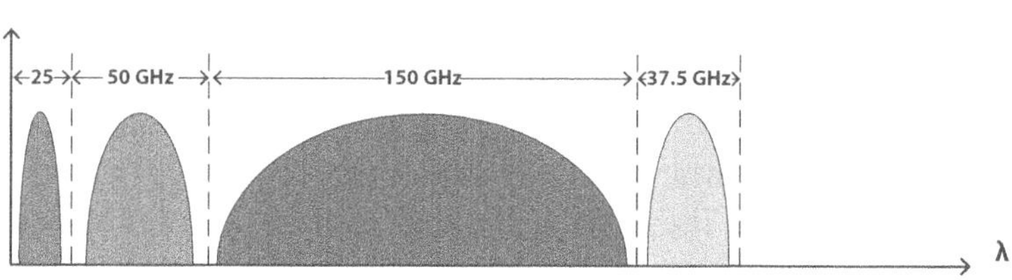

Figure 11.3 The flexible grid elastic optical network with every color representing a different connection.

By grouping a number of such adjacent slices, frequency channels are created, narrower and wider than those in the WDM OTNs. Such a technique enables the allocation of the channels tailored to the volume of the incoming transmissions [9]. Specifically, in the EONs, the lightpaths which carry the super-channels are routed via the network over the single-mode fiber (SMF) links over aptly allocated spectrum channels switched within the nodes of the network by wavelength selective switches (WSSs) [10]. Also, given a traffic demands set, the lightpath(s) routing mandates a contention-free assignment of the spectrum over every link that belongs to every connection's route that realizes any demand. The aforementioned is known as the routing and spectrum allocation (RSA) problem, which necessitates, for the end-to-end demands competing for spectrum, evaluation of the lightpath connections, which are tailored according to the actual width of the super-channels [11]. This RSA problem exists during the network planning as well as the network operational phase [6]. However, even though the EONs can support higher bit-rate traffic requirements much more efficiently than the fixed-grid WDM OTNs, it has now emerged that the capacity of EONs will soon reach the upper-bound to the conventional SMF's non-linear Shannon limit [12].

In view of further extending the capacity limit in the OTNs (including WDMs and EONs) recently; the space division multiplexing (SDM) technology has been introduced in which the capacity extension is achieved by exploiting a number of spatial resources and their simultaneous application to realize the transmissions [13, 14]. The SDM technology is a prime candidate for provisioning the future demands requiring enormous bandwidth as it enables the simultaneous transmission of multiple 'spaces (i.e., modes, cores, or fibers)' through the appropriate fiber solutions. The space domain could correspond to multiple cores inside the same cladding of the multi-core fibers (MCFs); multiple modes inside the same core of the few-mode fibers (FMFs); or parallel SMFs within the same bundle [15, 16]. The rationale behind enabling the EONs with the SDM technology, termed as 'SDM-EONs,' will result in major benefits such as, increased network capacity and more flexibility in resource(s) (i.e., spectrum and spaces) usage and sharing of the already network-integrated devices. However, SDM technology's integration within the EONs also introduces many challenges such as, SDM-EON optimization and formulation of efficient resource assignment strategies and algorithms for SDM-EONs, which need timely solutions. The SDM technology for flexi-grid OTNs is shown in Figure 11.4.

With the introduction of the spatial domain, in addition to the optical carriers of the super-channels being carried by continuous frequency slots,

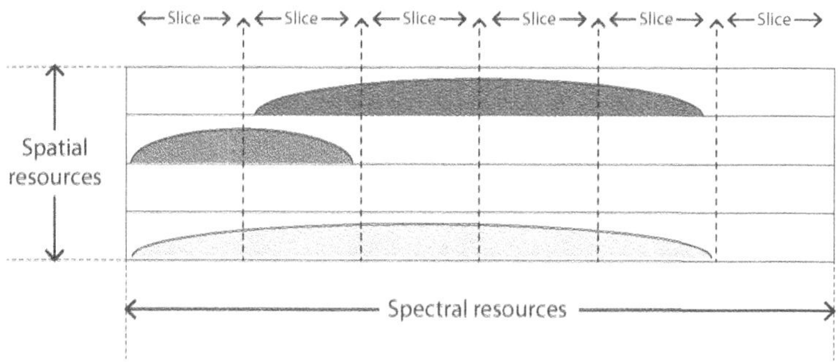

Figure 11.4 Space division multiplexing for flexi-grid optical networks (each channel, assigned with a different color, represents a set of frequency slices that is allocated on a particular spatial resource).

these optical carriers can also be distributed over multiple spatial modes [17]. Hence, the SDM-EONs belong to the class of spectral-spatial flexible OTNs in which the RSA technique (applicable in the EONs) transforms to the routing, space, and spectrum allocation (RSSA) technique due to the presence of the spatial mode, which in turn requires mathematical optimization methods to select the appropriate spatial mode which will be used to transmit the signals [18]. Hence, compared to the RSA technique, the RSSA method in the SDM-EONs assumed higher complexity due to the addition of the spatial mode, which increases the decision variables within the optimization problem [19]. Also, multiple SDM-EON scenarios can exist which require various fiber solutions and switching methods [20].

The integration of the SDM technology within the EONs is envisioned to offer (i) an enormous increase in the capacity, (ii) enhanced management of the resources owing to the presence of the spatial domain, and (iii) savings in-network cost due to the opportunity of being able to share the resources and the devices such as, switches, amplifiers, or transponders [21–23]. However, the aforementioned also introduces numerous challenges, including whether the SDM-EONs will provide long-haul transmission and whether the SDM-EONs be integrated within the existing legacy OTNs [24]. Specifically, simultaneous transmission of many spatial modes within the appropriate fiber solutions and the implementation of flexible spatial and spectral techniques are envisioned to increase the SDM-EON performance. This, in turn, also requires appropriate resource assignment schemes, which include spectral, spatial, or spectral-spatial transmission. Further, the efficient resources (spectrum and space) management includes the lightpath allocation in order to provision the demands and formulation of novel algorithms to solve the routing, spatial mode, and spectrum allocation. The aforementioned may have a severe impact on transmission quality due to interference among the modes, and this phenomenon must be accounted for within the SDM-EON planning and operational phases.

This chapter surveys the various SDM-EON technologies that combine different state-of-art technologies to form multiple SDM-EON scenarios. This includes the fiber solution and the switching technique used for obtaining the maximum capacity. Next, we also review the various RSSA techniques proposed in the literature, which offer optimized solutions in regard to the resource assignment in the spectrally-spatially flexible OTNs. Lastly, we also identify the research gaps which will aid in the planning and operation of the SDM-EONs.

11.2 Review of SDM-EON Physical Models

This section discusses the various fiber solutions and spectral and spatial switching techniques resulting in various spectral and spatial superchannels. Lastly, we also detail the various SDM-EON scenarios which are possible with a combination of the fiber solution and the switching methods.

11.2.1 Optical Fibers for SDM-EON

Several optical fibers have been proposed in the literature, which apply to the SDM-EONs [24–26]. The most popularly used fibers are as followed:

- Single-mode MCF (SM-MCF) – This type of fiber comprises many single-mode cores within a fiber cladding wherein every core guides one mode. For the MCFs, many core placement structures have been proposed, which include one ring, two rings, linear-array, two-pitched, or hexagonal structures. Further, the spacing can be even or uneven, and the packing of the core can be close or wide.
- FMF – This type of fiber is a type of multiple mode fiber (MMF) which, for SDM transmission, uses very few numbers of transverse modes that traverse within a high index core. Compared to the MMFs, the FMFs can transmit for longer distances as the physical layer impairments in the FMFs can be mitigated much easier than in the MMFs as the FMFs have a lesser amount of modes that propagate simultaneously.

- Few mode MCF (FM MCF) – This type of fiber comprises of the advantages of the MCFs and the FMFs, and in this fiber, the transverse modes propagate within the cores with a higher refractive index.
- Fiber bundle (FB) – In this type of fiber, multiple fibers exist in the form of a bundle. The bundle could be of SMF (SMFB) in which every fiber is capable of supporting an individual mode, or it could be of FMF (FMFB) in which every fiber is capable of guiding few modes. In view of instant migration from the existing deployed legacy networks to the SDM-EONs, the SMFB are most likely to be the first commercially available fibers.

Further, the fiber solution for SDM-EONs can be distinguished based on how the spatial modes interfere with the co-propagating modes. In the strongly coupled mode type, the spatial modes interfere; however, the data remains within the modes set. At the receiver, multiple-input multiple-output (MIMO) digital signal processing (DSP) must be used to retrieve the information [14]. In comparison, the weakly coupled type allows for the spatial modes to be routed and switched individually over the network, and hence, the MIMO DSP is not required, which minimizes the cost [20].

The literature review found that the MCFs are the most popular choice for realizing the SDM-EONs as most articles have considered the use of MCFs for the SDM-EON implementation (see Table 11.1). In the majority of the studies in the literature, the MCF core is placed in an evenly wide hexagonal shape with cores amount ranging between 7 and 12 being considered for the analysis. Further, the SMFB is the next most popular fiber solution, which is considered in the literature. Also, fibers such as OAM [41] and HC-PBGFs [42] find their less prominent mention in the literature.

Table 11.1 SDM optical fibers within the survey.

Type of fiber	Reference
SMFB	[27, 28, 51]
MCF	[27–35]
FMF	[36–38]
FMFB	[36]
FM MMF	[32, 39, 40]

It must be noted that the most popular fiber solution for SDM-EONs viz., the MCFs are affected by signal impairments due to the inter-cross-talk (XT) that occurs between the neighboring cores whenever the signals are transmitted over spectrum that overlaps [43]. Further, XT denotes disturbance level due to the signal power that leaks from the adjacent cores on the core in which the desired signal propagates. The XT is known to increase with the transmission distance [44]; it depends on the frequency and is affected by random variations in time [45, 46].

11.2.2 Switching Techniques for SDM-EON

The SDM-EONs need to be deployable simultaneously, ensuring scalability, switching flexibility, and cost-effectiveness. In this regard, multiple SDM-EON networking techniques are proposed in the literature that require state-of-art switching and networking [20, 21]. These methods offer various dimensions of spectral and spatial switching flexibility and can function with different fiber solutions. The existing switching solutions proposed in the literature are as follows:

- Independent Switching (InSw): In this type of switching technique, the spatial or the spectral channel can be switched independently [47]. InSw offers the finest granularity in the switching method as all the frequency slots, and the spatial modes can be directed to the desired output port in an independent manner. However, this technique incurs increased complexity of hardware [21].
- Mode Switching (MoSw): In this type of switching, also known as spatial switching (SpSw), all the spatial modes can be switched across all the channels, i.e., in every mode, the complete band of communication can be switched in a joined manner [47]. Thus, MoSw provides the switching of spatial modes across all the wavelength channels [48].
- Joint Switching (JoSw): In this type of switching, all the spatial modes are considered as one entity, and the frequency slots are switched by the WSS [47]. Thus, JoSw provides switching of spectral modes across all the spatial modes [48].
- Fractional Joint Switching (FrJoSw): In this type of hybrid switching, spatial modes groups are treated as a single entity, and the frequency slots can be switched to the output ports independently [47]. Thus, FrJoSw switches the spectral modes across sub-groups of the spatial modes [48].

Table 11.2 Switching techniques within the survey.

Switching technique	Reference
InSw	[32–35, 38, 47–51]
JoSw	[38, 48–51]
FrJoSw	[27, 38, 49, 50]
MoSw	[30, 49]
Hybrid	[28, 30, 49]

Further, the aforementioned switching techniques can also be included in hybrid techniques detailed in [28, 30, 49]. In Table 11.2, we present the survey of the studies which have adopted a particular switching technique. From the table, it can be clearly observed that among the various techniques, the InSw technique is the most considered by the existing studies. The next most considered techniques are the JoSw and the FrJoSw methods whereas the MoSw and the hybrid methods are less prominently considered.

Further, the SDM-EONs enable the transmission of various types of super-channels which comprise of many carriers that split into the spectral and/or spatial domains. In this regard, following the studies in [21, 47], the following SDM supporting transponders exist, which can generate or terminate any super-channel:

- Spectral transponders: Such transponders generate the spectral super-channels, which consist of many carriers required to be routed simultaneously with the use of one spatial mode in any fiber solution. Further, in regard to improving the spectral efficiency, it is ensured that the carriers follow the Nyquist condition. There are no guardbands between the carriers, whereas a guardband is required between the adjacent super-channels.
- Spatial transponders: Such transponders enable the simultaneous transmission over any adopted fiber solution in which the aggregate carriers of any super-channel utilizes similar frequency; however, the carriers are allotted varied spatial modes. The spatial transponders have been demonstrated to (i) share one LASER as the source and (ii) use the SDM enabled multiplexer/de-multiplexer for the parallel

transmission of many signals. However, such transponders result in the spatial super-channel requiring guard-bands over every individual mode of space.

- Spatial spectral transponders: Such transponders combine the idea of the spatial transponders and the spectral transponders and hence, enable the generation and reception of the spatial and spectral super-channels with many spatial modes used for the transmission of the spectral super-channels. Such transponders also allow few devices such as, LASER sources to be shared and require guard-bands for every mode. It must be noted that the super-channels transmission over the network necessitates the appropriate assignment of the resources (both spatial and spectral), and the switches need to be configured appropriately. In Table 11.3, we present the survey of the studies which have adopted a particular super-channel technique. It can be observed from the table that the majority of the existing studies have considered the spectral super-channels in the analysis.

The use of spatial super-channels or the combination viz., spatial-spectral super-channels is less frequently considered in the literature. Further, from the survey, it has emerged that, under similar capacity, the use of the spatial super-channels results is low spectral-efficiency compared to the consideration of the spectral super-channels. The aforementioned occurs since, in the spatial super-channel case, guardbands for switching are required to be inserted over each spatial dimension to separate the adjacent super-channels. However, as specified in [52], in view of improved network performance according to the bandwidth blocking probability, the width of the guardbands that separate the adjacent spatial super-channels can be minimized.

Thus far, the discussed fiber solutions, switching techniques, and transponders can be combined to result in various SDM-EON scenarios. We classify the different SDM-EON scenarios which find the most mention in

Table 11.3 Super-channel techniques within the survey.

Super-channel technique	Reference
Spectral	[32–35, 38, 47–52]
Spatial	[38, 47–50]
Spatial-Spectral	[27, 38, 49]

the literature. Firstly, it must be noted that all combinations of the various solutions are impossible as every solution has its limitations. As detailed in [48], for complete exploitation of the considered spatial-spectral switching techniques, it is mandatory to utilize the apt fiber solution. For example, few fibers (SMFBs or MCFs) are compatible with all switching techniques, whereas fibers such as FMFs and FM MCFs demonstrate efficiency only when the JoSw technique is implemented [27]. Further, it has been shown in the literature that there is a correlation between the super-channel type, fiber solution, and the switching technique which is adopted [21]. For instance, the spectral super-channels do not show efficiency when used in an MCF solution, and JoSw technique scenario since spectral super-channels occupy individual modes; however, combination (spatial-spectral) super-channels can be used with any fiber solution implementing InSw or JoSw technique [18].

Further, the SDM-EON scenarios can be characterized by an appropriate resource assignment method resulting from the SDM technology implemented. For example, the JoSw technique requires the assignment of aggregate spatial modes within a provided range of spectrum over the routes' entre links for the lightpaths. Additionally, few constraints are distinguished, such as the space continuity constraint and the spectrum continuity constraint, which relate to the utilized devices' capabilities [27]. The space continuity constraint and the spectrum continuity constraints imply that for every demand, overall, the links of the route, allotment of the same spatial route and spectrum is ensured [18]. Additionally, space contiguity constraint and the spectrum contiguity constraints may also be applied, which ensures for a demand, allotment of a contiguous spectrum in terms of frequency slots of contiguous spatial routes. Lastly, the demands' non-overlapping constraint could also apply, i.e., over the links, it is ensured that a set of FSs is allotted to only one demand [27]. However, it is found from the survey that the majority of the existing works relax the space continuity constraint and optimize the network under the spectrum continuity constraint. Thus, in these studies, space mode conversion is allowed by implementing the InSw or the MoSw techniques [32].

From our survey of the SDM scenarios existing in literature, we have observed that in comparison to the spatial super-channels, the spectral super-channels demonstrate superior network performance owing to the higher spectral-efficiency; however, owing to sharing of various network devices by all spatial modes being permitted in the spectral super-channels, such super-channels provide higher savings in the cost when compared to the spectral super-channels [27, 47]. Further, the study in [48] has shown that the various switching techniques' performance depends on the granularity of the spectral switching and the amount and size of traffic demands. In [53], the authors have shown that InSw, JoSw, and FrJoSw techniques demonstrate almost the same performance when the network is offered higher loads of traffic. The authors have compared the SDM-EON performance considering the FMFB and SMFB solutions and the MCF and SMFB solutions in [36] and [54], respectively. The results show that the capacity of the network scales efficiently when the fiber solution considered is SMFB.

Lastly, in Table 11.4, we present the survey of the combined classification of the surveyed SDM network scenarios. Also, with the use of the equipment (switches, transponders, etc.) and the spatial and/or spectral modes discussed in this section, every super-channel results in a different resource assignment technique, which is detailed in the next section.

Table 11.4 SDM-EON scenario within the survey.

SDM-EON scenario	Switching technique	Super-channel technique	Reference
SMFB	InSw	Spectral	[27, 32, 47, 50]
		Spatial	[27, 28, 32, 50]
		Spatial-spectral	[55]
	JoSw	Spatial	[27, 28, 36, 52]
		Spatial-spectral	[55]
	Hybrid	Spatial	[50]
MCF	InSw	Spectral	[31, 32, 54, 56,]
		Spatial-spectral	[32, 54]
	JoSw	Spatial	[52, 54]
	MoSw	Spectral	[57]
FMF	JoSw	Spatial	[36]
		Spatial-spectral	[38]

11.3 Review of SDM-EON Resource Assignment Techniques

Using various equipment such as, switches, transponders, etc., and the spatial and/or spectral modes result in every super-channel using a different resource assignment technique. In this section of the chapter, we segregate the various resource assignment algorithms which can be used to optimize the resources in the SDM-EONs.

In general, the problems which are addressed include the SDM-EON design and optimized resource assignment. These problems can be static or dynamic depending on whether the connection demands are known a-priori or appear stochastically. In the static case, the resource assignment does not incur any constraint of strict time of processing as the decisions are made offline, and hence, the mathematical programming and meta-heuristic methods can be used for optimization [48]. On the other hand, in the dynamic case, resource assignment is required to be conducted in an online mode, and the decisions are made depending on the currently available resources within the SDM-EON [47].

Our survey has revealed that most of the existing studies have adopted the dynamic version of the resource assignment problem compared to the static version. This is because the static optimization of SDM-EONs fails to use the energy and network resources efficiently [27, 28]. Further, it has been established that it is simpler to obtain feasible solutions for the dynamic resource assignment problems when compared to the static version, which requires the use of complex optimization techniques [36, 52]. Overall, for the SDM-EON optimization, considering all the demands, the complexity is high due to multiple decision variables and the inclusion of many constraints required in choosing the appropriate path, spatial mode, and spectrum.

In the literature, considering the dynamic optimization cases, the studies have focused on providing the lightpath problem [28]. The studies based on static optimization have focused on the lightpath planning issue [38]. Few studies have also focused on other advanced issues such as, planning of network equipment [30, 38], MIMO processing [36], and reducing the overall SDM-EON cost [58]. In regard to the static version of optimization, the majority of the studies have assumed a fixed set for the demands except [30], which has accounted for a quasi-static model of the traffic which assumes that the demands, which are once established within the SDM-EON, remain within the network till they are completely provisioned.

In regard to the algorithmic techniques, a common method used by studies in the literature is based on mathematical modeling in which, depending on the variables, the formulation is either an integer linear program, an integer program, or a mixed-integer program. The solutions to such aforementioned problems can be obtained through solvers such as, CPLEX optimizer software, GUROBI optimizer, etc., which efficiently implement the algorithms. However, for large cases of complex problems, such mathematical models demonstrate low scalability as they cannot find the desired solutions within reasonable times. On the other hand, these models provide exact solutions to the problems. Considering the aforementioned trade-off, studies in the literature have also resorted to using heuristic and meta-heuristic techniques to obtain the solutions. Such techniques can provide approximate solutions to large problem cases in reasonable execution times; however, the exactness of the solution is compromised. In the majority of the studies which implement the heuristic techniques to find the solutions to the resource assignment problem, the greedy mechanism is used in which, within one simulation run, the decisions are made [59, 60]. Our survey observed that the most implemented mathematical model is the integer linear program [30, 31, 38, 55, 57], followed by greed approach-based heuristics [27, 30, 48]. The meta-heuristics-enabled resource assignment algorithms are less frequently used in the literature [48, 61]. As an exception, the study in [62] has implemented an integer program including the constraints which are non-linear and related to the various physical layer impairments.

Additionally, it must also be pointed out that the meta-heuristic techniques have remained an effective tool to find optimal solutions to the various network problems [63]. However, it has been shown in [64] that the use of meta-heuristic techniques to optimize the SDM-EONs incurs scalability issues owing to the high complexity of the RSSA problem. Hence, one of the key open areas of research is to formulate novel problem-solving methods to ensure efficient and high-speed algorithm performance. Further, the primarily used integer linear program models for SDM-EONs use the link-route or the node-link routing models in which, compared to the node-link model, the link-route model incurs less time to obtain the solutions [64]. In regard to the spectrum assignment model, the slice-based [30, 38, 57] or the channel-based [65, 66] methods have been used by the studies in the literature. As for the fiber solution, the majority of the integer linear program models for the SDM-EONs are proposed under the deployment assumption of the MCFs [14].

Specifically, for the dynamic SDM-EONs, existing studies have implemented simpler heuristic techniques to assign the resources for the demands which arrive randomly. This is due to the fact that for the dynamic RSSA, decisions are to be made quickly for individual demands, which in turn makes the resource assignment decision in the dynamic case simpler when compared to the static optimization case. Therefore, in the majority of the studies available in the literature, the first fit method is implemented by the adopted heuristic so that the demand is established over the very initial wavelength path available, spatial mode, and the frequency slot [33–35]. Also, few studies use a cost-based metric to choose the route, spatial mode, and frequency slot [30, 57]. Lastly, it must also be noted that the dynamic resource assignment methods can also be applied to the static optimization case if the demands are processed according to a pre-selected sequence.

Next, we address the details of the RSSA algorithms, which are detailed in the literature. SDM brings the added dimension of 'space' within the EONs. Hence, in the SDM-EONs optimization, every demand must be allocated to one or more available spatial resources, resulting in the optimization problem being that of routing and assigning the spatial mode and spectrum appropriately [28, 48, 50]. Further, if the fiber solution adopted is enabled by MCF, the RSSA is extended to the routing, core, and spectrum allocation (RCSA) problem [32, 47]. Also, if modulation format is considered during the optimization of SDM-EON, RSSA extends to routing, modulation format, spatial mode, and spectrum allocation (RMSSA) [30, 31, 57]. The existing studies have shown that the RSSA technique is *NP*-hard [11] and can be solved in two ways: (i) joint RSSA, in which the resources and variables are jointly optimized [30], and (ii) decomposed RSSA, in which the routing problem is solved first followed by the spatial and spectrum assignment problem [33]. Specifically, in regard to the second technique, which is more popular, the main aim is to choose the appropriate spatial and spectral resources for the lightpaths. In the decomposed RSSA method, when a lightpath connection request arrives, the solution to the spatial and spectrum assignment problem is found by allotting an available spectral and spatial resource over every link of the already found route. Compared to the SDM-EONs, which are lightly loaded, in the SDM-EONs, which are heavily loaded, the adopted spatial and spectrum assignment technique has a major effect on the network performance. In the majority of the existing studies, the spatial and spectral resource allotment is done using the matrix structure technique wherein each element of the matrix denotes the availability of the frequency slot over the corresponding spatial mode [67].

Compared to the EONs, which use the first-fit spectrum assignment policy [68], in the SDM-EONs, the resource assignment varies depending on the order of choice of the spectrum and the spatial domain. From our survey, we found that the articles which implement the spectral super-channels, the spatial and spectrum assignment technique may be any one of the following:

- Selection of spectral super-channel with the least frequency slot followed by the selection with the least core index.
- Selection of spectral super-channel with the least core index followed by the selection with the least frequency slot.
- Selection of a previously computed region within the spectrum which corresponds to the bandwidth required by the connection.
- Selection of a spectral super-channel based on the least cost metric value where, the cost value could denote the cost to deploy a specific node architecture or the effect of crosstalk.

It must be noted that all the aforementioned strategies can be applied to both the spatial super-channels and the spectral super-channels. However, the SDM-EON performance owing to the use of the various strategies depends largely on a scenario of the network assumed, specification of the traffic, and assumed physical model of the network [32]. In regard to the switching techniques, studies have demonstrated that for efficient use of the spectrum allotment strategies, InSw is to be used; whereas, for the spatial allotment strategies, JoSw or FrJoSw needs to be used [20, 21].

Considering the practical scenario of dynamic SDM-EONs wherein the established lightpaths do not remain within the network for long and are to be released after a fixed interval of time, resource fragmentation issues start to emerge. This occurs owing to the variation in the network state frequently simultaneously with multiple demands of the bandwidth. In fact, studies have demonstrated that fragmentation of resources (spatial, spectral, or both) results in the inefficient use of resources and a drastic increase of the blocking probability of the demands [18]. Few studies in the literature present solutions to the fragmentation problem. In [29, 32], the authors have proposed a core classification method considering the MCF case and the spectral super-channels. The authors in [69] have studied the concept of cross-core virtual concatenation for the SDM-EONs, considering the relaxation of the spectrum contiguity constraint. Further, in [70], a fragmentation-aware method is proposed, which aims at congestion avoidance. Also, similar to the EONs case [68], in the SDM-EONs also, the resource management methods with fragmentation-awareness can be either proactive or reactive.

Lastly, we also present the results of our survey in terms of the performance metrics applicable for the SDM-EONs. These metrics aid in assessing the SDM-EON performance and optimization. From the survey, we observed that majority of the existing studies have aimed at reducing the use of spectrum considering the following functions: (i) maximum spectrum, defined as the largest frequency slices amount which is required to serve all demands in SDM-EON, (ii) aggregate spectrum, which expresses the aggregate frequency slices amount assigned over all the SDM-EON links, and (iii) mean spectrum, which represents the mean spectrum use in every SDM-EON link over each fiber [61]. Further, few studies also exist that use the overall network cost as a metric and aim to minimize it [71], whereas the study in [72] has the reduction of the switching modules amount as the objective. The authors in [38] have included the cost of the transceiver required for installing the SDM-EON within the objective function network. In regard to the performance metrics, the connection blocking probability is the most popular [27, 48], whereas few studies have also used the bandwidth blocking probability [33–35].

11.4 Research Challenges in SDM-EONs

Through our extensive survey in this chapter, we were able to identify the major open research avenues related to SDM-EONs, which require timely solutions. We identified that there exists a large technological chasm between the transport solution and SDM network implementation. The subject of SDM-EON is still in its infancy and has not yet been extensively researched in literature, specifically in comparison to corresponding relevant research on fixed-grid WDM OTNs and flexible-grid EONs. Specifically, the appropriate dedicated solutions to the 'classical' OTN issues and cases, which have been addressed regarding WDM networks and EONs, are not completely researched in the context of the SDM-b-EONs. Hence, we list the following key research avenues as a scope for future research in the SDM-EON domain:

- The SDM-EON performance analysis considering the various fiber solutions and different switching and super-channel technologies.
- Formulation of new RSSA algorithms which can efficiently optimize SDM-EON performance and which are also scalable.

- Comparison of various optimization problems solving ILP models (e.g. integer linear programming models, meta-heuristics, etc.) for various SDM-EON scenarios considering path, space and spectrum resources.
- Detailed analysis of the effect of cross-talk on the SDM-EON performance under various scenarios.
- Analyzing the SDM-EON performance considering the multiple traffic grooming techniques which are used for high bit-rate transmission.
- Application of the fault and protection techniques within the SDM-EONs.
- Develop the fragmentation techniques for the SDM-EONs, accounting for the fragmentation of the spatial and spectral resources.

11.5 Conclusion

In this chapter, we presented a survey on the various aspects to be considered for the planning and operation of the various SDM-EON scenarios. In this regard, firstly, we presented the various fiber solutions available for implementation in the SDM-EONs, followed by a detailed discussion on the various switching technologies applicable in the SDM-EONs. Further, we described the various transponders types and super-channels which SDM-EONs support. The aforementioned, as a combination, are observed to result in various SDM-EON scenarios. Lastly, we also discussed the various resource allocation schemes and algorithms that can be formulated for the various SDM-EON scenarios and address the RSSA problem, including crosstalk and resource fragmentation.

Our survey makes it clear that (i) SMFB or MCF are the most considered fiber solutions, (ii) Ind-Sw is the most considered switching technique, and (iii) spectral super-channels are the most considered in the existing literature. We also observed that multiple studies had compared SDM-EON performance considering various schemes for resource assignment. Finally, the literature mentions static and dynamic resource assignment issues through various heuristic and meta-heuristic approaches simultaneously considering various performance and optimization metrics.

References

1. Cisco visual networking index: Forecast and methodology, 2016–2021, white paper, 2017, https://www.cisco.com/c/en/us/solutions/collateral/service-provider/visualnetworking-index-vni/complete-white-paper-c11-481360.pdf. https://virtualization.network/Resources/Whitepapers/0b75cf2e-0c53-4891-918e-b542a5d364c5_white-paper-c11-738085.pdf
2. Singh, S.P., Iyer, S., Kar, S., Jain, V.K., Study on mitigation of transmission impairments and issues and challenges with PLIA-RWA in optical WDM networks. *J. Opt. Commun., Gruyter*, 33, 2, 83–101, 2012.
3. Singh, S.P., Sengar, S., Bajpai, R., Iyer, S., Next-generation variable line-rate optical WDM networks: Issues and challenges. *J. Opt. Commun., Gruyter*, 34, 331–50, 2013.
4. Iyer, S. and Singh, S.P., Spectral and power-efficiency investigation in single and multi-line-rate optical wavelength division multiplexed (WDM) networks. *Photonic Netw. Commun.*, Springer, 33, 1, 39–51, 2017.
5. Iyer, S. and Singh, S.P., Investigation of launch power and regenerator placement effect on the design of mixed-line-rate (MLR) optical WDM networks. *Photonic Netw. Commun.*, Springer, 40, 1, 14–30, 2020.
6. Gerstel, O., Jinno, M., Lord, A., Yoo, S.J.B., Elastic optical networking: A new dawn for the optical layer? *IEEE Commun. Mag.*, 50, 2, 12–20, 2012.
7. ITU-T recommendation G.694.1. Spectral grids for WDM applications: DWDM frequency grid, 1–14, Jun. 2002. accessed at: T-REC-G.694.1-200206-S!!PDF-E.pdf
8. Iyer, S., Investigation of cost and spectrum utilization in internet protocol-over-elastic optical networks. *J. Opt.*, Springer, 49, 3, 279–290, 2020.
9. Iyer, S., Traffic grooming with survivability and power-efficiency in software defined elastic optical networks. *J. Opt.*, Springer, 47, 3, 351–365, 2018.
10. Jinno, M., Takara, H., Kozicki, B., Tsukishima, Y., Sone, Y., Matsuoka., S., Spectrum-efficient and scalable elastic optical path network: Architecture, benefits, and enabling technologies. *IEEE Commun. Mag.*, 47, 11, 66–73, 2009.
11. Agrell, E., Karlsson, M., Chraplyvy, A., Richardson, D., Krummrich, P., Winzer, P., Roberts, K., Fischer, J., Savory, S., Eggleton, B., Secondini, M., Kschischang, F., Lord, A., Prat, J., Tomkos, I., Bowers, J., Srinivasan, S., Brandt-Pearce, M., Gisin, N., Roadmap of optical communications. *J. Opt.*, 18, 1–40, 2016.
12. Winzer, P.J., Spatial multiplexing in fiber optics: The 10x scaling of metro/core capacities. *Bell Labs Technol. J.*, 19, 22–30, 2014.
13. Li, G., Bai, N., Zhao, N., Xia, C., Space-division multiplexing: The next frontier in optical communication. *Adv. Opt. Photonics*, 6, 413–487, 2014.
14. Saridis, G.M., Alexandropoulos, D., Zervas, G., Simeonidou, D., Survey and evaluation of space division multiplexing: From technologies to optical networks. *IEEE Commun. Surv. Tutorials*, 17, 4, 2136–2156, 2015.
15. Essiambre, R.J., Ryf, R., Fontaine, N.K., Randel, S., Breakthroughs in photonics 2012: Space-division multiplexing in multimode and multicore fibers

for high-capacity optical communication. *IEEE Photonics J.*, 5, 2, article is: 0701307-0701307, 2013. https://ieeexplore.ieee.org/document/6508867
16. Morioka, T., Awaji, Y., Ryf, R., Winzer, P., Richardson, D., Poletti., F., Enhancing optical communications with brand new fibers. *IEEE Commun. Mag.*, 50, 2, s31–s42, 2012.
17. Xia, T.J., Fevrier, H., Wang, T., Morioka, T., Introduction of spectrally and spatially flexible optical networks. *IEEE Commun. Mag.*, 53, 2, 24–33, 2015.
18. Klonidis, D., Cugini, F., Gerstel, O., Jinno, M., Lopez, V., Palkopoulou, E., Sekiya, M., Siracusa, D., Thouenon, G., Betoule, C., Spectrally and spatially flexible optical network planning and operations. *IEEE Commun. Mag.*, 53, 2, 69–78, 2015.
19. Ryf, R., Chandrasekhar, S., Randel, S., Neilson, D.T., Fontaine, N.K., Feuer., M., Physical layer transmission and switching solutions in support of spectrally and spatially flexible optical networks. *IEEE Commun. Mag.*, 53, 2, 52–59, 2015.
20. Marom, D.M., Colbourne, P.D., D'Errico, A., Fontaine, N.K., Ikuma, Y., Proietti, R., Zong, L., Rivas-Moscoso, J.M., Tomkos, I., Survey of photonic switching architectures and technologies in support of spatially and spectrally flexible optical networking. *IEEE/OSA J. Opt. Commun. Netw.*, 9, 1, 1–26, 2017.
21. Marom, D.M. and Blau, M., Switching solutions for WDM-SDM optical networks. *IEEE Commun. Mag.*, 53, 2, 60–68, 2015.
22. Krummrich, P.M., Optical amplifiers for multi mode/multi core transmission, in: *Proceedings of OFC*, Los Angeles, USA, 2012.
23. Randel, S., Space division multiplexed transmission, in: *Proceedings of OFC*, Anaheim, USA, 2013.
24. Richardson, D.J., Fini, J.M., Nelson, L.E., Space division multiplexing in optical fibres. *Nat. Photonics*, 7, 354–362, 2013.
25. Nakajima, K., Sillard, P., Richardson, D., Li, M.-J., Essiambre, R.-J., Matsuo, S., Transmission media for an SDM-based optical communication system. *IEEE Commun. Mag.*, 53, 2, 44–51, 2015.
26. Mizuno, T., Takara, H., Sano, A., Miyamoto, Y., Dense space-division multiplexed transmission systems using multi-core and multi-mode fiber. *IEEE J. Light. Technol.*, 34, 2, 582–592, 2016.
27. Khodashenas, P.S., Rivas-Moscoso, J.M., Siracusa, D., Pederzolli, F., Shariati, B., Klonidis, D., Salvadori, E., Tomkos, I., Comparison of spectral and spatial superchannel allocation schemes for SDM networks. *IEEE J. Light. Technol.*, 34, 11, 2710–2716, 2016.
28. Pederzolli, F., Siracusa, D., Shariati, B., Rivas-Moscoso, J.M., Salvadori, E., Tomkos, I., Improving performance of spatially joint-switched space division multiplexing optical networks via spatial group sharing. *IEEE/OSA J. Opt. Commun. Netw.*, 9, 3, B1–B11, 2017.
29. Fujii, S., Hirota, Y., Tode, H., Murakami, K., On-demand spectrum and core allocation for reducing crosstalk in multicore fibers in elastic optical networks. *IEEE/OSA J. Opt. Commun. Netw.*, 6, 12, 1059–1071, 2014.

30. Muhammad, A., Zervas, G., Forchheimer, R., Resource allocation for space division multiplexing: Optical white box vs. optical black box networking. *IEEE J. Light. Technol.*, 33, 23, 4928–4941, 2015.
31. Perello, J., Gene, J.M., Pages, A., Lazaro, J.A., Spadaro, S., Flex-grid/SDM backbone network design with inter-core XT-limited transmission reach. *IEEE/OSA J. Opt. Commun. Netw.*, 8, 540–552, 2016.
32. Tode, H. and Hirota, Y., Routing, spectrum, and core and/or mode assignment on space division multiplexing optical networks. *IEEE/OSA J. Opt. Commun. Netw.*, 9, A99–A113, 2017.
33. Iyer, S., On routing, modulation format, space and spectrum allocation with protection in space division multiplexing based elastic optical networks. *J. Inf. Telecommun.*, Taylor & Francis, 4, 1, 105–117, 2020.
34. Iyer, S., An online routing algorithm for space division multiplexing based elastic optical networks. *Int. J. Commun. Networks Distrib. Syst., Inderscience*, 24, 2, 167–185, 2020.
35. Iyer, S., On the minimization of resource utilization for cost reduction in space division multiplexing based elastic optical networks. *Int. J. Internet Protoc. Technol.*, Inderscience, 15, 1, 29–40, 2022.
36. Diamantopoulos, N.P., Shariati, B., Tomkos, I., On the power consumption of MIMO processing and its impact on the performance of SDM networks, in: *Proceedings of OFC*, Los Angeles, USA, pp. 1–3, 2017.
37. Huang, H., Huang, S., Yin, S., Zhang, M., Zhang, J., Gu, W., Virtual network provisioning over space division multiplexed optical networks using few-mode fibers. *IEEE/OSA J. Opt. Commun. Netw.*, 8, 10, 726–733, 2016.
38. Rottondi, C., Boffi, P., Martelli, P., Tornatore, M., Routing, modulation format, baud rate and spectrum allocation in optical metro rings with flexible grid and few-mode transmission. *IEEE J. Light. Technol.*, 1, 1, 61–70, 2017.
39. Yang, Y., Chen, X., Yan, H., Hua, B., Li, J., Hao, Y., Chen, Z., He, Y., A scattered spectrum-scan routing and spectrum allocation scheme for spatial-division-multiplexing optical networks based on blocking OXCs, in: *Proceedings of ACP*, Wuhan, China, 2016.
40. Li, Y., Hua, N., Zheng, X., A capacity analysis for space division multiplexing optical networks with MIMO equalization, in: *Proceedings of OFC*, Los Angeles, USA, pp. 1–3, 2017.
41. Boffi, P., Martelli, P., Gatto, A., Martinelli, M., Mode-division multiplexing in fibre optic communications based on angular momentum. *J. Opt.*, 15, 7, 075403, 2013.
42. Xu, J., Lyngso, J.K., Leick, L., Carpenter, J., Wilkinson, T.D., Peucheret, C., Mode division multiplexing exploring hollow-core photonic bandgap fibers, in: *Proceedings of IEEE ICTON*, Cartagena, Spain, 2013.

43. Koshiba, M., Design aspects of multicore optical fibers for high-capacity long-haul transmission, in: *Proceedings of IEEE Topical Meeting MWP*, Sendai, Japan, pp. 318–323, 2014.
44. Hayashi, T., Taru, T., Shimakawa, O., Sasaki, T., Sasaoka, E., Design and fabrication of ultra-low crosstalk and low-loss multi-core fiber. *Opt. Express*, 19, 17, 16576, 2011.
45. Ye, F., Tu, J., Saitoh, K., Takenaga, K., Matsuo, S., Takara, H., Morioka, T., Wavelength dependence of inter-core crosstalk in homogeneous multi-core fibers. *IEEE Phot. Technol. Lett.*, 28, 1, 27–30, 2016.
46. Luis, R.S., Puttnam, B.J., Cartaxo, A.V.T., Klaus, W., Mendinueta, J.M.D., Awaji, Y., Wada, N., Nakanishi, T., Hayashi, T., Sasaki, T., Time and modulation frequency dependence of crosstalk in homogeneous multi-core fibers. *IEEE J. Light. Technol.*, 34, 2, 441–447, 2016.
47. Siracusa, D., Pederzolli, F., Klonidis, D., Lopez, V., Salvadori, E., Resource allocation policies in SDM optical networks, in: *Proceedings of ONDM*, Pisa, Italy, pp. 168–173, 2015.
48. Shariati, B., Klonidis, D., Rivas-Moscoso, J.M., Tomkos, I., Evaluation of the impact of spatial and spectral granularities on the performance of spatial superchannel switching schemes, in: *Proceedings of IEEE ICTON*, Trento, Italy, pp. 1–4, 2016.
49. Zhao, Y., Tian, R., Yu, X., Zhang, J., Zhang, J., An auxiliary graph based dynamic traffic grooming algorithm in spatial division multiplexing enabled elastic optical networks with multi-core fibers. *Opt. Fiber Technol.*, 34, 52–58, 2017.
50. Pederzolli, F., Gerola, M., Zanardi, A., Forns, X., Ferran, J., Siracusa, D., YAMATO the first SDN control plane for independent, joint and fractional-joint switched sdm optical networks. *IEEE J. Light. Technol.*, 35, 8, 1335–1341, 2017.
51. Yang, M., Wu, Q., Zhang, Y., Joint assignment of spatial granularity, routing, modulation, and spectrum in SDM-EONs: Minimizing the network CAPEX considering spectrum, WSS, and laser resources. *IEEE J. Light. Technol.*, 36, 18, 4153–4166, 2018.
52. Rumipamba-Zambrano, R., Perelló, J., Gené, J.M., Spadaro, S., On the scalability of dynamic flex-grid/SDM optical core networks. *Comput. Networks*, 142, 208–222, 2018.
53. Shariati, B., Khodashenas, P.S., Rivas-Moscoso, J.M., Ben-Ezra, S., Klonidis, D., Jimenez, F., Velasco, L., Tomkos, I., Evaluation of the impact of different SDM switching strategies in a network planning scenario, in: *Proceedings of OFC*, San Diego, USA, pp. 1–3, 2016.
54. Rumipamba-Zambrano, R., Perelló, J., Gené, J.M., S., Effective capacity quantification of joint-switching-enabled flex-grid/SDM optical backbone networks, in: *Proceedings of OFC*, Los Angeles, USA, pp. 1–3, 2017.

55. Walkowiak, K., Klinkowski, M., Lechowicz, P., Sen, A., ILP modeling of flex-grid SDM optical networks, in: *Proceedings of Networks*, Montreal, Canada, pp. 1–3, 2016.
56. Zhao, Y. and Zhang, J., Crosstalk-aware cross-core virtual concatenation in spatial division multiplexing elastic optical networks. *IEEE Electron. Lett.*, 52, 20, 1701–1703, 2016.
57. Muhammad, A., Furdek, M., Zervas, G., Wosinska, L., Filterless networks based on optical white boxes and SDM, in: *Proceedings of ECOC*, Dusseldorf, Germany, pp. 1–3, 2016.
58. Rivas-Moscoso, J.M., Shariati, B., Mastropaolo, A., Klonidis, D., Tomkos, I., Cost benefit quantification of SDM network implementations based on spatially integrated network elements, in: *Proceedings of ECOC*, Dusseldorf, Germany, pp. 1–3, 2016.
59. Pióro, M. and Medhi, D., *Routing, flow, and capacity design in communication and computer networks*, Morgan Kaufmann, San Francisco, CA, USA, 2004.
60. Simmons, J., *Optical network design and planning*, 2nd edition, Springer Publishing Company Incorporated, United States, 2014.
61. Walkowiak, K., *Modeling and optimization of cloud-ready and content-oriented networks*, vol. 56 of Studies in Systems, Decision and Control, Springer, Switzerland, 2016.
62. Li, Y., Hua, N., Zheng, X., A capacity analysis for space division multiplexing optical networks with MIMO equalization, in: *Proceedings of OFC*, Los Angeles, USA, pp. 1–3, 2017.
63. Pióro, M. and Medhi, D., *Routing, flow, and capacity design in communication and computer networks*, Morgan Kaufmann, San Francisco, CA, USA, 2004.
64. Iyer, S., Multiple core fibers versus multiple fibers enabled space division multiplexing based elastic optical networks: A performance comparison. *Int. J. Comput. Netw. Inf. Secur.*, 8, 13–23, 2019.
65. Li, G., Bai, N., Zhao, N., Xia, C., Space-division multiplexing: The next frontier in optical communication. *Adv. Opt. Photonics*, 6, 4, 413–487, 2014.
66. Li, Y., Li, Y., Hua, N., Zheng, X., Shared backup path protection in multi-core fiber networks with MIMO-based crosstalk suppression, in: *Proceedings of OFC*, San Diego, USA, pp. 1–3, 2016.
67. Shi, Z., Zhao, Y., Yu, X., Li, Y., Zhang, J., Liu, C., Zhang, G., Liu, Z., Contaminated area based RSCA algorithm for super-channel in flex-grid enabled SDM networks, in: *Proceedings of ACP*, Wuhan, China, 2016.
68. Talebi, S., Alam, F., Katib, I., Rouskas, G.N., Spectrum management techniques for elastic optical networks: A survey. *Opt. Switch. Netw.*, 13, 34–48, 2014.
69. Zhao, Y. and Zhang, J., Crosstalk-aware cross-core virtual concatenation in spatial division multiplexing elastic optical networks. *IEEE Electron. Lett.*, 20, 1701–1703, 2016.

70. Proietti, R., Liu, L., Scott, R.P., Guan, B., Qin, C., Su, T., Giannone, F., Yoo, S.J.B., 3D elastic optical networking in the temporal, spectral, and spatial domains. *IEEE Commun. Mag.*, 53, 2, 79–87, 2015.
71. Rivas-Moscoso, J.M., Shariati, B., Mastropaolo, A., Klonidis, D., Tomkos, I., Cost benefit quantification of SDM network implementations based on spatially integrated network elements, in: *Proceedings of ECOC*, Dusseldorf, Germany, pp. 1–3, 2016.
72. Muhammad, A., Zervas, G., Saridis, G., Salas, E.H., Simeonidou, D., Forchheimer, R., Flexible and synthetic SDM networks with multi-core-fibers implemented by programmable ROADMs, in: *Proceedings of ECOC*, Cannes, France, pp. 1–3, 2014.

12

Role of Wavelength Division Multiplexing in Optical Communication

P. Gunasekaran*, A. Azhagu Jaisudhan Pazhani, A. Rameshbabu and B. Kannan

Department of ECE, Ramco Institute of Technology, Rajapalyam, Tamil Nadu, India

Abstract

Optical communication, also known as optical telecommunication, is a method of long-distance communication in which light is used to carry data. Wavelength Division Multiplexing (WDM) is used for fast data transmission. WDM (wavelength division multiplexing) is a fiber-optic communications device that uses different wavelengths (or colors) of laser light to multiplex a range of optical carrier signals into a single optical fiber. This technique, also known as wavelength-division duplexing, allows bidirectional communication over a single strand of cable. WDM describes an optical carrier that is traditionally specified by wavelength, while frequency-division multiplexing describes a radio carrier that is typically defined by frequency. This is strictly conventional since wavelength and frequency convey the same information. The frequency (in Hertz, which is cycles per second) multiplied by wavelength determines the carrier wave's velocity (the physical duration of one cycle). The lower case letter c is used to represent the speed of light in a vacuum.

In glass fiber, it moves at a much slower pace, about 0.7 times c. The data rate, which should ideally be at the carrier level, is always a fraction of the carrier frequency in practical devices. Normal WDM, Coarse WDM, and Dense WDM are the three wavelength patterns used in WDM systems. The data transmission speed increases as a result of the WDM concept.

***Keywords*:** Wavelength Division Multiplexing (WDM), Dense Wavelength Division Multiplexing (DWDM), on-off keying (OOK), Polarization-Mode Dispersion (PMD)

**Corresponding author*: gunasekaran@ritrjpm.ac.in

Chandra Singh, Rathishchandra R Gatti, K.V.S.S.S.S. Sairam and Ashish Singh (eds.) Modeling and Optimization of Optical Communication Networks, (217–234) © 2023 Scrivener Publishing LLC

12.1 Introduction

Light is employed as the information carrier in optical communication systems. The optical fiber is the most common medium for optical communication channels, hence the system is known as a fiber-optic communication system. A laser produces light at the transmitter, and the information, which was at first an electrical sign, is balanced interested in the optical transporter. The tweaked optical sign is guided by a fiber optic connection with optical speakers to the recipient, which ought to have the option to recuperate the first electrical information by identifying and demodulating the optical sign [1].

The utilization of tweak designs former than on-off keying (OOK) in optical fiber correspondence frameworks, similar stage and recurrence shift keying was broadly explored preceding the presentation of the optical speaker, basically because of their high recipient affectability. The main issues were the system's complexity and the requisite laser line width. The optical amplifier was created not solitary to alleviate the problem of receiver compassion, nevertheless also to allow the effective routine of numerous wavelengths. As a result, OOK has been employed as the most cost-effective modulation style for over a decade, and the channel data rate has expanded exponentially above 100Gbit/s. Several linear and nonlinear phenomena restrict the optical broadcast of a single channel at high channel data rates.

Transmission systems employing traditional OOK signals in one or the other non-return-to-zero (NRZ) or return-to-zero (RZ) format and running at channel data speeds more than 40 Gbit/s are extremely difficult to build. Since of their improved signal tolerance to fiber nonlinearities, chromatic scattering, polarization-mode diffusion, and great spectral competence, advanced modulation formats have been proposed as an unusual to OOK in wavelength-division-multiplexing (WDM) schemes functional at data rates superior as 10Gbit/s per channel. Advanced modulation formats are divided into two categories. The first employs coherent detection at the receiver, which necessitates the employment of a local oscillator and digital signal processing to retrieve the data. The second employs straight discovery without a local oscillator and digital signal processing, making receiver design much easier.

A cascade of identically equipped portions generally makes up the link or optical transmission line. A regular single-mode fiber (SMF), a dispersion reimbursing fiber (DCF), and two optical amplifiers are included in each segment to compensate for power losses during fiber transmission.

The cumulative amplified unstructured emission noise created by the optical amplifiers, polarization-mode dispersion (PMD), and fiber spreading and nonlinearity are the major sources of performance deterioration for these systems, which are commonly quantified in terms of bit-error rate (BER). When light travels through a fiber, amplified unstructured emission noise interrelates with the optical signal due to nonlinearity in the fiber, resulting in nonlinear phase noise.

These factors, composed with random riddling, make determining the BER accurately the maximum difficult component of the receiver model. After reviewing the BER, it's possible that you'll want to improve the system's performance. The system routine may have several native and universal extrema due to various in lines and nonlinear components along the connection, which presents a difficulty for simple optimization strategies. The role of wavelength division multiplexing in optical communication is described in this concept.

12.2 Modules of an Optical Communication System

Fiber optic connection is the best alternative for gigabit and beyond gigabit data transport. Voice, video, telemetry, and data are all sent across great distances over LAN or workstation networks using this kind of communication. Light wave technology is used to transport data through fiber by converting electrical information into light in a fiber optic communication system. Massive bandwidth, slighter diameter, lightweight, long-distance signal transmission, short attenuation, transmission safety, and other unique characteristics of this type of communication technology make it a critical component of any communications structure. The following info on fiber optic communication systems focuses on its distinguishing characteristics, fundamental parts, and other specifics.

Fiber optic correspondence is a strategy for conveying information starting with one area then onto the next by sending infrared light heartbeats across an optical fiber. The light is in the method of a transporter signal, which is adapted to store the records in this case. When great distance, high bandwidth, and electromagnetic interference resistance are required, fiber optic cables replace electrical wires [2].

Fiber-optic communication is typically utilized in LANs to convey audio, audio-visual, and telemetry. Optical fiber is used by plentiful telecommunication businesses to passage telephone, cable TV, and Internet message signals. Bell Labs researchers researched fiber-optic communication and set a new record for BW distance of 100 petabit kilometers per second [19].

12.2.1 How a Fiber Optic Communication Works?

It's nothing like copper wire transmission, which depends solely on electrical driving forces successful over the link has fiber optics transmission included the exchange of light transmissions starting with one spot then onto the next. A fiber optic correspondence network likewise incorporates transmission and getting hardware, a light source, and locator gadgets, for example, those found in Figure 12.1.

The transmitter hardware turns the info information, which is as electrical driving forces, into a light sign with the guide of a light source. For productive transmission, the sufficiency, recurrence, and periods of this LED source should stay steady and liberated from change. A fiber optic link conveys the light pillar from the source to the objective hardware, where the data is changed over back to an electrical sign by a collector circuit [7].

A light finder and a reasonable electronic circuit make up the recipient circuit, which can gauge the adequacy, recurrence, and period of the optic field. This strategy for correspondence uses frequencies simply over the visual reach, moving toward the infrared area. Contingent upon the application, together LED and Laser can be utilized as light sources.

12.2.2 Codes of Fiber Optic Communication System

A fiber optic communication system is made up of three primary components. They really are.

- Dense Light Source
- Low loss Optical Fiber
- Photo Detector

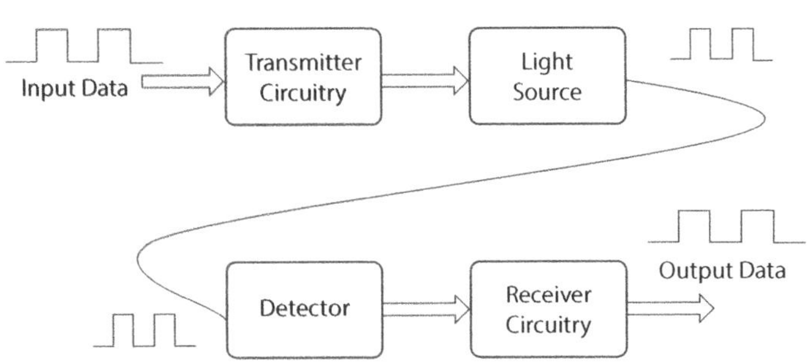

Figure 12.1 Working of fiber optical communication.

12.2.2.1 Dense Light Source

The light source requirements differ depending on the request, such as local area networks and long-haul communication systems. Power, speed, spectral linewidth, noise, durability, cost, temperature and other factors are among the foundations supplies. Light-emitting diodes (LEDs) and laser diodes are both employed as light sources. The following Figure 12.2 shows the dense light sources [3].

Because of their short bandwidth and power limitations, light-emitting diodes are employed for tiny range and low data rate uses. Surface and Edge Emitting Systems are two examples of LED architectures. Surface-emitting diodes are modest to build and dependable, while edge-emitting diodes are more commonly utilized due to their wider linewidth and modulation frequency restriction. Edge emitting diodes feature a high power output and a shorter linewidth.

Laser Diodes are favored for lengthier spaces and great data rate transmission owing to their high power, high speed, and shorter spectral linewidth features. These, on the other hand, are intrinsically non-linear and supplementary susceptible to temperature changes.

Many advances and innovations have made these sources more dependable in recent years. Below are a few examples of such comparisons between these two sources. Direct and external modulation methods are used to modify each of these sources.

12.2.2.2 Low Loss Optical Fiber

Optical fiber is a type of tow that is also recognized as a low-loss tube-shaped dielectric waveguide. Optical fiber also takes into account factors such as the working environment, ductile strength, toughness, and stiffness. Fiber optic cable is flexible and composed of superior extruded glass

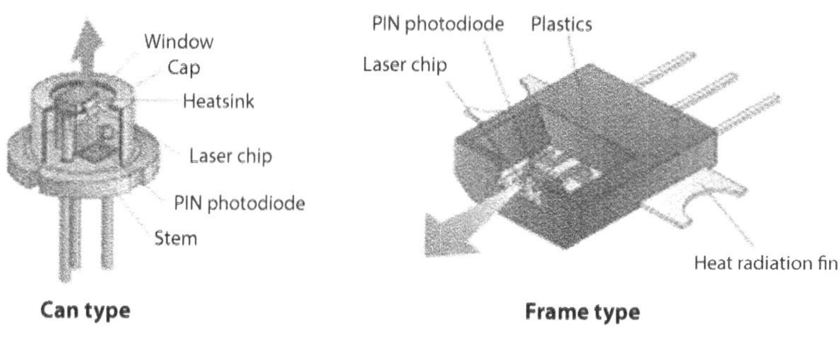

Figure 12.2 Dense light sources.

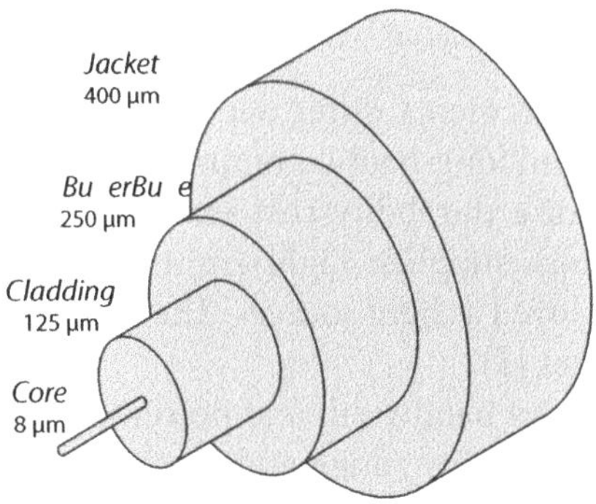

Figure 12.3 Fiber structure.

or plastic. The fiber optic cable's diameter ranges from 0.25 to 0.5mm. The following Figure 12.3 shows the fiber structure [3].

A Fiber Optic Cable comprises of following parts [5].

- Core
- Cladding
- Buffer
- Jacket

Core
A fiber cable's core is a chamber of artificial plastic that spans the length of the cable and provides safety through cladding. The core's width is determined by the application. Light moving inside the core replicates after the core, the cladding border, owing to interior reflection. For the majority of applications, the core cross-section must be round.

Cladding
The core is protected by cladding, which is an exterior optical substance. The cladding's primary role is to reflect light rear into the core. After light passes over the core and enters the cladding, it deviates angle and returns to the core.

Buffer
The buffer's key purpose is to preserve the fiber and several fibers grouped into optical cables from damage. The outer coating of the cable, known as the jacket, protects these bundles.

Types of Optical Fibers [6]
Single-Mode Fibers: Single-mode fibers, which remain employed in telephone and television sets, convey just one signal per fiber. The cores of single-mode fibers are tiny.

Multi-Mode Fibers: Multimode fibers remain employed to transport numerous signals per fiber; these communications are employed in bigger core PC and local area networks.

12.2.3 Photo Detectors

Photodetectors are utilized to change over light signals once again into electrical signs. For optical recipients in optical correspondence frameworks, two sorts of photodetectors are regularly utilized: PN photodiode and torrential slide photodiode. The material arrangement of these gadgets fluctuates relying upon the frequencies utilized in the application. Silicon, germanium, InGaAs, and different materials fall inside this class.

12.3 Wavelength-Division Multiplexing (WDM)

The technique of choice for transferring huge volumes of data between locations has traditionally been wavelength division multiplexing, or WDM. It boosts bandwidth by allowing several data streams to be delivered across a single optical fiber network at the same time. WDM enhances the utility of fiber and aids in the optimization of network investments in this way. Carriers and service providers have traditionally used WDM systems. Large-scale systems developed for "national infrastructures" make private network use excessively expensive and difficult. Things have altered in recent years. Furthermore, technology is quickly developing [4].

WDM networking solutions for business corporations, governmental institutions, and privately held datacenters are now available. Solutions that is less complicated and less expensive than standard carrier-grade solutions. And in what way it can be utilized to get the most out of fiber networks and maximize network investments.

The ability to transport diverse data types through fiber networks in the form of light is at the heart of WDM. A single virtual fiber grid is established by assisting several light channels, each with its own wavelength, to be broadcast simultaneously via an optical fiber network. Relatively than needing plentiful fibers for each service, a single fiber can be shared by

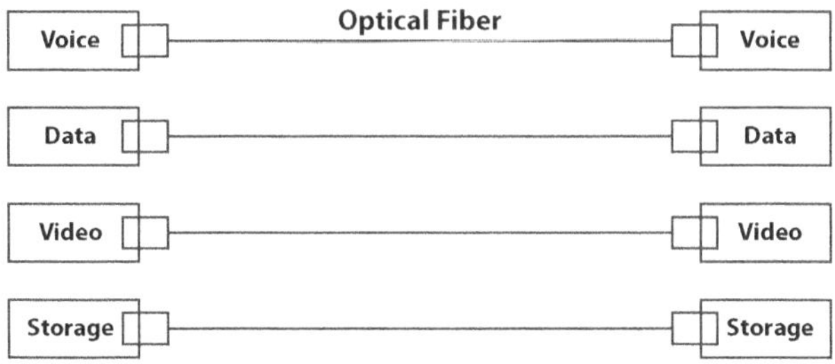

Figure 12.4 Individual fiber channel.

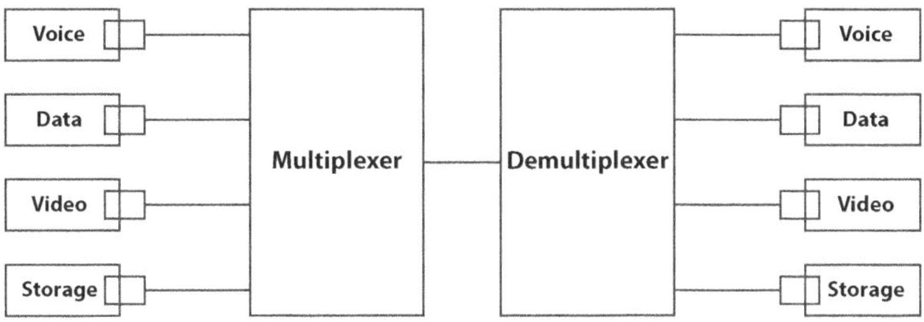

Figure 12.5 Multiplexing combines multiple channels in a single fiber.

many services. Figures 12.4 and 12.5 shows the individual fiber channel and Multiplexing combines multiple channels in a single fiber [3].

WDM boosts bandwidth and enhances the utility of fiber in this way. Fiber leasing or purchase accounts is for a sizable portion of networking expenditures. As a result, leveraging an existing fiber to transmit various traffic channels may save a lot of money.

WDM systems are made up of four components in their most basic form is shown below Figure 12.6.

12.3.1 Transceivers – Transmitting Data as Light

Information from SAN and IP switches are converted to light signals that may be transferred into the fiber via transceivers, which are wavelength-specific lasers. Each data stream is transformed into a signal with a distinct color wavelength. Channels cannot interact with each other due to the physical qualities of light. As a result, all WDM wavelengths are independent. By using this method to create virtual fiber channels, the amount of fibers required is diminished by the factor of the wavelengths employed. It also enables for the addition of new channels as needed without interfering with existing traffic services [10].

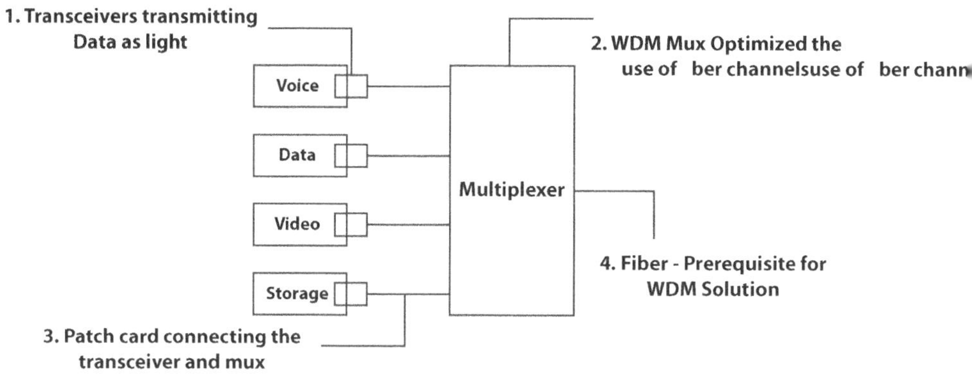

Figure 12.6 WDM System component.

Any mix of different services may be carried concurrently over a single fiber or fiber pair since each channel is see-through to the speed and type of information.

12.3.2 Multiplexers Enhancing the Use of Fiber Channels

The WDM multiplexer, also known as the passive multiplexer, is essential for optimizing or maximizing the fiber's utilization. The multiplexer is at the heart of the process, bringing all of the information streams together to be sent across a single fiber at the same time. The streams are demultiplexed, or divided into different channels, at the opposite end of the cable.

Two bi-directional channels might be transported via a pair of fibers in early WDM systems. The number of channels and the amount of data delivered per channel have both risen as technology has progressed. At any given moment, up to 80 channels can be concurrently sent via a fiber [8].

Multiplexers are sometimes known as terminal muxes since they are frequently found at the network's termination points. A multiplexer is placed at each site to create a point-to-point connection when linking two locations. Many networks have extra locations where some sort of connectivity is necessary, but not all forms of stream of traffic. Optical add drop multiplexers (OADMs) are castoff to excerpt the wavelengths mandatory for a given location while avoiding traffic kinds that aren't required. This allows for the creation of more adaptable ring, distribution, and access networks.

12.3.3 Categories of WDM

WDM might be ordered into two gatherings dependent on frequency [9]:

Course WDM (CWDM)
CWDM ordinarily has eight channels with a 20 nm division between them. It utilizes less energy and is less exorbitant than DWDM. The limit of the connections, just as the distance upheld, are, notwithstanding, restricted.

Dense WDM (DWDM)
The quantity of multiplexed diverts in DWDM is generously higher than in CWDM. It's either 40 GHz separating at 100GHz or 80 GHz dividing at 50GHz. Subsequently, they can move enormous measures of information through a solitary fiber connect. DWDM is normally utilized in media communications and link organizations' center organizations. It's likewise utilized for IaaS administrations in cloud server farms.

12.4 Modulation Formats in WDM Systems

Fiber nonlinear special effects, intermodal dispersal, polarization mode dispersal and increased impulsive emission noise all cause problems in high-capacity WDM systems. In WDM systems, these variables restrict the transmission capacity and distance. A broad range of strategies have been proposed to increase the transmission performance of WDM systems:

- Advanced modulation are utilized to compromise commotion strength, fiber Propagating qualities,
- Fiber types lessening nonlinear signal spins and empowering sophisticated signal dispatch power,
- Methods for scattering, scattering slant, and PMD compensation
- Administration of nonlinear impedances
- Signal adjustment
- Forward error correction.

The most efficient way to deal with transmission problems is to employ sophisticated modulation schemes. Because dissimilar data formats have distinct waveforms and bands, they provide varying signal superiority at the receiver side for a particular broadcast channel. Links with varied system factors may require different optimum data formats at the same time. For long-haul, high-speed WDM broadcast connections, the optimum modulation format is one with a dense range, low sensitivity to fiber nonlinear effects, substantial scattering tolerance, and simple and economical generation and recognition setups.

The most common modulation formats, such as NRZ, RZ, CRZ, CSRZ, DPSK, and CSRZ-DPSK will be discussed and contrasted in this chapter. We'll start by propagating a single channel using well-known modulation methods. Using various types of optical fibers, components, and amplifiers, we steadily expand the numeral of channels, bitrates, and system complication. Different fiber types and modulation formats may have an impact on system performance. As a result, several modulation formats must be compared. We'll also conduct a thorough research of optical in lines and nonlinear special effects for optimum system routine, i.e., the lowest possible bit error ratio (BER). The eye-diagram is being used to track how well the system is working.

12.4.1 Optical Modulator

Direct modulation and external modulation are the two most used modulator technologies.

12.4.1.1 Direct Modulation

Direct modulation of semiconductor lasers remained the preferred method for low-capacity logical optical systems with tiny transmission distances in the 1980s and 1990s. They are cost-effective and practical for short distances. Direct modulation, on the other hand, creates chirping, which causes the carrier frequency of a signal to change over time, resulting in signal pulse broadening or dispersion. Furthermore, laser phase noise and generated jitter hinder direct modulation's progress toward increased capacity and bit rate transmission. It can't handle bit rates more than 2.5 Gigabits per second. Furthermore, direct modulation, particularly SPM, causes nonlinearity [12].

12.4.1.2 External Modulation

The external modulation approach can be used to circumvent the constraints of direct modulation. Nonlinearities and excessive chirp are avoided by External Modulation. Electro-optic modulators (EOM) modulators can be used to accomplish external modulation.

The EOM work on the basis of electro optic effect principles. In a solid state, polymeric, or semiconductor material, the change in refractive index is relational to the pragmatic electric field. Popular terms of squeak, extermination ratio, and modulation swiftness, they perform better. Electro optic modulator waveguides have mostly been incorporated on the factual

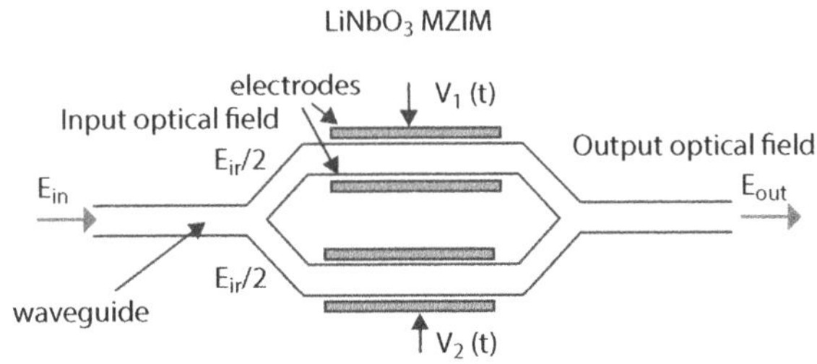

Figure 12.7 Mach-Zehnder interferometer structure.

platform of lithium niobate ($LiNbO_3$), which is high efficiency, low loss, and easy to produce over the years. It becomes increasingly common after the introduction of the Erbium doped optical fiber amplifier (EDFA) in the late 1980s. EOMs are used to change the phase or intensity of a lightwave carrier [17].

The phase of optical carrier signals is manipulated by an electro optic phase modulator beneath the stimulus of an electric field formed via the supplied voltage. When an RF driving voltage is provided to the electrode, the refractive index varies, causing variations in the amount of delays of the propagating lightwave. Because the delays correlate to the phase changes, an electro optic phase modulator is utilized to convey the phase modulation of the optical carrier [15].

The mechanism of optical intensity modulation is built on the interference of the optical fields of the two lightwave components. The most often used modulator is the LiNbO3-based Mach-Zehnder interferometer (MZIM). It's extensively utilized in communication networks with speeds of 2.5, 10, and 40 gigabits per second.

Figure 12.7 depicts the construction of a $LiNbO_3$-based Mach-Zehnder interferometer [18]. When entering the modulator, the inward light is separated into two artilleries. The optical signals' power is separated by the power splitter. An electro optic phase modulator is used in each arm of the $LiNbO_3$ modulator to adjust the phase of the optical carrier if necessary. The light waves of the two artilleries phase modulators are connected and hindered with further at the MZIM's output. The input voltages $V_1(t)$ and $V_2(t)$ are launched into the modulator's arms.

12.4.2 Modulation Formats

The modulation format specifies how information is encoded on an optical signal. The optical signal's basic parameters and state of polarization (SOP) may all be modified. Contingent on which of the optical features of the carrier's electric field relates to which of the four modulation formats, they may be divided into four categories:

$$E(t) = \hat{e}Ae^{j(wt+\phi)}$$

A: Amplitude Shift Keying
ω: Phase Shift Keying
ϕ: Frequency Shift Keying
ê: Polarization Shift Keying

Depending on whether the sign to be sent is a mark ("1") or a space ("0"), ASK codes messages by switching on or off the brightness of light at a level proportional to the communication frequency. Each binary symbol ("1" or "0") in this modulation style is signified by the occurrence or non-appearance of light. Non-return to-zero (NRZ), return-to-zero (RZ), and duo-binary plans are all included. The RZ format is also available in a variety of versions. Simple RZ, carrier suppressed RZ (CSRZ), chirped RZ (CRZ), vestigial sideband RZ (VSB-RZ), and dispersion-managed soliton-based RZ are some of the types of RZ (DMS-RZ) [13].

PSK modulates the phase of light to encode data. Each binary symbol ("1" or "0") in this modulation scheme is characterized by a light phase of "0" or "π". Differential PSK (DPSK), RZ-DPSK, CSRZ-DPSK, differential quaternary PSK (DQPSK), and their pulse imprinted versions are all included.

FSK, which comprises FSK, continuous phase FSK (CPFSK) and minimum shift keying (MSK), encrypts data through modulating the frequency of light. PolSK modulates the polarization of light to encode data. The modulator creates two orthogonal polarization states in polarization shift keying, which correspond to "1" and "0" bits.

12.4.2.1 Non Return to Zero (NRZ)

Because it is simple to make, detect, and process, non return to zero (NRZ) consumes the dominant modulation plan in intensity modulated-direct detection (IM/DD) fiber-optical communication over the last several years. Figure 12.8 shows the representation of NRZ code [11].

The NRZ transmitter is shown in Figure 12.9 [11]. The pragmatic electric field, whose voltage fluctuates through a defined role, modulates the greatness of the carrier light wave. An electrical NRZ signal is used to drive

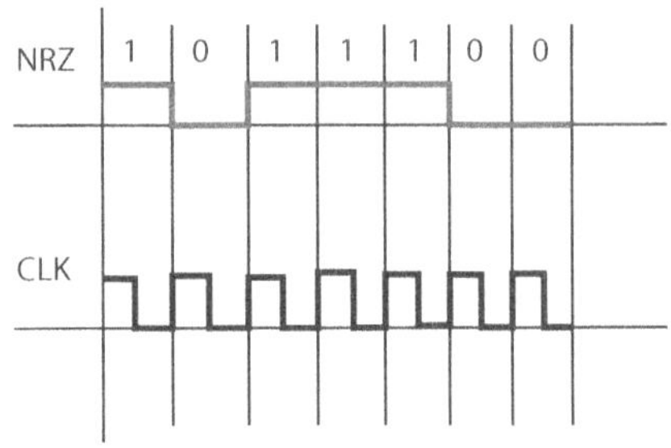

Figure 12.8 Representation of the NRZ code.

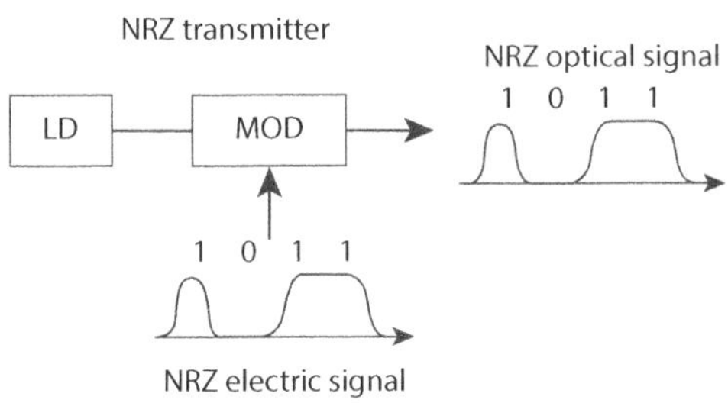

Figure 12.9 NRZ transmitter diagram.

the Mach- Zehnder modulator (MZM) at the quadrature point of the modulator power transfer function.

Because of the lower on-off transitions, the optical spectrum of NRZ pulses is narrow. Abridged spectral width progresses dispersion tolerance and allows for greater spectral efficiency, but it also impacts inter-symbol interference (ISI) between pulses. When compared to its NR equivalent, the NRZ modulated optical signal is more resistant to fiber nonlinear consequence. NRZ, on the contrary, has the meekest transceiver design. In comparison to RZ, it necessitates a comparatively small electrical bandwidth for transceivers. NRZ uses around partial the bandwidth of RZ, making it cooler to deploy and fewer expensive.

12.4.2.2 Return to Zero (RZ)

Recent studies and examinations revealed that RZ systems are superior to traditional NRZ systems. The power level for the logical 1 bit in RZ format returns to 0 after partial of the time, but the power level for the 0 bit remains constant. During the full bit period, binary 0 is signified by the lack of an optical pulse. Figure 12.10 shows the representation of the RZ code [11].

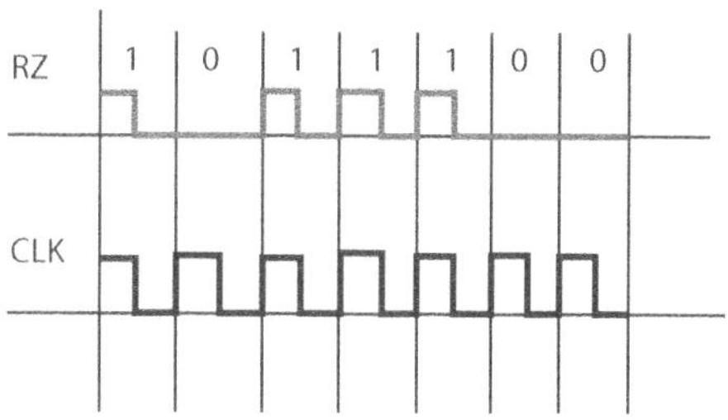

Figure 12.10 Representation of the RZ code.

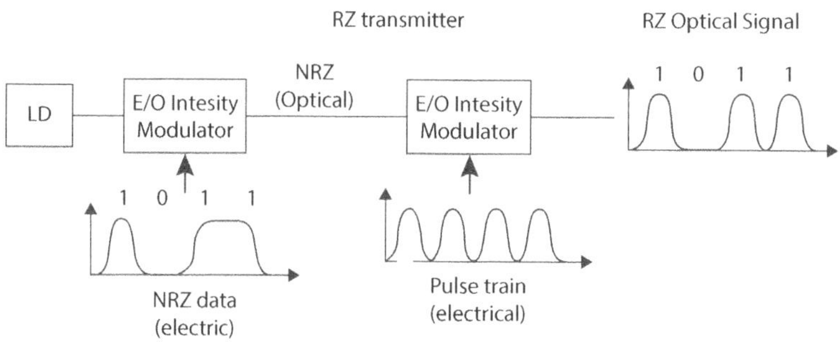

Figure 12.11 RZ transmitter.

The RZ transmitter is seen in Figure 12.11 [11]. An external intensity modulator generates the NRZ optical signal first. Then, using another intensity modulator, it is modulated through a synchronized pulse train by the similar data rate by means of the electrical signal.

12.4.2.3 Chirped RZ (CRZ)

Chirped RZ (CRZ) is one of the RZ format's variants. The CRZ transmitter is shown in Figure 12.12 [3]. The data is modulated into the laser source to form an NRZ signal, which is then fed into the RZ modulator to generate RZ pulses. Lastly, a phase modulator is utilized to synchronously modulate the RZ pulse in order to chirp the yield, with the highest chirp occurring in the middle of the pulse. A phase modulation of 1.5 radians is ideal.

The signal bandwidth in CRZ is spectrally broadened by bit-synchronous periodic chirp. Even if this limits the setup's applicability for great spectral efficiency WDM schemes, it improves its resilience to fiber nonlinearity in general.

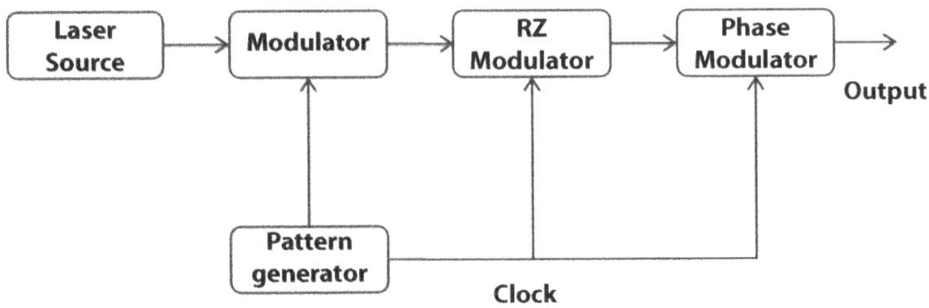

Figure 12.12 Chirped RZ transmitter.

12.4.2.4 Carrier Suppressed RZ (CSRZ)

CSRZ is a kind of RZ in which the carrier is disabled. In high data rate systems, the CSRZ format decreases nonlinear impairments in the channel and enhances spectral efficiency. The CSRZ signal differs from standard RZ in that it features a phase shift amongst consecutive bits. Fashionable optical domain, this phase change yields no DC element; hence there is no carrier module for CSRZ. The vital frequency modules are reduced to partial of the data rate by phase alternating across neighboring bit slots. Because of its lower optical power, CSRZ is more tolerant of chromatic dispersion, permitting for supplementary channels to be multiplexed in broadcast. Furthermore, carrier conquest affects FWM's efficiency in WDM systems.

The CSRZ transmitter's block diagram is exposed in Figure 12.13 [3]. As demonstrated in this diagram, generating a CSRZ optical signal necessitates the use of two electro-optic modulators. The NRZ data is programmed by the first MZ modulator. The additional MZ modulator then modulates the resulting NRZ optical signal to form a CSRZ optical signal.

12.4.2.5 Differential Phase Shift Key (DPSK)

Prompt optical power levels through optical intensity modulation can be used to represent digital signals. Similarly, the phase of an optical carrier may

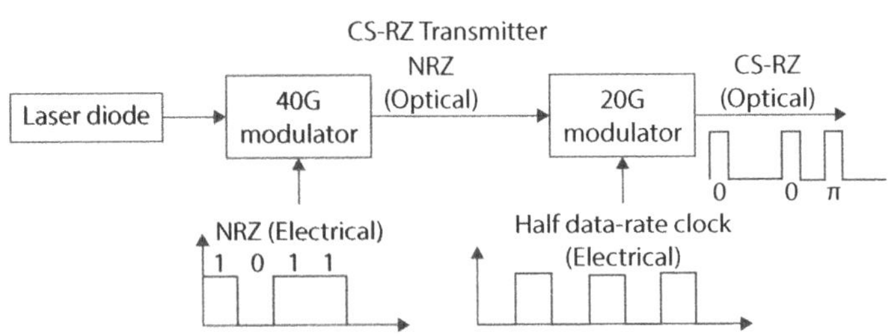

Figure 12.13 CSRZ transmitter.

be used to represent a digital signal, which is known as optical phase shift key (PSK). Because semiconductor laser sources were immature in the initial days of optical infrastructures, the optical phase was not steady abundant for phase based modulation methods. PSK has been viable in actual optical systems in recent years, thanks to the fast development of single frequency laser sources and the custom dynamic optical phase lodging. The differential phase shift key (DPSK) format is the most frequently utilized [14].

The encoding setup DPSK modulation tracks changes in the binary stream. The binary phase alternate between consecutive bits is encoded in DPSK: a 1-bit is represented by a phase change, however a 0-bit is characterized by the lack of a phase variation. DPSK may be employed in RZ and NRZ formats, much as intensity modulation. The major benefit of adopting DPSK over intensity modulation is a 3-dB increase in receiver sensitivity. Nonlinear effects are also tolerated better by DPSK. When compared to intensity modulation formats, it is more resistant to XPM and FWM. RZ-DPSK and CSRZ-DPSK have also been shown to have better transmission performance than ordinary DPSK.

12.4.3 Uses of Wavelength Division Multiplexing

- A fiber optic communications system's effective bandwidth is multiplied by WDM.
- The erbium amplifier, a fiber optic repeater device, may mark WDM an economical and enduring solution.
- WDM lowers the cost of data transmission while increasing the capacity of the cable.
- Multiple wavelengths (colors of light) are used to convey signals across a single cable in Wavelength Division Multiplexing (WDM) [16].
- At the receiving end, WDM employs optical prisms to separate the distinct colors, and optical prisms do not require a power supply.
- Temperature stabilized lasers were utilized in WDM systems to give the required channel count.

References

1. Mynbaev, D.K. and Scheiner, L.L., *Fiber-optic communications technology*, Pearson, Jan 2002.
2. Senior, J.M., Pearson Education Limited, Edinburgh Gate Harlow Essex CM20 2JE, *Optical fiber communications principles and practice*, Second Edition, England, 1992.

3. Keiser, G.E., A review of WDM technology and applications. *Opt. Fiber Technol.*, 5, 3–39, 1999.
4. Mukherjee, B., WDM optical communication networks: Progress and challenges. *IEEE J. Sel. Areas Commun.*, 18, 10, 1810–1824, 2000.
5. Agrawal, G.P., *Fiber optic communication systems*, Third edition, John Wiley & Sons, Inc., 605 Third Avenue, New York, 2002.
6. Mynbaev, D.K., Gupta, S.C., Scheiner, L.L., *Fiber optic communications*, Pearson Education, India, 2005.
7. Petermann, K., *Optical communication theory and techniques*, Springer Science + Business Media, Inc, Boston, 2005.
8. Winzer, P.J. and Essiambre, R.-J., Advanced optical modulation formats. *Proc. IEEE*, 94, 5, 952–985, May 2006.
9. Goldfarb, G., Li, G., Taylor, M.G., Orthogonal wavelength-division multiplexing using coherent detection. *IEEE Photonics Technol. Lett.*, 19, 24, 2015–2017, 2007.
10. Scheutzow, M., Reisslein, M., Maier, M., Seeling, P., Multicast capacity of packet-switched ring WDM networks. *IEEE Trans. Inf. Theory*, 54, 2, 623–644, 2008.
11. Proakis, J.G. and Salehi, M., 1221 Avenue of the Americas, *Digital communications*, 5th edition, New York, NY 10020. Copyright © 2008 by The McGraw-Hill Companie, 2008.
12. Binh, L.N., *Digital optical communications*, CRC Press, Taylor & Francis Group, Boca Raton London New York, 2009.
13. Gond, V.J. and Goel, A., Performance evaluation of wavelength routed optical network with wavelength conversion. *J. Telcommun.*, 2, 1, April 2010.
14. Ji, P.N. and Patel, A.N., Flexible wavelength division multiplexing (FWDM) networks. *International Conference on Optical Communications and Networks*, pp. 1–2, 2011.
15. Minato, N., Kobayashi, S., Sasaki, K., Kashima, M., Design of hybrid WDM/OCDM add/drop filters and its experimental demonstration for passive routing in metropolitan and access integrated network. *J. Lightwave Technol.*, 32, 6, 1120–1131, 2014.
16. Montero, S. *et al.*, Recent advances in wavelength division multiplexing plastic optical fiber technologies, INTECH, Advances in Optical Fiber Technology: Fundamental Optical Phenomena and Applications, 2014, Available at: http://dx.doi.org/10.5772/59518.
17. Mahloo, M. *et al.*, Toward reliable hybrid WDM/TDM passive optical networks. *IEEE Opt. Commun. Mag.*, 52, 2, 14–23, 2014.
18. Yuan, D., Dong, Y., Liu, Y., Li, T., Mach-Zehnder interferometer biochemical sensor based on silicon-on-insulator rib waveguide with large cross section. *Sensors*, MDPI, 15, 21500–21517, 2015, doi:10.3390/s150921500.
19. Agrawal, G.P., Optical communication: Its history and recent progress, in: *Optics in Our Time*, M. Al-Amri, M. El-Gomati, M. Zubairy (Eds.), Springer, Cham, 2016.

13

Optical Ultra-Sensitive Nanoscale Biosensor Design for Water Analysis

Shaikh Afzal[1]* and Manju Devi[2]†

[1]*Dept. E&C, Anjuman-I-Islam's Kalsekar Technical Campus, Mumbai, India*
[2]*Dept. of E&C, The Oxford College of Engineering, Bengaluru, India*

Abstract

The water-borne pathogen causes different types of poisoning to humans, such as emesis and diarrhea. Food contamination due to *B. Cereus* is of great concern. Detection, differentiation of the water-borne microorganism such as *B. Cereus* spores using the conventional medical test are costly, involves a complex manual process. Further, a microbe is related to a phenotypic trait or the existence of organism-related genes. Many researchers attempted to segregate the closely related *B. Cereus* spores' members, a simple and low-cost approach yet to be realized. The inspiration for the current paper was to investigate a fast identification approach for water analysis. The present research is on the variation of material index of refraction between the normal water and water infected with *B. Cereus* spores. This technique relies on the unique index of refraction as a spectral signature for the *B. Cereus* spores detection. The designed biosensor is Circular and Triangular resonator structures using Photonic Crystal. Analysis of the observation presents a variation in the spectrum between water without infection and infected water. From the obtained result Q factor of 7068.18 and sensitivity 832.17 nm/RIU is observed for Triangular resonator structure. Q factor of 5553.57 and sensitivity 776.69 nm/RIU is observed circular resonator structure. The obtained Q factor and sensitivity is higher than the published Q factor 2066.24 and sensitivity of 546.72 nm/RIU, for a PHC sensor. Thus, we concluded that a triangular resonator is an ultrasensitive sensor and is the best suitable for detecting the presence of *B. Cereus*.

***Keywords*:** *B. Cereus* spores; bio-sensor, photonic crystal, index of refraction, ring resonator structure, wavelength, sensitivity, Opti-FDTD, bacteria, optical sensor

**Corresponding author*: afzal.aiktc@gmail.com
†*Corresponding author*: manju3devi@gmail.com

Chandra Singh, Rathishchandra R Gatti, K.V.S.S.S.S. Sairam and Ashish Singh (eds.) Modeling and Optimization of Optical Communication Networks, (235–250) © 2023 Scrivener Publishing LLC

13.1 Introduction

A sensor that senses bio analytes is termed a biosensor. This is an instrument for analyzing and detecting bio elements such as mucus, saliva, tear, urine, blood, and sweat. Several biomaterials are extremely receptive, to name a few, tissues, virus, bacteria, many more, or an element of a larger biometric that binds with the analysis that is under observation for study. One such types are *B. Cereus* spores that give results to a gram-positive stain that has spores in the bacteria body which is in the shape of a rod named rod-shaped. Bacillus bacteria are seen in the atmosphere. *B. Cereus* is omnipresent and mostly available in the soil, where it germinates a saprophytic life cycle [1]. While *B. Cereus* is destroyed in the course of day-to-day cooking, spores are hard to get destroyed. Spores in water become more prone to infect the intestines and generates toxins causing diarrhea. Spores need greater than 5 minutes at 121°C (250°F) or highly cold weather to get killed. *B. Cereus* infects two types of water contaminated health hazard – emetic (vomiting) and diarrheal syndromes. As per published literature, *B. Cereus* infected diarrhea is due to the erosion observed on in the small intestine [2].

The spores are of great challenge to the hospitality and beverage industries because of their longevity in adverse conditions. Several steps are recommended to destroy the vegetative cells, and prevent germination and growth in water, thereby reducing their shelf-life [3]. The pasteurization process destroys most of the vegetative cells and prevents the formation of a suitable environment to germinate and grow [4]. In the poultry sector, the raptured & infected eggs with unhygienic environment, causes the spread of *B. Cereus* [5]. Biofilms of *B. Cereus* likely are a potential threat of contamination in water. This results in spread of infection in water and food supply chain. Over the years many researchers tried to get a cheap and easy solution to prevent the spread of contamination because of biofilm. Still, there is no viable solution available, and every year we could get news of food poisoning on a large scale. Especially in a rural area during public food festivals or social occasions [6].

The current work involves the use of Photonic Crystal (PHC) to detect *B. Cereus* spores for water quality analysis. The objective is to find a cheap and easy solution that can be deployed at every vulnerable place without the need for trained medical staff. The proposed sensor is aimed at the identification of *B. Cereus* strains rapidly. Quick detection is the need of the hour as any delay in detection may lead to an increase in the number of food poisoning incidences.

13.2 Related Work or Literature Survey

13.2.1 *B. Cereus* Spores' Study for Water Quality

B. Cereus spores causes serious, life endangering infections in immune deficient patients. A microorganism's capacity to form biofilm on water is the real cause for infections [7]. Appropriate antibiotic treatment is required to stop the progression of the infection and eventually death. Recent work on the analysis of *B. Cereus* using PHC is available in reference [8].

Although there are significant developments in the dairy industry, unintentional or accidental contamination of milk because of *B. Cereus* spores still happening and is a problem that results in high scale wastage of milk products and subsequent financial losses. There is a requirement to identify spores from water. This is because *B. Cereus* spores grow in water only in the sporulated form. The presence of *B. Cereus* spores in milk is very minimal in quantity which indicates that an effective technique with greater sensitivity is needed for dairy products screening to detect spores. The Figure 13.1 is referred to the work of Vidic *et al.* which gives the benefits and drawbacks of standard techniques used to detect spores in dairy products using the state-of-the-art nanotechnology. The drive to get nano technology involved is because of the low cost and ease of integration in compact device. For example, integration of the sensor in a lactometer [9].

Figure 13.2 shows the image of *B. cereus* spores. The coat is made up of protein, lipids, and carbohydrates. This helps to protect the bacteria from an oxidizing agent, chemicals, heat, and gamma radiation. The *B. cereus*

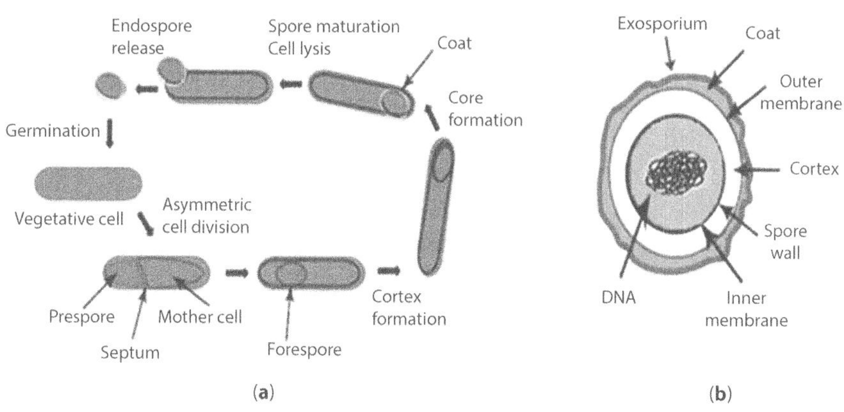

Figure 13.1 (a) Spores germination by endospore-forming *B. cereus*. (b) Schematic diagram of bacterial spore structure. DOI:10.3390/bios10030015

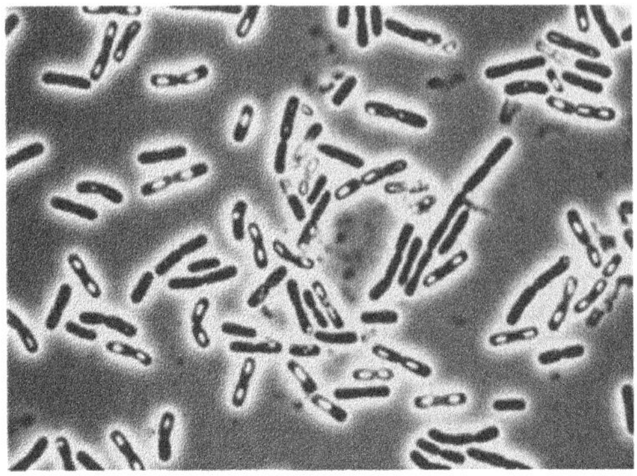

Figure 13.2 *B. cereus* spores – the white spots in the rod-shaped bacteria are spores. www.eurekalert.org

spores can digest many compounds like carbohydrates, protein, peptides, and amino acids for their growth. India being a sub-tropical country shows highly prevalent cases of *B. cereus* Spores infection due to contaminated water [10].

13.2.2 History Use of Optical Property for Biosensing

Research on the use of optical effect due to the interaction of light with *B. cereus* spores as a discriminator of biological particles was conducted at the University of Massachusetts, USA. To model an ensemble of spores, the researchers used experimental parameters appropriate to *B. cereus*. The results show that it is possible to use for the identification of aerosols made up largely of single biological particles [11].

Evanescent field trapping of *B. cereus* with the help of PHC sensor is another work that demonstrates optical trapping of single *B. cereus* spores. This work included the enhanced field of a cavity at resonance in a planar silicon PHC. The researcher concluded that planar PHC provides a compatible platform for the sensing of bio-particles [12]. In another work, the researcher presented a Kramers–Krönig analyses based on spectrum (transmitted and reflected) to obtain the real (n) and imaginary (k) parts of the complex refractive index (RI) as shown in Equation 13.1.

$$N = n + ik \tag{13.1}$$

Solid film samples considered for the experiment, pressed pellets, and suspensions in water and glycerol. The intended experiment is on the early adopters of optical method. It was observed that the optical properties of spores in water were different from the spores in glycerol. Further observation revealed, spores with a history of exposure to water had different optical properties compared to the spores that had not been in contact to water [13]. This work is one of the successful research projects which involved detection of *B. cereus* using variation in optical properties. By virtue of change in optical property caused by the presence of *B. cereus* or their variants, it is possible to detect their presence in a nondestructive way. That means without using any reagent which may change the characteristics of the organism.

13.2.3 Photonic Crystal

PHC's are a periodic structure of regularly shaped lattices having separate optical properties. A PHC configuration is multi-dimensional such as 1D, 2D, and 3D depends on the structure of the crystal as in Figure 13.3. More on PHC structures are available in published reference [14]. PHC is functionally analogous to the semiconductor. In the case of semiconductor movement of electrons is controlled by the structure of the material. In PHC movement of photons is governed by the PHC structure. It allows the selected wavelength of light to travel through the medium and stop other wavelengths.

In one-dimensional photonic crystal's, the periodic variation of RI is there in a single direction, and the RI is distributed uniformly for the remaining area. The photonic bandgap is there due to RI variation according to the periodicity direction [15]. Two-dimensional structured photonic crystal has the same periodicity in two directions only. 2D photonic crystals have abruptly occurring PBG only when the crystal has a higher RI.

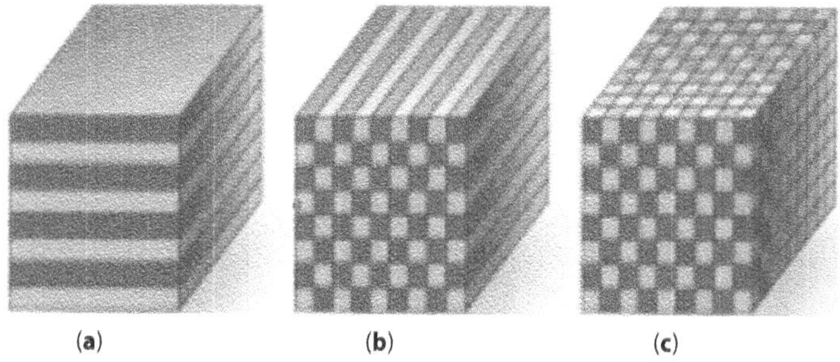

Figure 13.3 (a) One-Dimensional, (b) two-dimensional, and (c) three-dimensional photonic crystal.

Three-dimensional PHCs have defined variation of permittivity along the three planes of electromagnetic propagation. The 3D structure has higher dielectric contrast. Defect engineering is a critical aspect of PHC based sensor design. The defects are introduced in the crystal to create a bandgap. Bandgap selects what frequencies to allow for the light that passes through the medium of PHC. The design becomes complex depending on the dimension selection for the crystal [16].

13.3 Tools and Techniques

13.3.1 Opti FDTD

Finite Difference Time-Domain (FDTD) is a computational modeling implementation using software that facilitates the computer-Aided Design (CAD) and simulation for nonlinear photonic integrated circuits. FDTD helps to design analysis and simulation for wave propagation and the nonlinear optical interaction with matter. FDTD tools from Optiwave with a Uniaxial Perfectly Matched Layer (UPML) solving some of the complex design issues in nonlinear optics. The software tool provides a solution to electromagnetic wave propagation. In the temporal and spatial domain, it uses the full vector differential form of Maxwell curl equations. The full vector Maxwell equations is an essential tool. We are using both temporal and spatial domains to solve the design issue for the present investigation. It also encounters to design any arbitrary model geometries. This is highly flexible tool for simulation and there are no limitations on design for any optical element for nonlinear optics [17]. Opti FDTD provides a handy toolset for researches for nanophotonic device design and simulation. Opti FDTD output of current work is in the result section [18].

13.3.2 EM Wave Equation

The Maxwell equations are essential to solve the problem of electromagnetic propagation in PHC's [19]. The macroscopic Maxwell equations are represented by Equation 13.2 to Equation 13.4.

$$\nabla x E = -\frac{1}{c}\frac{\partial B}{\partial t} \tag{13.2}$$

$$\nabla x H = -\frac{1}{c}\frac{\partial D}{\partial T} + \frac{4\pi}{c} J \tag{13.3}$$

$$\nabla \cdot D = 4\pi\rho \tag{13.4}$$

$$\nabla \cdot B = 0 \tag{13.5}$$

Considering ñ = 0 (vacuum), and J = 0,
The Equations 13.2–13.5 is represented as given below [20]:

$$\nabla . \mathrm{E} = 0 \tag{13.6}$$

$$\nabla . \mathrm{B} = 0 \tag{13.7}$$

$$c = \frac{1}{\sqrt{\mu_0 \epsilon_0}} = 2.99792458 \times 10^8 \text{ m/s} \tag{13.8}$$

$$v_p = \frac{1}{\sqrt{\mu_0 \mu_r \epsilon_0 \epsilon_r}} \tag{13.9}$$

μ_0 = Magnetic permeability = 4π x 10^{-7} H/m
μ_r = Relative permeability
ϵ_0 = Free space permittivity = 8.854 x 10^{-12} F/m
ϵ_r = Relative permittivity
c = Light's velocity in free space
v_p = Velocity in medium

13.3.3 Optical Ring Resonator

Optical resonators consist of a set of waveguides where light pass through a confined space and can interact with the matter before exiting through the opposite end of entry [21].

In Figure 13.4 a schematic of the 3D model design of a ring resonator is given. Input light incident on Port A of the Bus waveguide, transmitted light output through Port B. The Dropping waveguide Port C and D detects the backward and forward transfer of reflected light [22]. The light confinement is happening in the ring area, and due to this, the resonator is a ring resonator.

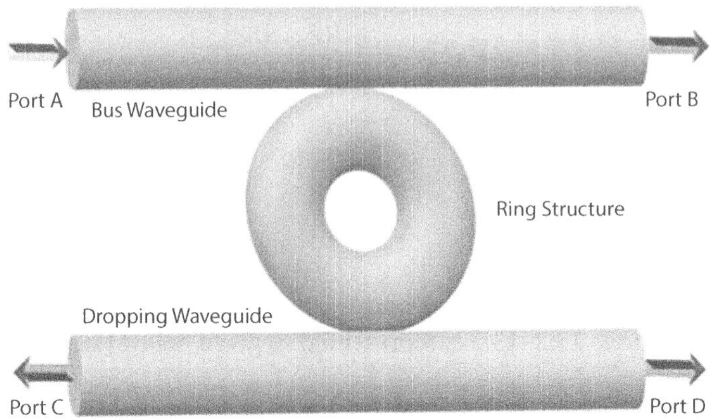

Figure 13.4 Schematic 3D model of a ring resonator.

13.3.4 Output Power Computation

The power is computed using real part of the equations:

$$P_{TE} = \frac{1}{2}\int Re\left(\dot{\vec{E}} \times \vec{H}^{*}\right) \cdot ds$$
$$= \int [\dot{E}_y \times Re(H_x^{*} + H_z^{*})] \cdot dxdy; \quad dy = 1 \tag{13.10}$$

$$P_{TM} = \frac{1}{2}\int Re\left(\dot{\vec{E}} \times \vec{H}^{*}\right) \cdot ds$$
$$= \int [Re(\dot{E}_x + \dot{E}_z) \times (H_z^{*})] \cdot dxdy; \quad dy = 1 \tag{13.11}$$

Where, H and E are components of transverse input field components. Also, the input field is normalized by:

$$\frac{\sqrt{P_\mu}}{\sqrt{P_m}} F_m \tag{13.12}$$

Here; P_μ is value of power entered, P_m is either P_{TE} or P_{TM} depending on whether the simulation is TE or TM. F_m represents transverse input field.

13.4 Proposed Design

For the proposed biosensor design two types of resonators considered- Circular type and Triangular. The design specifications are as given below:

- Lattice configuration = Rods in air
- Lattice Structure (a) = 1µm
- Rod's Radius (r) =0.20µm
- Wavelength = 1551nm
- Material type = S_i (Silicon)
- RI value of *B. cereus* spores = 1.503

Each structure consists of four variants A- without waveguide, B- with Bus waveguide, C- with dropping waveguide and D- with both bus waveguide and dropping waveguide.

13.4.1 Circular Resonator PHC Biosensor

FDTD analysis is done using Opti-FDTD for the four circular structures A, B, C, and D as shown in Figure 13.5.

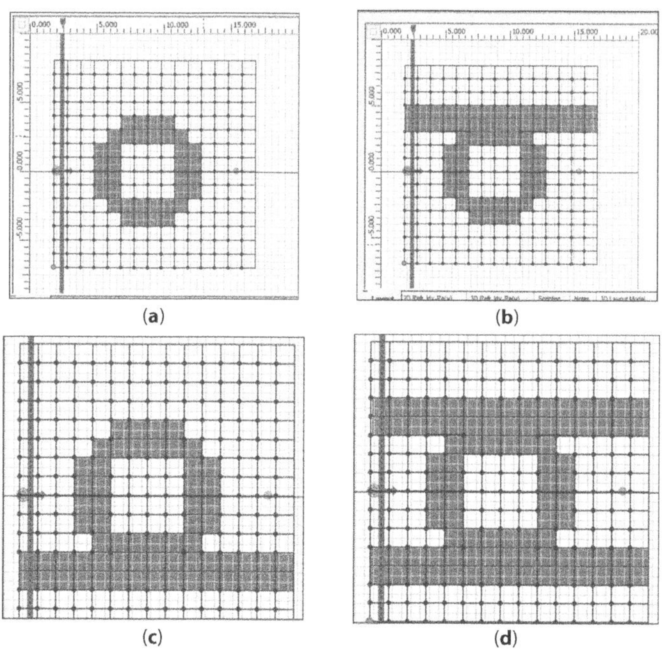

Figure 13.5 Photonic crystal circular structure.

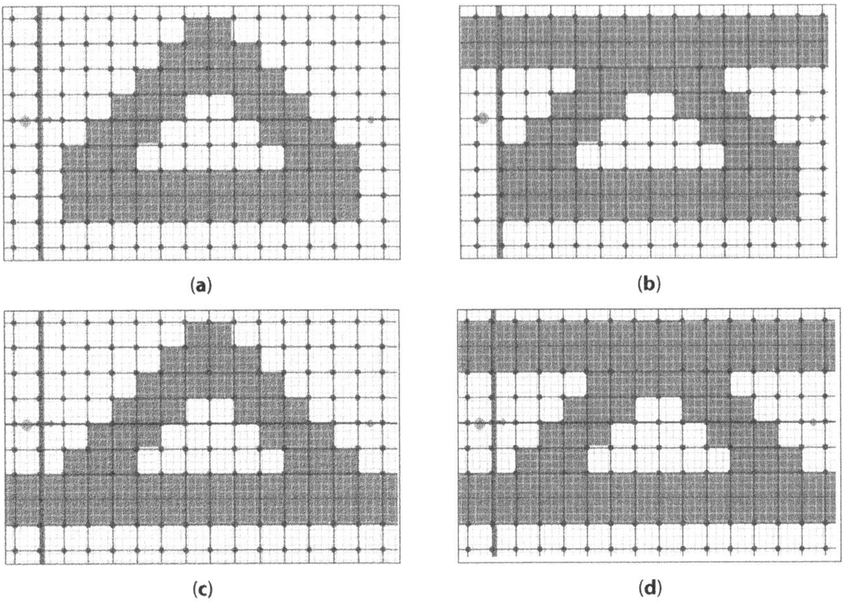

Figure 13.6 Photonic crystal triangular structure.

13.4.2 Triangular Structure PHC Biosensor

FDTD analysis is done using Opti-FDTD for the four triangular structures A, B, C, and D as shown in Figure 13.6.

13.5 Simulation

Figure 13.7 represents the simulation for Gaussian source [23]. FDTD analysis is done using Opti-FDTD for the four different structures for detecting the presence of *B. cereus* spores. Table 13.1 shows the optical parameters used during simulation [24].

13.6 Result and Analysis

Figure 13.8 represents the power spectrum with *B. cereus* spores by varying different combination of PHC circular structures.

Results obtained from Opti-FDTD analyzer [25]. The amplitude of structure A was 0.206 units, while that of structures B, C, and D were 0.232, 0.238, and 0.276 units respectively.

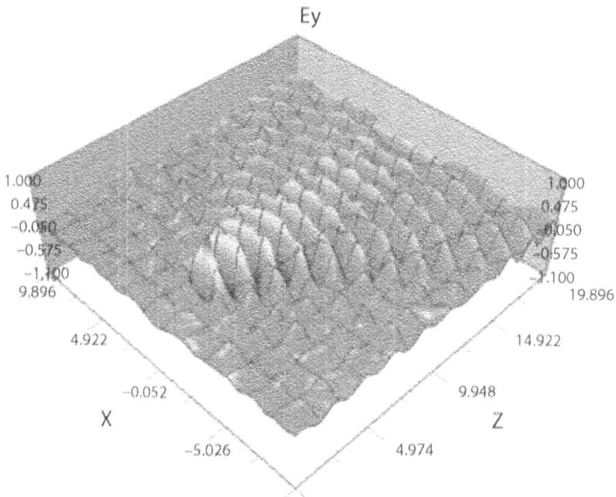

Figure 13.7 Simulation for Gaussian source.

Table 13.1 Simulation parameters.

S no	Refractive index	Lattice (a in μm)	Peak wavelength (μm)
1	1.503	1.0	1.551

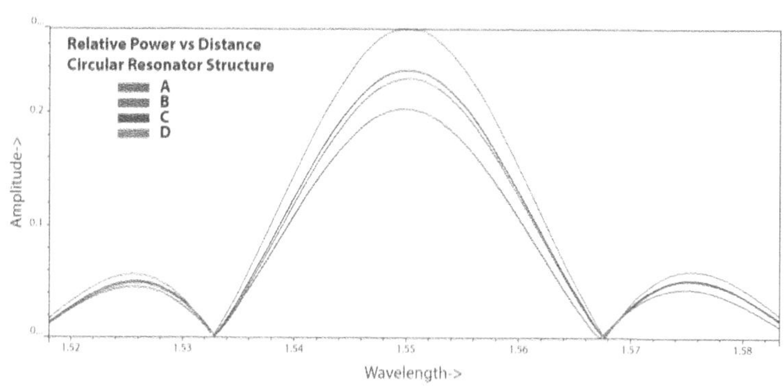

Figure 13.8 Combined spectrum of circular structure.

From the above graph, it is visible that the amplitude of the response obtained for structure D i.e., ring resonator with line defect above and below it is highest. Thus, structure D is best suitable for detecting infected spores in water.

Figure 13.9 represents the power spectrum with spores by varying separate combinations of Photonic Crystal triangular structures.

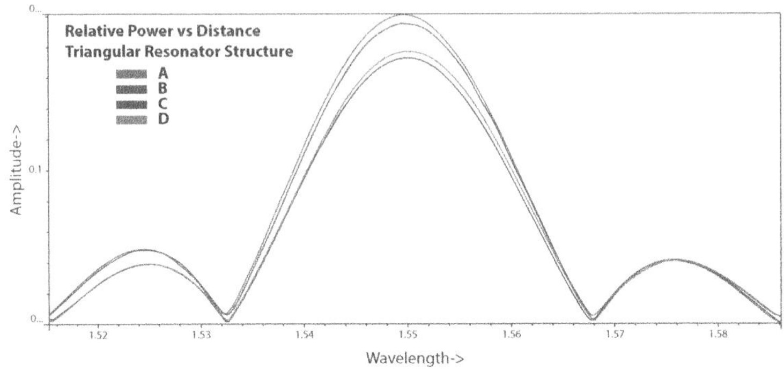

Figure 13.9 Combined spectrum of triangular structure.

The amplitude of structure A was 0.2007 units, while that of structures B, C, and D were 0.1953, 0.1729, and 0.774 units respectively.

From the above graph, it is visible that the amplitude of the response obtained for structure A i.e., triangular structure is highest. Thus, structure A is best suitable for detecting the presence of spores in contaminated water.

Table 13.2 shows the comparison of different structures plotted in Figure 13.8 and Figure 13.9.

Table 13.2 Result comparison.

Plot reference	Structure	Wavelength shift (μm)	Amplitude shift
Figure 13.8-A	Circular only	0.4	0.00027627
Figure 13.8-B	Circular with Bus waveguide		0.00132497
Figure 13.8-C	Circular with Dropping waveguide		0.00127745
Figure 13.8-D	Circular with Bus and Dropping waveguide		0.00234241
Figure 13.9-A	Triangle only	0.3	0.00440054
Figure 13.9-B	Triangle with Bus waveguide		0.00443482
Figure 13.9-C	Triangle with Dropping waveguide		0.00403937
Figure 13.9-D	Triangle with Bus and Dropping waveguide		0.00417915

From Table 13.2 shows that shift in the wavelength is better for circular resonance structure while the amplitude shift is better for Triangle with Bus and Dropping waveguide.

Table 13.3 shows the computed Q factor for both the structures. Triangle structure shows better Q factor compared to the circular resonator structure.

From Table 13.4 it is evident that the circular structure with bus and dropping waveguide has the highest transmitted power of 27.6%.

Table 13.3 Simulation parameters.

S no	Structure	Peak Wavelength (μm)	Wavelength Shift (μm)	Q factor
1	Circular with Bus and Dropping waveguide	1.555	0.4	3887.50
2	Triangle with Bus and Dropping waveguide	1.555	0.3	5183.33

Table 13.4 Transmitted power for each structure.

Plot reference	Structure	Transmitted power (%)	Power in db.
Figure 13.8-A	Circular only	20.60	-13.8816
Figure 13.8-B	Circular with Bus waveguide	23.2	-14.2744
Figure 13.8-C	Circular with Dropping waveguide	23.8	-12.3579
Figure 13.8-D	Circular with Bus and Dropping waveguide	27.6	-14.9155
Figure 13.9-A	Triangle only	20.07	-14.8022
Figure 13.9-B	Triangle with Bus waveguide	17.29	-13.5312
Figure 13.9-C	Triangle with Dropping waveguide	19.53	-14.7536
Figure 13.9-D	Triangle with Bus and Dropping waveguide	17.74	-13.4439

Table 13.5 Performance parameters.

S no	Structure	Sensitivity	Q factor
1	Circular with Bus and Dropping waveguide	776.69	5553.57
2	Triangle with Bus and Dropping waveguide	832.17	7068.18

Table 13.5 shows the computed Q factor and Sensitivity for both the structures. The Triangular Resonator structure shows better performance compared to the Circular Resonator structure.

13.7 Conclusion and Future Scope

In the purposed work, a PHC based biosensor with two different structure sets is proposed, designed, and simulated for the identification of spores in contaminated water. The proposed structures are designed and analyzed using Opti-FDTD and measurements like spectrum shift, change in the amplitude, etc. are obtained for different structures. The presence of the spores is detected by observing the shift in the transmission spectrum by observing the change in the RI value of the analyte. From the above result Q factor of 7068.18 and sensitivity 832.17 nm/RIU is observed for Triangular Resonator structure. Q factor of 5553.57 and sensitivity 776.69 nm/RIU is observed in Circular Resonator structure. The obtained Q factor and sensitivity is higher than the published Q factor 2066.24 and sensitivity of 546.72 nm/RIU, for a PHC sensor [26]. Thus, we concluded that a Triangular Resonator with bus and dropping waveguide is an ultra-sensitive biosensor and is the best option for detecting the presence of *B. cereus* spores in water compared to other proposed structures. Further, given that the structure is simple and can be fabricate for the real-time application of bacteria detection in contaminated water. It will benefit services involving portable water because of the easy and fast detection of *B. cereus* spores. With the advancement of fabrication technologies, the development of a contamination detection system using the proposed design presented in this work is possible. In the long run, any portable system built with the proposed technology likely be beneficial for the food and restaurant industry immensely.

References

1. Vilain, S., Luo, Y., Hildreth, M., Brozel, V., Analysis of the life cycle of the soil saprophyte Bacillus cereus in liquid soil extract and in soil. *Appl. Environ. Microbiol.*, 72, 4970–4977, 2006.
2. Fagerlund, A., Lindback, T., Storset, A.K., Granum, P.E., Hardy, S.P., Bacillus cereus Nhe is a pore-forming toxin with structural and functional properties similar to the ClyA (HIyE, SheA) family of haemolysins, able to induce osmotic lysis in epithelia. *Microbiology*, 154, 693–704, 2008.
3. Granum, P.E. and Lund, T., Bacillus cereus and its water poisoning toxins. *FEMS Microbiol. Lett.*, 157, 223–228, 1997.
4. Daelman, J., Vermeulen, A. *et al.*, Growth/no growth models for heat-treated psychrotrophic Bacillus cereus spores under cold storage. *Int. J. Water Microbiol.*, 161, 7–15, 2013.
5. Andersson, A., Ronner, U., Granum, P.E., What problems does the water industry have with the spore-forming pathogens Bacillus cereus and Clostridium perfringens? *Int. J. Water Microbiol.*, 28, 145–55, 1995.
6. Waites, W.M., Harding, S.E., Fowler, D.R., Jones, S.H., Shaw, D., Martin, M., The destruction of spores of Bacillus subtilis by the combined effects of hydrogen peroxide and ultraviolet light. *Lett. Appl. Microbiol.*, 7, 139–40, 1988.
7. Bahl, M.I. and Rosenberg, K., High abundance and diversity of Bacillus anthracis plasmid pXO1-like replicons in municipal wastewater. *FEMS Microbiol. Ecol.*, 74, 1, 241–247, 2010.
8. Priyadarshini, R. and Narayanappa, C.K., The study of bacillus cereus on a photonic crystal biosensor. *Int. J. Eng. Technol.*, 7, 3.4, 231–234, 2018.
9. Vidic, J., Chaix, C., Manzano, M., Heyndrickx, M., Water sensing: Detection of Bacillus cereus Spores in *Dairy Products. Biosensors*, 10, 15, 2020, doi: 10.3390/bios10030015.
10. Rana, N., Panda, A.K., Pathak, N. *et al.*, Bacillus cereus: Public health burden associated with ready-to-eat Waters in Himachal Pradesh, India. *J. Water Sci. Technol.*, 57, 2293-2302, 2020, https://doi.org/10.1007/s13197-020-04267-y.
11. Druger, S., Czege, J., Li, Z., Li, Z., Bronk, B.V., *Calculations of light scattering measurements predicting sensitivity of depolarization to shape changes of spores and bacteria, report*, Edgewood Chemical Biological Center, USA, 2008.
12. Leest, T. and Caro, J., Evanescent field trapping of bacterial spores using photonic crystal cavities. *Proc. SPIE 8570, Frontiers in Biological Detection: From Nanosensors to Systems*, 5 March 2013, vol. 857006, https://doi.org/10.1117/12.2002046.

13. Tuminello, P.S., Arakawa, E.T., Khare, B.N., Wrobel, J.M., Querry, M.R., Milham, M.E., Optical properties of Bacillus subtilis spores from 0.2 to 2.5 μm. *Appl. Opt.*, 36, 2818–2824, 1997.
14. John, D.J., Steven, G.J., Joshua, N.W., Robert, D.M., *Photonic crystals molding the flow of light*, Second edition, Princeton University Press, United States, 2008.
15. Sharma, P. and Sharan, P., Photonic crystal based ring resonator sensor for detection of glucose concentration for biomedical applications. *Int. J. Emerg. Technol. Adv. Eng. (IJETAE)*, 4, 3, 2250–2459, 2014.
16. Sharma, P. and Sharan, P., Design of PC based biosensor for detection of glucose concentration in urine. *IEEE Sens. J.*, 15, 2, 1035–1042, 2015.
17. Susuma, N., Recent progresses and future prospects of two- and three-dimensional photonic crystals. *J. Light. Technol.*, 24, 1, 4554–4567, 2006.
18. Roy, S.K. and Sharan, P., Photonic crystal based sensor for DNA analysis of cancer detection, in: *Silicon Photonics & High-Performance Computing. Advances in Intelligent Systems and Computing*, vol. 718, A. Mishra, A. Basu, V. Tyagi (Eds.), Springer, Singapore, 2018.
19. Robinson, S. and Nakkeran, R., Photonic crystal ring resonator based optical filters. *Adv. Photonic Crystals*, Feb. 2013, doi: 10.5772/54533.
20. Nandhini, V.L., Suresh B., K., Roy, S.K., Pandit, K., Photonic crystal based micro interferometer biochip (PC-IMRR) for early stage detection of melanoma. *Pertanika J. Sci. Technol.*, 26, 3, 1505–1512, 2018.
21. Chetty, G. and Yamin, M., A distributed smart fusion framework based on hard and soft sensors. *Int. J. Inf. Technol.*, 9, 19, 2017, https://doi.org/10.1007/s41870-017-0008-9.
22. Sharma, P., Roy, S.K., Sharan, P., Design and simulation of photonic crystal based biosensor for detection of different blood components, *2014 IEEE Region 10 Symposium*, pp. 171–176, 2014.
23. Sharma, P. and Sharan, P., Design of photonic crystal based ring resonator for detection of different blood constituents. *Opt. Commun.*, 348, 19–23, 2015.
24. Nandhini, V.L., Suresh Babu, K., Roy, S.K., Sharan, P., Multichannel biosensor for skin type analysis, in: *Advances in Machine Learning and Computational Intelligence. Algorithms for Intelligent Systems*, S. Patnaik, X.S. Yang, I. Sethi (Eds.), Springer, Singapore, 2021, https://doi.org/10.1007/978-981-15-5243-4_57.
25. Roy, S.K. and Sharan, P., Design of ultra-high sensitive biosensor to detect E. Coli in water. *Int. J. Inf. Technol.*, 12, 775–780, 2020. https://doi.org/10.1007/s41870-019-00327-5.
26. Maache, M., Fazea, Y., Bile Hassan, I., Alkahtani, A.A., Ud Din, I., High-sensitivity capsule-shaped sensor based on 2D photonic crystals. *Symmetry*, 12, 1480, 2020.

14

A Study on Connected Cars–V2V Communication

Chandra Singh*, Sachin C. N. Shetty, Manjunatha Badiger and Nischitha

Sahyadri College of Engineering and Management, Mangalore, India

Abstract

Connected Cars–Vehicle-to-Vehicle (V2V) communication comprises a wireless network where cars send messages and receive message about real time data of road condition and impending collision. The basic idea of this is to transmit the safety information. It uses a VANET (Vehicular ad hoc networks) for communication purpose. It works 360 degrees at any direction. Exchange of information takes place by adapting a suitable protocol. The real time status of the vehicle and road is updated to server instantly. In case of road blocks the location details is updated to server and alert message is sent to driver if there is any collision. It helps to take a major decision about the travel path to reach destination and avoid impending collision. The sensor continuously monitors its environment for changes.

Keywords: VANET, Ad Hoc wireless networks, V2V

14.1 Introduction

Traffic and accidents are the major problems that are faced with the increase in the number of vehicles in roadways. Majority of the accidents are due to collision. Traffic and accidents can be avoided to some extent by providing prior information about the state of the road and warning about the impending collision [10, 11]. Today technology has reached to such an extent which enables communication between Internet and cars. Present day cars are equipped with many types of connections that may be internal or external. Passengers or drivers always wish to experience all the comfort while driving. Internet

**Corresponding author*: chandrasingh146@gmail.com

Chandra Singh, Rathishchandra R Gatti, K.V.S.S.S.S. Sairam and Ashish Singh (eds.) Modeling and Optimization of Optical Communication Networks, (251–266) © 2023 Scrivener Publishing LLC

of Things is playing a major role in today's automobile technology which has put a good effort in relieving drivers from stressful operation. The associated vehicle is a car that is proficient to get to the Internet and can speak with keen gadgets, different cars and gather real time information from various sources is playing a significant job in the Internet of Things. In this present age vehicle makers, software and equipment designers have grasped the test of giving creative answers for new age vehicles [12, 13]. The present cars are designed such that it soothes drivers from the most distressing activities required while driving by giving them entertainment functions. Meanwhile, they must know about the safety, security and reliability. The target of this undertaking is to give an outline of the conceivable outcomes offered by associated functionalities in cars and the related technical problems and issues, just as to enumerate the at present accessible equipment and programming arrangements and to evaluate the status for use of vehicle-to-vehicle (V2V) communication [14]. This is used to transmit basic data between vehicles to encourage alerts to drivers about impending accidents and roadblocks. The transport division is usually subordinate to a few issues, for example traffic blockage and mishaps. The primary aim of this is to build road security by endeavoring to maintain a strategic distance from the peril and to avoid vehicles entering in to blocked streets. Vehicle-to-Vehicle (V2V) and Vehicle-to-Infrastructure (V2I) innovations means to give correspondence models that can be actualized in vehicles in various applications. The subsequent foundation is a specially made to organize were hubs vehicles and furthermore cell phones outfitted with remote modules [15]. It is based on Ad-hoc mesh network. The cooperation with the numerous associated modules empowers data exchange by appropriate communication protocol. The primary aim of this chapter is to make use of relevant communication protocol, application and system for upcoming road infrastructure that can be made use by vehicles [16].

14.2 Literature Survey

The paper titled "Position-Aware Ad Hoc Wireless Networks for Inter-Vehicle Communications: the Fleet net Project" by Hannes Hartenstein NEC Europe Ltd., Bernd Bochow GMD Focus, Andre Ebner University of Hamburg-Harburg [1] presented the project Fleet net which was developed and demonstrated the position aware wireless ad-hoc network for inter-vehicle communication. In this they use multi-hop communication which will help in improving safety features by creating awareness to the drivers by quickly sending and distributing the sensed data to the nearby vehicles allowing them to take quick appropriate actions. Applications like security distance warning and overtaking assistance is done by using IP addressing and current position of the vehicle which enables position based routing. This also provides traffic flow information and emergency notifications. For knowing the current position of the node GPS is used. This paper supports applications based on the information present in the web using Internet. Each node is capable of collecting information from other vehicles and generating information so it can be considered as server which will provide updated information. The paper titled "Vehicle-to-Vehicle Communication Technology" by Albert Demba Department of Informatics and Deitmar P.F. Moller Department of Stochastics and OR, Germany [2] suggested a V2V system which would operate without interference and physical security. This method does not rely on third party unlike other methods. Because the ad hoc network is used it has a range of 1000m with 360-degree view. It uses a wireless communication known as DSRC combined with GPS technology. This paper mainly focuses on the security. After making the comparative study on the various encryption methods asymmetric public key infrastructure (PKI) method was selected. It will verify the identities by digital signatures and creates certificates for each device. A technology used along with this is hash algorithms which provides authenticity, protects integrity and confidentiality. The paper titled "A proposed System for Vehicle-to-Vehicle Communication: Low Cost and Network Free Approach" by Mohammed Ateeq Alanzi college of Computing and IT, Saudi Arabia [3] presented a system which provides safe driving and an alert to improve performance. In this model LCD display is installed at the back of the car.

To display the necessary messages, a voice and touch enabled devices and wireless transceiver are present inside the car so that the driver can input or receive information. This method does not need any network architecture or protocols for communications. Various messages are generated by the sensors present in the vehicle. The proposed system uses a Wi-Fi based transmitter. The applications enable the driver to input towards microcontroller, whenever the driver is approaching any abnormal circumstances then he can touch the exact warning messages and it will be displayed. The different warning messages are put previously. A threshold for distance between the vehicles is already set. In the proposed method any real time data is not used and since fewer components are used the cost is low. The paper titled "Vehicle Tracking System using GSM and GPS Technologies" by Dr. Bharati Wukkadada and Allan Fernandes K.J Somaiya Institute of Management Studies and Research, India [4] says that the tracking system is a device which makes it easier to locate position, time and mobility. The system consists of GPS receiver and GSM modem for monitoring the movement of vehicles and updates the status. Once the devices are connected continuous streaming of data takes place. Modem shows the location and time based on longitudinal and latitudinal points given by the satellites. This helps in anti-theft system with help of GPS. The system can also be used for animal tracking, driving behavior and delivery services. Interaction between the two components is carried by the microcontroller where the processing of the data takes place. For the node to be visible it requires four satellites in visible horizon. The paper "Smart Road Accident Detection and communication System" [5] discusses on means to facilitate communication to the relatives or the emergency needs in case of a vehicle collision or accidents. The accidents are characterized by vibrations during the drive, if the vibration exceeds a preset threshold value set as a danger limit the driver has a time window of 10 seconds where he has to press the reset button. If the button isn't activated in the speculated time, the control unit senses this as an accident or a condition where the driver is no longer in full control of the vehicle and sends a warning message along with the co-ordinates of the accident location (GPS module) to the pre-determined contacts via gsm module. In other cases, where the vehicle may topple or tilt rather than collide, the gyro sensor will notify the accident if the angle of tilt is greater than the threshold value. Even in this case the driver has the 10 seconds window to activate the reset button.

The paper titled "Accident prevention system for vehicles using v2v communication [6]" describes on some of the measures for the implementation of vehicle collision detection systems. The proposed method uses ZigBee for wireless communications between cars in order to alert the cars in case of impromptu breaking or obstacles on the road. The system has collision avoidance system which helps in alerting the vehicle behind to not overtake or a so called "Donot-overtake" Warning. In addition to this, there is an intersection assist method which helps in avoiding T-Bone accidents which are caused at road intersections. Blind spot warning helps drivers during the night drives where the visibility extent is limited, the warning message alerts the driver of any further blockages or stopped cars in the front.

The paper titled "Design and Implementation of Real Time Wireless System for Vehicle Safety and Vehicle to Vehicle Communication [7]" proposes use of ZigBee protocol to help in avoidance of accidents by using sensors and GPS modules in place of conventional speedometers. The main reason is to tackle the inaccurate reaction during the driver intervention in scenarios where responses are required in a matter of moments. Here the cars communicate with each other using its altitudes (positions) to avoid probable accidents. The proposed system has a main unit which acts as a transceiver to send information or alert messages to other cars for lane clearance or clear the slow moving traffic in conditions where there is a need for an emergency vehicle to pass. The other important identities used for the calculation of the drive are speed, distance between successive cars, and proximity to obstacles, temperature, and humidity. Arduino is used as the central unit which combines all these factors and gives an optimum output for the drive in the road.

The paper titled "Obstacle Detection of vehicles under Fog" by Greeshma G, Lalithamani Nithyanandham [8] propose a method for detection of objects during fog or cloud covered roads for prevention of collision. Since the visibility factor varies for fog and cloud or moist, it is important to differentiate between these factors to calculate the distance and the blockage in order to assist a safe driving. The detection and classification of fog is done via Gray level co-occurrence matrix features and support vector machine (SVM). SVM is used to classify the input images into the set of predefined images and conditions. Then the algorithm is made to measure the distance between the moving obstacles. This method can be used in future applications in order to employ self-driving vehicles.

The paper titled "Cloud-assisted Real-Time Road Condition Monitoring System for Vehicles" proposes [9] real time feed of road conditions to cloud using machine learning. The vehicle is fitted with an accelerometer whose values are monitored continuously. The values are predetermined for smooth road, speed breakers. When the output values are different it symbolizes that the road condition isn't smooth and this information is fed to the cloud and all the similar vehicles which are connected to the server are being alerted about the condition of the drive through that road. The GPS module send out the co-ordinates of the road hence showing which part of the road will not provide a pleasure driving experience.

14.3 Software Description

Thing Speak is an open-source Internet of Things (IoT) application and API to store and retrieve data from things. It uses HTTP and MQTT protocol over the Internet or via a Local Area Network. It requires a Things Api account which is free of cost for a small project. Things Api makes you create a channel which is the identification of your project. It can either be a public channel that can be shared to all the users or a private channel which can be used for a specific group of people generally used before the announcement of the project to the public. The channel has to be given a description which includes fields to be used for analysis of input data from sensors i.e. Temperature, Pressure, distance. We can visualize new data or interact with web services, social media platforms and with various devices. This also provides us with tools to perform actions such as react to data and queuing the commands which needs to be executed. API Keys or Application Programming Interface is the only way to access your channel. It is like a password to enter your project. API Keys are used for two actions: To update data in channel, to retrieve data in channel. To update data in channel, API write key is used. To retrieve data in channel, API read key is used. To access the channel:http://api.thingspeak.com/update?api_key=YOUR-API&field1=VAR-1 and field2=VAR-2 is used and YOUR-API is substituted to your channel API key, VAR-1 and VAR-2 is used for the data fields you're using. To read Channel: http://api.thingspeak.com/channels/YOUR-CHANNEL-ID/fields/FIELD.json?results=NOS-OF-RESULTS and api_key= YOUR- API is used where you've to add your channel ID and fields.

Arduino IDE is an open source software which runs on Windows, Mac OS and Linux. We can write the code easily and upload it to the board. This application is written with the help of functions from the C and C++. It connects to the hardware and helps in communication. Programs written using this software are called sketches. Library manager can be accessed to import new libraries. Port option contains all the devices connected. Function provided by library is we can manipulate data. The sketch is generated in the form of hex file which is uploaded to the microcontroller. IDE contains two parts that is editor and compiler. The inbuilt functions play a major role for various purpose like debugging, editing and compiling. IDE has three main sections. the menu bar includes file, sketch, edit, tools and help. The check mark is used to verify the code. Arrow button is used to upload or transfer the codes. Upward arrow is used for opening existing project. Save button is present to save the code. Serial monitor is used for sending or receiving the data. We need to select baud rate of the Arduino board accordingly. The bottom of main screen is the output pane which provides the status of running code, if any errors we can fix them [17].

14.4 Methodology

Connected vehicle is a car. The vehicle has capability to access the internet anytime. Access to the Internet may be using built in device that has modern applications. V2V acts as a real time routing map. It works on wireless network where car sends a message about the road condition. The system results in a mesh network where each vehicle can send and capture the information signal. It is made possible with the help of IoT. The devices like GPS receivers are used for location detection, road sensors for collecting the real time data about the road condition. The system is designed to ensure it is capable of getting all the information for safe and better navigation. If there is any road block the message is sent to server. The other cars which has internet connectivity can get access to the same server and can collect the real time data about the road condition. If there is road block, the other cars can avoid entry to the blocked road and can find some alternate path for the destination. It helps in controlling further traffic. The other working is collision detection. As soon as the collision is detected the warning is given using alert message to the driver. It helps driver to take necessary action to avoid impending collision. Avoiding collision will prevent unnecessary road block and traffic. There are many kinds of sensors made available for different applications. The sensor used in this project will be continuously monitoring the surroundings. Sensor used has the capability to identify collision and collision avoidance. If there is any accident that has taken place, the location details will be sent to the nearest hospital. All the functionalities together make the connected car. Internet and communication between cars are playing a major role. The main aim is to deal with traffic efficiency by route navigation and alert message to the driver.

14.5 Working

In this system, the vehicle will be installed with an ultrasonic sensor. The sensor will be continuously monitoring the surroundings and it can detect the road-blocks due to accident or tree fall. When any kind of road-blocks are detected by the vehicle, the system will fetch the GPS co-ordinates and the data will be sent to an internet application and road status with location will be updated. The website will also show which of the device has sent the message. The information about the co-ordinates will also be shown in the LCD display. This holds good for all the vehicles connected to the internet. The switch array is made use in order to update the road real time data. The other vehicles which is coming on the same path can communicate with the server by linking to the same web-application and will be able to monitor the status of the roads. This will prevent other vehicles entering the blocked roads and help in clearing the traffic accordingly. Using this information, the driver can find alternate route through which destination can be reached. It helps in mobility management. If there is any need for emergency the location can also be sent to ambulance. The other instance is where the accident has already occurred due to negligence. The driver can press one of the buttons which will send the co-ordinates of the location to the website and also the emergency numbers via GSM module. In case the driver is no longer in his full conscious or isn't in a condition to push the button, the car following can alert in its own system. This is powered by a 2A adaptor. Ultrasonic sensor will also sense any impending collision and warning will be provided to driver so that driver should take necessary actions to avoid accident. This helps in foggy areas where visibility factor reduces to slim values. The buzzer alerts with a beep and the display shows the distance at which the obstacle is present at. It is programmed such a way that when two vehicles come too close or crosses the minimum distance to be maintained in order to avoid collision the warning will be sent to driver. The ultrasonic sensor is fitted to all the corners of the car. A 16x2 LCD is used to display the status messages and a Wi-Fi Transceiver ESP8266 is used to get connected to an internet device through TCP-IP protocols. In this way, the status of all the vehicles and roads are being updated to the server instantly. Buzzer is made use to facilitate with the warning sound incase the driver fails to see the LCD display and can take necessary action to avoid collision. For continuous communication internet is must. The co-ordinates sent by the GPS module can be traced out to exact location via simple applications present in the digital market The hardware's can be either built in or brought in. Internet connection is from smartphone. Configuration of the device must be done for individual data connection separately by changing the SSID and the passwords. With the help of all this hardware and technology the car can be made aware of its surroundings and communicate bidirectional. API keys are made use in order to update the information to the server as it provides authentication token and a unique identifier. It supports in providing security to the updated data and avoids conveying information by hackers. It rejects or do not provide access to any request that is not enabled in the particular API. The vehicle which

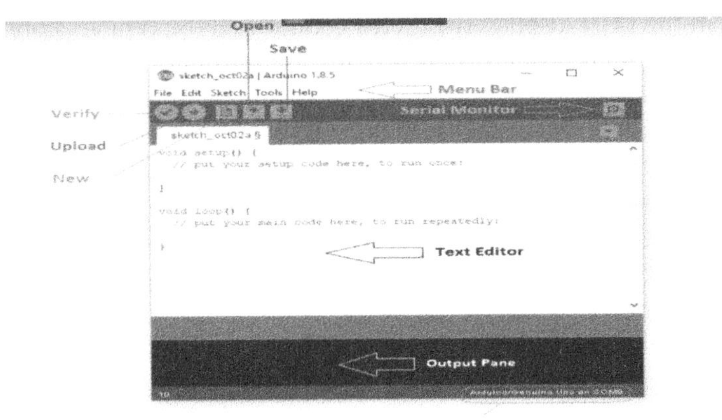

Figure 14.1 Arduino IDE environment.

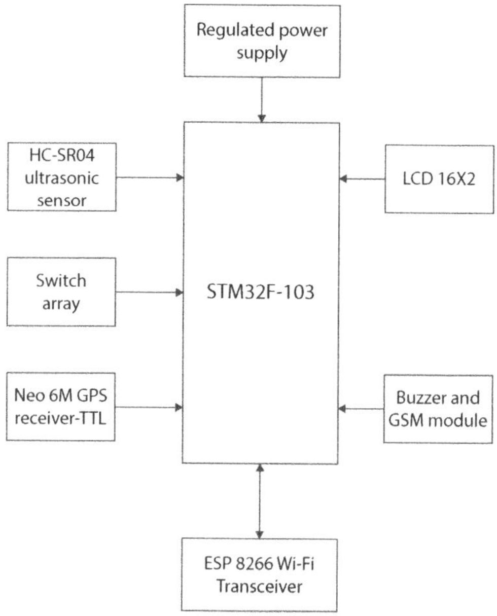

Figure 14.2 Block diagram of vehicle 1.

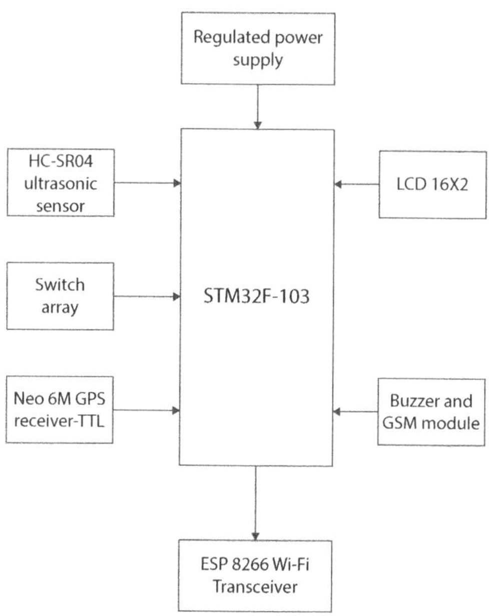

Figure 14.3 Block diagram of vehicle 2.

sends information acts as a sender and all other vehicles pinging to server will act as receivers. If the information is irrelevant to the receiver the message is ignored or discarded is depicted in Figures 14.1 to 14.6.

As shown in the image Figure 14.7 and Figure 14.8, the necessary pin connections have been made between the STM controller and the above described sensors. Figure 14.9 shows the Thing Speak channel dedicated to this project with channel ID 1053894. The input values will be updated in the latitude and longitude section and this determines the clearance of road or blockage on the way. Figure 14.10 shows the status of the vehicle running, with latitude and longitude information given of the vehicle location. The switching 'ON' of Road %2520% Clearance field shows that the road ahead is clear with no obstruction. If any obstruction is found, the Accident Status field is switched 'ON' and the location of the vehicle is uploaded to the channel.

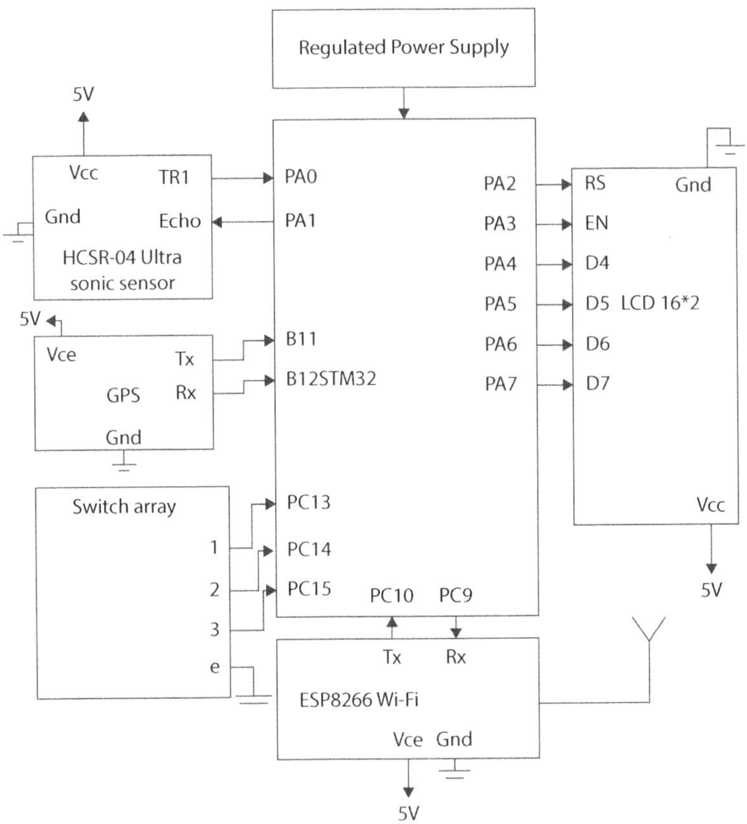

Figure 14.4 Pin description for vehicle.

Figure 14.5 Hardware connection of ultra sonic sensor.

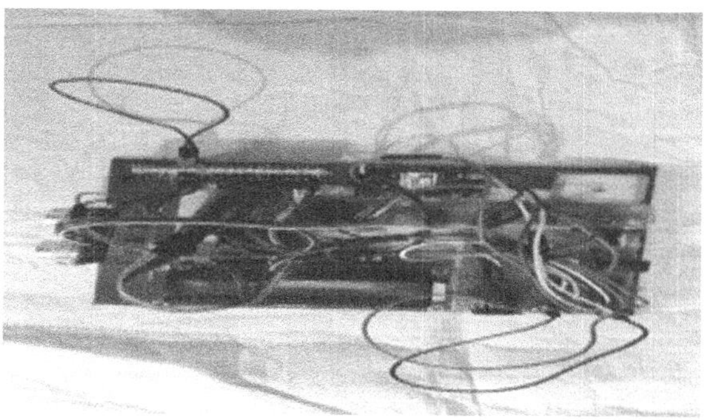

Figure 14.6 Connected car module.

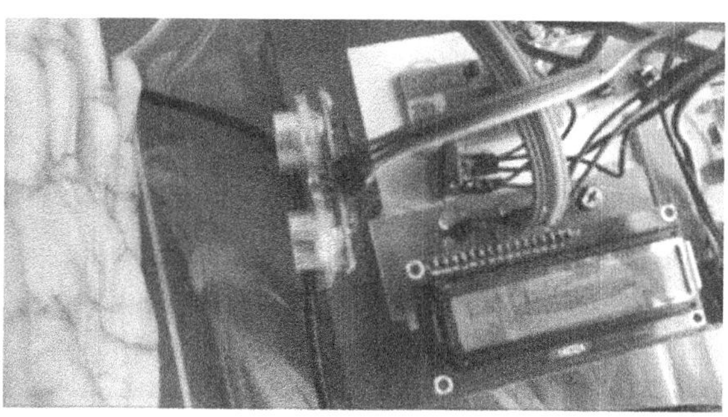

Figure 14.7 Collision detection.

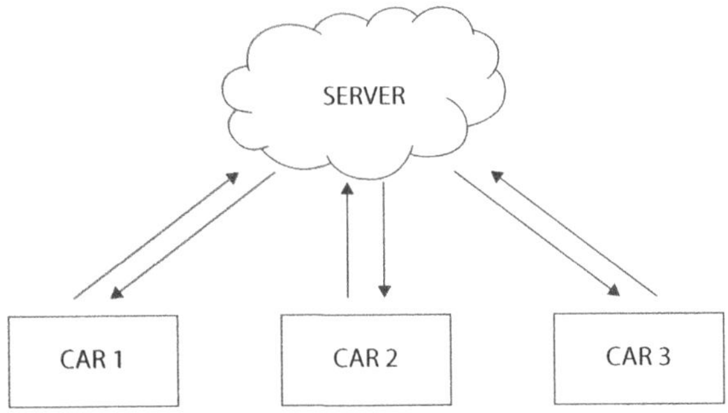

Figure 14.8 Car connected to server.

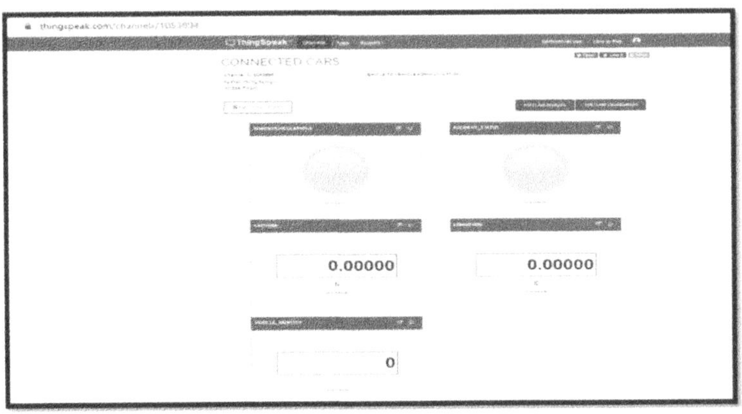

Figure 14.9 Things speak channel.

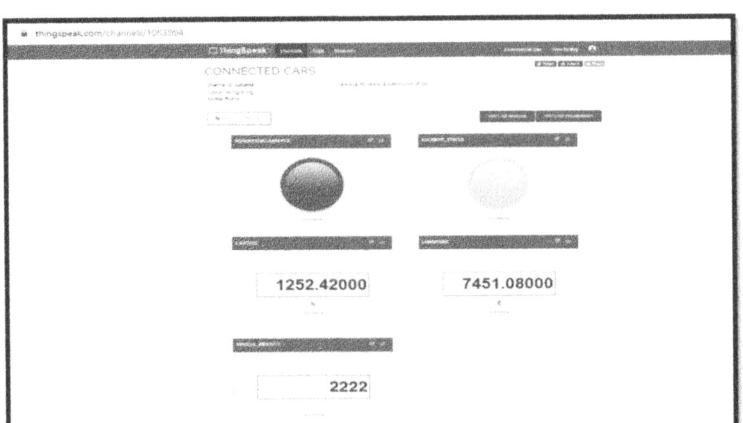

Figure 14.10 Channel showing channel clearance.

14.6 Advantages and Applications

1. V2V communication reduces the entry of the cars to the blocked path by providing prior information.
2. Prior information about the state of the road provides an option for driver to find the alternate path for destination and hence the time can be saved.
3. Real time road information can be accessed.
4. Ratio of car accidents can be decreased.
5. Helps in traffic congestion management.
6. Alert message about the impending crashes to driver can save many life and prevents unnecessary traffic.
7. Quality of safety and security of roadway is being provided.
8. Results in economic saving of individual and nation.
9. It can be implemented in other vehicles as well since the functionalities remains the same.
10. Continuous communication between the vehicles is provided in the presence of internet connectivity.
11. Ad-hoc network helps in providing anywhere, anytime access environment.

Applications

1. Safety application: In avoiding collision.
2. Incident report: If there is any road block, prior information is being sent to the server.
3. Convenience: It provides convenience to the drivers while driving by providing real time information of the surrounding and relives drivers from stressful operation.
4. Traffic Congestion: Reduces unnecessary traffic and entry to the blocked roads.

14.7 Conclusion and Future Scope

With safety a priority and concerning issue in the automobile world, this project proves to be a very reliable and inexpensive way to prevent crashes and traffic congestion. V2V network is also versatile as any automobile can use this technology without any special modification except the range of the sensors used. This important factor makes the technology affordable and very accurate. This chapter will demonstrate how a cloud based server is used for the communication of the vehicles using this network.

Future Scope

There are many attractive perspectives in hardware and software development. The model can be modified to predict the unsafe driving. With the advancement in technology the model can be merged with autonomous driving technology.

References

1. Hartenstein, H., Bochow, B. *et al.*, Position-aware ad hoc wireless networks for inter-vehicle communication: The fleetnet project. *MobiHOC 2001*, Long Beach, CA, USA, © ACM 2001 1-58113-390-1/01/10.
2. Demba, A. and Möller, D.P.F., Vehicle-to-Vehicle communication technology, in: *2018 IEEE International Conference on Electro/Information Technology (EIT)*, Rochester, MI, USA, pp. 0459–0464, 2018.
3. Alanezi, M.A., A proposed system for vehicle-to-vehicle communication: Low cost and network free approach. *Indian J. Sci. Technol.*, 11, 12, 1–10, March 2018, DOI: 10.17485/ijst/2018/v11i12/121337.
4. Wukkadada, B. and Fernandes, A., Vehicle tracking system using GSM and GPS technologies. *IOSR J. Comput. Eng. (IOSR-JCE)*, 1, 05–08, eISSN: 2278-0661, p-ISSN: 2278-8727, 2017.
5. Vatti, N.R., Vatti, P.L., Vatti, R., Garde, C., Smart road accident detection and communication system. *Proceeding of 2018, IEEE, International Conference on Current Trends toward Converging Technologies*, 978-1-5386-3702-9/18.
6. Adithya, S. *et al.*, Accident prevention system for vehicles using V2V communication. *Int. J. Adv. Res. Comput. Sci.*, 9, 3, May 2018.
7. Mallikarjuna, G.C.P. *et al.*, Design and implementation of real time wireless system for vehicle safety and vehicle to vehicle communication. *IEEE, 2017, International Conference on Electrical, Electronics, Communication, Computer and Optimization Techniques (ICEECCOT).*
8. Greeshma, G. *et al.*, Obstacle detection of vehicles under fog. *International Journal of Simulation: Systems, Science and Technology (IJSSST)*, 20, 1, 1473-804x online, 1473-8031 print, DOI 10.5013/IJSSST.a.20.01.27.
9. Ameddah, M.A., Das, B., Almhana, J., Cloud-assisted real-time road condition monitoring system for vehicles, IEEE, ©2018, 978-1-5386-4727-1/18/$31.00.
10. Darbha, S., Konduri, S., Pagilla, P.R., Benefits of V2V communication for autonomous and connected vehicles. *IEEE Trans. Intell. Transp. Syst.*, 20, 5, 1954–1963, May 2019, doi: 10.1109/TITS.2018.2859765.

11. Tian, D., Zhou, J., Wang, Y., Sheng, Z., Xia, H., Modeling chain collisions in vehicular networks with variable penetration rates. *Transp. Res. C Emerg. Technol.*, 69, 36–59, Aug. 2016.
12. Lioris, J., Pedarsani, R., Tascikaraoglu, F.Y., Varaiya, P., Platoons of connected vehicles can double throughput in urban roads. *Transp. Res. C Emerg. Technol.*, 77, 292–305, Apr. 2017.
13. Zadobrischi, E. and Dimian, M., Vehicular communications utility in road safety applications: A step toward self-aware intelligent traffic systems. *Symmetry*, 13, 3, 438, 2021, https://doi.org/10.3390/sym13030438.
14. Huang, T., Yuan, X., Yuan, J., Xiang, W., Optimization of data exchange in 5G vehicle-to-infrastructure edge networks. *IEEE Trans. Veh. Technol.*, 69, 9, 9376–9389, Sept. 2020, doi: 10.1109/TVT.2020.2971080.
15. Karkera, T. and Singh, C., Autonomous bot using machine learning and computer vision. *SN Comput. Sci.*, 2, 4, 1–9, 2021.
16. Li, Y., Luo, Q., Liu, J., Guo, H., Kato, N., TSP security in intelligent and connected vehicles: Challenges and solutions. *IEEE Wireless Commun.*, 26, 3, 125–131, Jun. 2019.
17. Singh, C., Sairam, K. V. S. S. S. S., Harish, M.B., Global fairness model estimation implementation in logical layer by using optical network survivability techniques. *International Conference on Intelligent Data Communication Technologies and Internet of Things*, Springer, Cham, 2018.

15

Broadband Wireless Network Era in Wireless Communication – Routing Theory and Practices

R. Prabha[1]*, G. A. Senthil[2], S. K. B. Sangeetha[3], S.U. Suganthi[1] and D. Roopa[1]

[1]Sri Sai Ram Institute of Technology, Tambaram, Chennai, India
[2]Agni College of Technology, Chennai, India
[3]SRM Institute of Science &Technology, Chennai, India

Abstract

Wireless technologies that enable people to link their home networks together as well as to the Internet via wireless community networks are in high demand. Despite the fact that many network architectures for broadband wireless networking have been studied and implemented in practice due to emerging technology and locales, Broadband Wireless Networking (BWN) has unquestionably been the most common architecture that has demonstrated strong dominance in various roles. Indeed, allowing such broadband networking and creating a group network has numerous advantages. This would boost the rural and isolated communities' quality of life, digital interaction, and collaborations. The design of efficient communication is a difficult problem for the next generation BWNs' performance in managing real-time and QoS-sensitive applications, as well as pleasing both service providers and consumers. Unfortunately, today's cutting-edge routing standards in BWNs are not ideally suited to tackle this task, as these standards are fundamentally complex and suffer from innate problems with regard to efficient communication-based implementations, according to the literature. As a result, the aim of this chapter is to provide BWNs with a straightforward roadmap of theoretical context so that they can manage various efficient routing mechanisms.

Keywords: Broadband Wireless Networking (BWN), Distributed-Data Interface (FDDI), Synchronous Optical Network (SON)

**Corresponding author*: r.prabha05@gmail.com

Chandra Singh, Rathishchandra R Gatti, K.V.S.S.S.S. Sairam and Ashish Singh (eds.) Modeling and Optimization of Optical Communication Networks, (267–280) © 2023 Scrivener Publishing LLC

15.1 Introduction

Remote broadband terms ought not be related with the normal word broadband availability or BISDN (Broadband Integrated Services Digital Network), which applies to various correspondence organizations (fiber or optical) utilized by ISPs and NSPs to arrive at Internet spine information speed of in excess of 155 Mbps. BISDN is the wire and connection that goes through dividers, under floors, from utility post to utility shaft, and under feet on a city street in layman's terms. BISDN is a perspective, a lot of developments, and a lot of norms for joining progressed correspondence organizations into a fiber optic and radio-based broadband network [1]. Data correspondence organization for fast data sent in wide detonates, the Fiber Distributed-Data Interface (FDDI), and the Synchronous Optical Network are exceptionally significant for BISDN (SONET). BISDN can impart data at speeds going from 2 Mbps to a lot more noteworthy. Remote broadband, on either hand, is a distant association structure that handles irrefutably the last issue by allowing us to interface disengaged customer premises to an ISP or carrier's establishment network without leasing customary T-1 and more significant copper or fiber stations from the local telecom provider. Fixed remote access that can be used by associations, endeavors, families, and remote workers who move beginning with one unequivocal spot then onto the following is implied as remote broadband. It doesn't resolve the issues of convenient customers making the rounds in its present structure [2]. Remote broadband is a utilitarian extension of the highlight point, remote LAN combination idea for conveying high velocity, high-limit pipes for voice, mixed media, and Internet network administrations. Albeit remote broadband is essentially used to interface LANs to the Network in straightforward applications, further developed executions permit you to associate various administrations (information, voice, and video) over a similar link. The last requires the establishment of multiplexing offices at the client's area or in a focal hub. Wireless Broadband detours actual phone networks as far as organization, and it is similarly as attainable in country regions all things considered in metropolitan regions. Seller arrangements that keep away from costly establishment, overhauling, and redesigns mean skirting 120 years of broadband improvement for geologies that haven't yet innovatively developed to fiber and cabling frameworks. Liberation in certain spaces has worked on the permitting system for Wireless Service Providers (WSPs) [3].

Remote broadband is faster to market, and clients are presented in stages rather than all at once, avoiding the costly upgrades that must be completed before wired subscribers can communicate. Wireless broadband is intended for fixed wireless connections; it does not meet mobility requirements, which are currently addressed by only 2.5 G and 3G networks. Broadband wireless radios may potentially be miniaturized and integrated into portable devices in the future. They would then be able to supplement 3G in mobile apps. However, none of the vendors have any getting experience in this field, so it's just a physics and circuitry possibility. Wireless broadband, of course, is intended to meet the needs of suburban Internet access by bypassing local telcos. Multiple research projects are underway to improve WOBAN, such as the management system and energy conservation [4]. As a result, the WOBAN architecture acts as a foundation for future discussion of the aforementioned topics. WOBAN is also known as FiWi BAN by some researchers (Fiber Wireless Broadband Access Networks). We use the same language as the source to ensure continuity between this survey and the references. As a result, when discussing related works, the word FiWi BAN can appear.

The architecture of BWA networks in general covers a wide range of topics. First and foremost, the access network's bandwidth must be increased. Optimal relay station positioning, bandwidth allocation scheme, and direct communication mode between subscriber stations have all been proposed. Other related research focuses on large-scale BWA network management systems and efficiency comparisons with multiplex strategies for 4G [5]. Another important consideration when installing BWA networks is energy and cost management. Reducing energy usage not only saves money on electricity bills, but it also helps to minimize CO2 emissions and is environmentally friendly. Several studies have been conducted on this subject, including an architectural study of energy efficiency, a mixed capacity access proposal, a study for minimizing emissions in long reach access, and a green WOBAN design. The enhancement of survivability and real application situations that can benefit from BWN networks are also important topics in BWN [6].

With this detailed introduction about BWN, Section 15.2 describes the outline of BWN, Section 15.3 discusses various routing mechanisms in BWN, Section 15.4 briefs about security issues in BWn followed by conclusion in Section 15.5.

15.2 Outline of Broadband Wireless Networking

15.2.1 Type of Broadband Wireless Networks

15.2.1.1 Fixed Networks

Rapid remote organizations that connect to fixed areas and are intended to help itinerant clients are known as fixed broadband remote innovations. Fixed remote innovations incorporate Wireless Fidelity (Wi-Fi) and Worldwide Interoperability for Microwave Access (WiMax). IEEE 802.11 and IEEE 802.16 guidelines control these two advancements, respectively [7, 8].

Wi-Fi was the first high-speed fixed wireless broadband technology to get into the industry. The first WLAN (802.11) was implemented in 1997 and could accommodate 2 Mbps, while IEEE accepted 802.11b in 1999. There are currently several wireless devices based on Wi-Fi technology, including the IEEE 802.11a, b, and g standards, as well as an 802.11n standard that has yet to be defined. It's difficult to say what effect 802.11n would have when it actually hits the market, but it's expected to offer up to 600 Mbps. In a 2006 survey, In-Stat predicted that more than half of all chipsets shipped in 2008 will be focused on 802.11n. Wi-Fi has seen widespread use as a high-speed wireless technology, most notably in hotspots around the world.

WiMAX

WiMax (Worldwide Interoperability for Microwave Access) is another proper broadband remote innovation that will give last-mile broadband inclusion across a more extensive topographical region than Wi-Fi. It should involve a space of one to six miles in measurement. WiMax inclusion scopes of this size are planned to give fixed and traveling remote broadband access without the requirement for a line-of-site (LOS) with a base station. WiMax can likewise have greater portability, higher-speed information applications, more prominent inclusion, and throughput than Wi-Fi. The execution of WiMax can give various advantages. Regardless, it considers higher throughput rates, quicker information speeds, and a more extensive working reach. Thus, the innovation is great for sending in troublesome landscape or in regions with deficient wired framework.

WiMax additionally upholds and effectively incorporates with other wired and remote technology. The greatest inconvenience of WiMax execution is the utilization of restrictive gear. To give ideal usefulness, WiMax hardware should have the option to productively use power. The yield power utilized by WiMax is controlled by a running technique that indicates the appropriate planning offset and force settings. Therefore, endorser station's transmissions are relied upon to show up at the base station simultaneously and with a similar force level. At the point when WiMax is utilized outside, it can encounter delays in non-view conditions, which can prompt intersymbol interference.

15.2.1.2 The Broadband Mobile Wireless Networks

The first (1G) associations, which depended on Frequency Division Multiple Access (FDMA) and were by and large used for voice correspondence, begun the advancement of convenient help. Second-age (2G) networks have for the most part supplanted the 1G organization, which is for the most part utilized for discourse applications. At a low speed, these 2G organizations conveyed circuit-exchanged information transmission administrations. The scramble to create and consolidate advanced frameworks brought about a plenty of divergent and contradictory specifications.2.5G is an improvement of 2G innovation that takes into account more information transmission capacity on 2G organizations. Hence, progresses like General Packet Radio Service (GPRS) and Enhanced Data Rates for Global Evolution (EDGE) were made (EDGE). The point of third-age (3G) networks was to give a typical general norm to rapid information and top-notch voice administrations. The point was for all clients all throughout the planet to utilize a typical norm, taking into account genuine worldwide meandering. Soon, 4G is a cutting-edge wave of broadband that will enhance and supplant 3G organizations. The principal provisions of 4G foundations are getting to data anyplace, anyplace, with a consistent connection to a wide assortment of data and benefits, and getting a lot of data, information, pictures, video, etc. Future 4G frameworks would be comprised of an assortment of organizations that utilization IP (Internet convention) as a standard convention, giving clients unlimited oversight about the applications and conditions they use.

In view of ebb and flow portable availability designs, 4G would highlight expanded limit, higher information rates, and smoother and quicker hand-off, with an accentuation on giving consistent inclusion through an assortment of remote frameworks and organizations. The main idea is to use new technology to integrate 4G features with all current mobile technologies. Some of the key aspects of 4G networks that consumers are interested in are application adaptability and high dynamism. These features imply that services can be provided and made available according to the preferences of individual customers, while still supporting traffic, air interfaces, radio environments, and service quality. The right and efficient link to network applications can be done in a variety of ways and at different levels [9].

15.2.2 BWN Network Structure

The BWN network is principally utilized as a highlight multipoint geography, with a cell conveyance of base stations that are totally associated with center organizations and in correspondence with fixed remote supporter stations. The endorser stations as a rule include housetop mounted receiving wires and radio units connected to indoor organization interface units, while the base stations are usually found on a pole. Multiple subnets in a star topology usually make up a completed BWN structure. In terms of structure, it resembles the historic slotted Aloha network proposed in the 1970s. The downstream of the vehicle system, from the central base station to the endorser stations, is sent, while the upstream, from the supporter stations to the central base station, utilizes time division various access (TDMA). The organization's fundamental parts are portrayed in Figure 15.1.

Thousands of customer premises equipment (CPE) can be found on the rooftops of various structures. Optical fiber can link the central base station with other server applications to the Internet backbone. The base station (BS) is a switching system that communicates directly with the subscriber station and is also referred to as a network bridge or router. The modem will serve as a data port on the subscriber station (SS), resuming data transmission over cable. The physical and data link layers of a BWA network are the two layers that differentiate it from other networks,

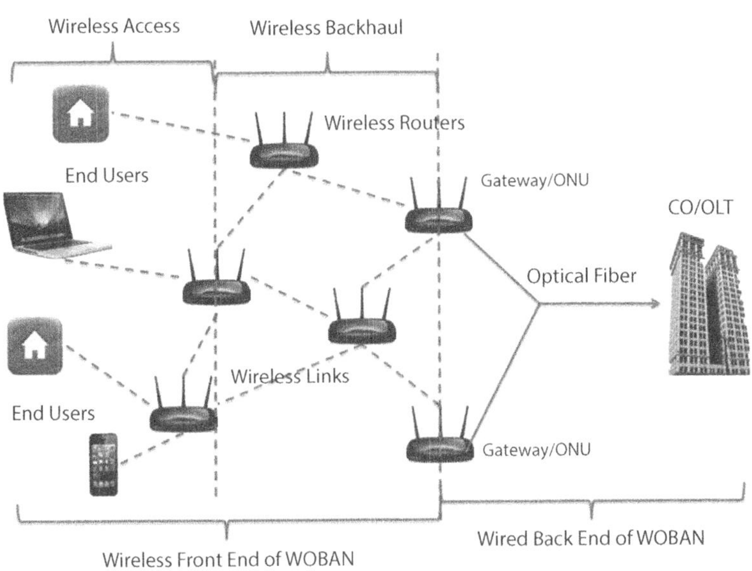

Figure 15.1 Structure of WBN.

according to the OSI model. They define the methods and specifications for communication between the Subscriber Station (SS) modem and the Base Station (BS). A BWA system's network protocol closely follows the DOCSIS (Data Over Cable Service Interface Specifications) format, with some modifications for wireless use [10].

15.2.3 Wireless Broadband Applications

Broadband communication refers to the transmission of digital audio (voice), data, and/or video at speeds faster than those of wideband communication (above 1 Mbps). Broadband connections enable the delivery of multiple services such as voice, data, and video over a single network [11].

Digital Telephone
Digital telephony is a communication device that represents and transmits analogue signals using digital data. Audio signals (acoustic sounds) or complex modem signals (other types of information) are examples of analogue signals. WiMax systems can use IP Telephony to provide telecommunications services (voice over Internet protocol VoIP). These IP networks use IP protocol to initiate, process, and receive voice or digital telephone communications. WiMax systems may use analogue telephone adapters (ATAs) or IP telephones to provide wireless telephone service. IP signals are converted into regular telephone (dial tone) formats by ATAs.

Broadband Data Connections
Broadband data networks are those that transmit digital data signals at speeds of 1 megabit per second or higher for consumer connections and 45 megabits per second or higher for LANs, MANs, and WANs. WiMax service providers can be referred to as wireless Internet service providers because they provide broadband data services that can link to the Internet (WISP).

Digital Television
The method or device that uses digital transmission to relay video images is known as digital television. Channels for digital video and audio are split into the digital transmission. Typically, these digital channels are compact.

15.2.4 Promising Approaches Beyond BWN

To meet future capacity demands, radio access strategies will need to be significantly improved. There is no easy alternative; instead, the solution would be a mixture of multiple options, depending on the time and place.

In networking, radio access, and modem integration, there are some promising strategies to take, which are briefly outlined below.

- Increases in capacity of 500-1000 times are needed.
- There is a scarcity of spectrum.
- Demands for energy conservation small cells are being used on a large scale. Super-broadband optical fiber is now available.
- Spectrum use that is adaptable

Every year, the market for broadband wireless communications grows dramatically across the world. The ever-increasing number of users who subscribe to broadband packages is a major factor in this growth. The movement toward flat-rate contracts has intensified this. Furthermore, new devices with strong multimedia capabilities, such as mobile phones and tablets, are entering the market, generating new demands for broadband wireless connectivity. Finally, new data services and apps are becoming available, which are critical success factors for mobile broadband. As both of these variables are combined, the data traffic in the wireless access system grows exponentially.

Over the next decade, this pattern is projected to continue in a similar manner. According to recent studies and extrapolations based on historical trends, overall traffic will rise by a factor of 50 to 10000 in the next period. Aside from the total traffic, the realistic throughput per user must be greatly improved. According to a rough estimate, the average and peak data rates would rise by at least ten times. Furthermore, important design standards that must be met more effectively than in today's systems include user fairness across the entire coverage area, latency to minimize response time, and better support for a variety of Quality of Service (QoS) requirements originating from various networks [12].

15.3 Routing Mechanisms

The directing convention used to track down the best course to the base stations is presently being investigated as one of the vital spaces of cross section organizations (or passageways). Clients who utilize this innovation can exploit their administrations all the more adequately and impart all the more proficiently, just as move their information stream through the remote correspondence climate. Directing is an assistance wherein the switch analyzes the entirety of the accessible ways for sending bundles to their objective and picks the right one. The rule of organization execution improvement is carried out by developing a steering tree choice that is characterized by topological properties that are independent of the organization's arrangement. The manner in which the tree is constructed and the hubs are masterminded takes into account a superior appropriation of hubs, which prompts better directing and collection. The likenesses between the tree's mathematical boundaries and the organization's proficiency should be assessed, and those with the best relationships ought to empower the best trees to be made, taking into consideration some directing and geography improvement.

There is right now a scope of steering conventions accessible, each with its own arrangement of benefits and detriments when applied to work organizations. There is no single directing convention that can be supposed to be awesome among such countless choices. The clarification for this is that they each have their own arrangement of attributes, and there is no single convention that is viewed as ideal in all situations. Every convention has a distinctive component that makes it proper for a particular application.

Adaptive routing algorithms can in turn be classified in two ways:

1. Distance Vector (DV): This became known as Routing Information Protocol (RIP) or Distributed Bellman-Ford Protocol (DBF) due to its application to bundle directing on the Internet (DBF). This calculation works by permitting every switch to keep a table (for example a vector) of the most brief distances to each known objective and figuring out which line to take to arrive. Steering is characterized as a metric unit in a distance vector that addresses the expense of a way between network hubs. The actual distance between hubs, the quantity of bounces (jumps), the transmission inactivity, hub clog, and different factors could all be remembered for this metric unit.
2. Link State (LS): This powerful calculation was made to tackle the issue of distance vector steering since it depended on the quantity of jumps to the objective, despite the fact that a bundle could arrive at its objective by voyaging a brief distance, for example with few bounces. Be that as it may, the association data transfer capacity can be limited, bringing about a more extended deferral. As a result, the connection state has emerged in order to find efficient routes, regardless of the total number of hops in the network is situated. Routing protocols for ad-hoc networks are divided into three categories: constructive, reactive, and hybrid. When a data packet needs to be sent, the proactive form requires us to maintain the route network for all possible destinations. The nodes in reactive protocols discover the destinations on request. The hybrid

protocols are those in which there is only one group of nodes that updates information on potential destinations on a regular basis. In BWN, QoS routing is a critical parameter for ensuring guaranteed QoS. In BWN, this problem has been thoroughly investigated. The point of QoS directing in these organizations is twofold: to track down the most ideal course for every approaching association within the sight of hidden connection, and to adjust the heap to augment network use.

Traditional routing protocols use the number of hops as their most common criterion. Notwithstanding, obviously these conventions are deficient for media applications that need QoS affirmations, for example, VoIP and video conferencing. Steering conventions with QoS should track down the most brief, yet additionally the best course that fulfills the start to finish QoS boundaries, independent of the measure of bounces or how the correspondence conventions should track down the best courses across various jumps. It is important that the most recent conventions and directing calculations consider boundaries and different measurements, for example, power utilization, the nearness of the spine network yield, and, specifically, the quality over amount association for clients and the nature of remote interchanges, while representing weakening, signal quality, and disturbance [13, 14].

15.4 Security Issues and Mechanisms in BWN

15.4.1 DoS Attack

Quite possibly the most major issues with a wide range of remote network, particularly broadband remote organizations, is forswearing of administration (DoS). A DoS break happens when enrolled clients are not furnished with a mentioned administration inside a characterized greatest holding up period. The most problematic and damaging assault can be directed on any layer of a broadband remote organization. DoS assaults expect to upset accessibility by impeding contact between network gadgets or keeping a solitary gadget from sending or getting traffic; accessibility guarantees that supported clients can get to information, administrations, and organization assets from any area whenever.

A DoS attack on the actual layer might be dispatched by utilizing a radio sticking framework or a wellspring of solid commotion to meddle with the actual channels, conceivably risking administration accessibility. Nonetheless, this kind of assault is remarkable on the grounds that it requires the utilization of specific equipment, and sticking assaults can be identified utilizing radio analyzers. It may cause major issues during the sharing of confidential information or during combat. None of the mechanisms are up to the task of dealing with a jamming assault on these BWN.

15.4.2 Distributed Flooding DoS

A distributed flooding DoS attack poses a significant threat to all wireless broadband networks, as it has the potential to bring an entire network down or consume a significant amount of network bandwidth. This type of attack begins by infiltrating a wireless network with a large number of innocent nodes known as Zombies, which are programmed by highly qualified programmers. These zombies submit data to pre-selected attack targets, clogging the network in the process. DDoS attacks are almost impossible to avoid in most situations, and they have the potential to flood and overflow networks.

15.4.3 Rogue and Selfish Backbone Devices

By breaching the core network equipment, the intruder will severely damage broadband wireless networks. A rogue BS, as defined by IEEE 802.16, is an intruder station that is used to confuse the network's mobile stations by appearing and acting like a legitimate BS. Sniffers are used by attackers to hack mesh routers or access points. A sniffer is a programmed that analyses network traffic passively as part of a passive traffic analysis attack. The BS is taken advantage of in IEEE 802.16 by reconstructing a PC with the equipment address of one more substantial gadget, which can be identified by capturing IEEE 802.1 administration messages with sniffers. A similar strategy can be utilized to think twice about switches and passages by utilizing the equipment address of another organization PC.

15.4.4 Authorization Flooding on Backbone Devices

Test demand outlines are utilized by WBN and IEEE 802.11 hubs to find a remote organization, and in the event that one exists, the AP reacts with a Probe reaction outline. Customers pick the AP that conveys the best message to them. The interloper might parody a surge of test demand outlines, representing an enormous number of hubs searching for a remote organization, making the AP or remote lattice switch be seriously overloaded. If the heap arrives at the limit esteem, the AP or remote cross section switch will quit reacting, possibly bringing about help blackout. Customer stations use testaments to validate and enlist with the BS in IEEE 802.16. The customer station's capacity to present an enormous number of enlistment solicitations to the BS which causes a DoS.

15.4.5 Node Deprivation Attack

The aggressors focus on a solitary hub in a hub hardship assault and keep it from taking part in normal organization tasks. In WBN and IEEE 802.11, hubs should initially confirm with the lattice switch or passageway, and afterward de-verify in the event that they at this point don't have any desire to utilize the organization administrations. The attackers will mimic. The attacker will impersonate the target node and spoof the de-authentication message to prevent it from accessing network resources [15].

15.5 Conclusion

Strong research efforts on emerging evolutionary and creative radio access network solutions are essential with the dramatic traffic increase expected for 2025 and beyond. On a global scale, these initiatives will require coordinated activities between industry and academia. Broadband wireless networking is establishing itself as a legal local access network for high-quality digital data, video, and voice services. It is hoped that through this chapter, which provides an introductory explanation of the BWN network, provides a complete outline of background, routing protocols and security issues. We may consider various issues relating to the security aspect of broadband technology based on the above review. There are many perspectives to consider when considering the security of wireless technologies. Safety encompasses a wide range of authentication, access control, and encryption technologies. While these are important and huge structure blocks for overall protection, they are not the subject of this section and will be considered in future work.

References

1. Craig, K.H., Hurley, S., Tjelta, T., Propagation studies for enhanced broadband wireless access, *URSI General Assembly, Maastricht*, 2022.
2. Ibikunle, F., Orunta, J., Dike, I., Broadband wireless access deployment approach to rural communities. *J. Comput. Netw.*, 1, 3, 38–45, 2013.

3. Guan, M. and Wang, L., Comparison of broadband wireless access technology for HAPS communication. *Sens. Transducers.*, 3, 166, 122–127, 2014.
4. Hwang, K. and Vemuri, R., Broadband wireless access (BWA) networks: A tutorial, 2022.
5. Kaydenko, M.M., Adaptive modulation and coding in broadband wireless access systems, pp. 275–276, 2013.
6. Lazov, I., Entropy analysis of broadband wireless access systems. *IEEE Syst. J.*, 11, 4, 2366–2373, Dec. 2017, 10.1109/JSYST.2015.2456941.
7. Lazov, I., An uncertainty quantification methodology for broadband wireless access systems. *Pervasive Mob. Comput.*, 42, 151–165, 2017, https://doi.org/10.1016/j.pmcj.2017.10.002.
8. Lin, H.-T., Lin, Y.-Y., Kang, H.-J., Adaptive network coding for broadband wireless access networks. *IEEE Trans. Parallel Distrib. Syst.*, 24, 4–18, 2013, 10.1109/TPDS.2012.101.
9. Liu, Y. and O'Leary, P., Broadband wireless access in an energy efficient environment, pp. 1–18, 2013, 10.1049/ic.2013.0097.
10. Nasser, N., Miller, R., Esmailpour, A., Taha, A.E., Bejaoui, T., Optimized bandwidth allocation in broadband wireless access networks. *Wireless Commun. Mobile Comput.*, 15, 2111–2124, 2015, 10.1002/wcm.2479.
11. Okello, D., Wasswa, W., Mukasa, P., Sebbaale, D., Kagarura, G., eInfrastructure: Next generation wireless broadband networks for Uganda 2020, 2015, 10.1109/ISTAFRICA.2015.7190548.
12. Osseiran, A., Braun, V., Hidekazu, T., Marsch, P., Schotten, H., Tullberg, H., Uusitalo, M., Schellmann, M., The foundation of the mobile and wireless communications system for 2020 and beyond: challenges, enablers and technology solutions, pp. 1–5, 2013, 10.1109/VTCSpring.2013.6692781.
13. Senthil, G.A., Prabha, R., Roopa, D., Babu, D.V., Suganthi, S., Improved cluster head selection for data aggregation in sensor networks. *2021 7th International Conference on Advanced Computing and Communication Systems (ICACCS)*, pp. 1356–1362, 2021, doi: 10.1109/ICACCS51430.2021.9442048.
14. Zander, J. and Mähönen, P., Riding the data tsunami in the cloud : Myths and challenges in future wireless access. *IEEE Commun. Mag.*, 51, 145–151, 2013, 10.1109/MCOM.2013.6476879.
15. Suganthi, S.U., Valarmathi, G., Subashini, V., Janaki, R., Prabha, R., Coal mine safety system for mining workers using LORA and WUSN. *Mater. Today: Proc.*, 46, Part 9, 3803–3808, 2021, ISSN 2214- 7853, https://doi.org/10.1016/j.matpr.2021.02.037.

16

Recent Trends in Optical Communication, Challenges and Opportunities

S. Kannadhasan[1*] **and R. Nagarajan**[2]

[1]*Department of Electronics and Communication Engineering, Study World College of Engineering, Coimbatore, Tamil Nadu, India*
[2]*Department of Electrical and Electronics Engineering, Gnanamani College of Technology, Tamil Nadu, India*

Abstract
Virtually all technologies have experienced a digital transformation as a result of technical advances in our daily lives and their convergence improvements in the communication network to allow for more seamless communication. As a result, a massive amount of digital data is shared through email, audio, and video chats, guaranteeing that people are constantly connected. As the market for digital information develops, the data presently transferred via optic fiber transmission technologies will become outdated due to the inherent nonlinear effects of digital information. By finding new methods and using current tools, machine learning remains a viable solution for coping with future device complexity. The goal of this chapter is to look at the nonlinearities that arise in optical fibers and the possible solution offered by machine learning techniques for increasing optical fiber communication capacity.

Keywords: Optical fiber, transmitter, receiver, nonlinearities, multiplexer and demultiplexer

16.1 Introduction

The growing interconnectivity of the global economy has a significant impact on the advancement of information and communication technologies. The world aims for continuous connectivity and uses digital technology such as smart phones and mobile phones, as well as communication

**Corresponding author*: kannadhasan.ece@gmail.com

Chandra Singh, Rathishchandra R Gatti, K.V.S.S.S.S. Sairam and Ashish Singh (eds.) Modeling and Optimization of Optical Communication Networks, (281–302) © 2023 Scrivener Publishing LLC

tools like as audio, video, and email, to communicate with people all over the world at any time and from any place. The data is sent via the internet using an optical fiber network design. Information is readily disseminated through the internet across nations and continents. Internet traffic is projected to surpass Exabyte this month. To transfer data at today's data rates and delays, optical fibers with high bandwidth and low loss are needed as a communication medium. The optical fiber is a light stream composed of a flexible, transparent strand of ultra-pure glass that transmits light between the fiber's two ends. The optical fiber is a light stream composed of a flexible, transparent strand of ultra-pure glass that transmits light between the fiber's two ends. A dielectric cladding layer surrounds a dielectric heart. The signals in the optical fire's center are confined by using a refractive index greater than the cladding. According to Wikipedia, it's a networking and telecommunications protocol. As light goes through the optic fiber, it bounces back and forth between the backbone and the cladding. The cladding of an optic fiber works on the concept of complete inward reflection, which means that no light from the center is absorbed [1–5].

In optical fiber cables, machine learning methods such as SVM, KNN, and RBFNN are used to decrease non-linearity, which causes service deterioration. The machine learning techniques utilize observable data to learn the characteristics of various non-linear impairments, and then combine the statistical representations of the impairments to be used later in either recompensing optical fiber damages or measuring the amount of distortions produced. For multiple-impairment tracking in optical networks, the ML seems to be a more feasible and cost-effective approach. Furthermore, the Bayesian approach is used for carrier synchronization and machine learning-based laser line width effect reduction. The article discusses how nonlinear effects in optical fiber communication may be reduced using machine-learning methods. According to a literature study, it goes on to explain how machine learning may be used in optical fibers for different applications, as well as well as a discussion of how to minimize nonlinear effects in optical fiber networks, as well as an overview of the different machine learning methods accessible. In the future, machine learning will be increasingly successful at reducing noise fluctuations in optical fiber links.

Due to its high data rate and long-distance communication capabilities, fiber optic infrastructure has dominated the industry for a long time. Metal cables have been proven to be inferior than fiber optic cables. Fiber optics offer less signal distortion and are less sensitive to interference. Many systems, such as finance algorithmic trading, cloud storage, and supercomputers, need low latency [6–10]. The bulk of communication networks utilize fiber optics to transfer data. Some of the areas where significant progress has been achieved in the last decade are as follows: While silica is still used in fiber processing, alternative materials such as chalcogenide glasses, fluoroaluminate crystalline compounds, and fluorozirionate materials are being employed to provide longer infrared wavelengths and improved communication capabilities. In the last decade, the field of fiber optic sensors has advanced considerably. Higher sensor efficiency ensures better data reception. New technical breakthroughs are being developed, such as free-space technology. Fiber optics is a fundamental component of our telecommunications network. The evolution of optical networking technologies has relied heavily on low-loss optical data transmission. For optical networks, the use of single mode fiber (SMFs) in passive optical splitters has been suggested. The next-generation network is the fiber-to-the-home network. In 1990, the first fiber optic amplifier was created. To stop regeneration and allow data impulses to reach hundreds of kilometers, fiber optic amplifiers are used. The aim of wavelength division multiplexing is to increase an optical network's transmitting capacity. In a number of circumstances, including deeper water, acoustic transmission has many flaws, and optical underwater communication is the most efficient way to overcome this shortcoming. The main reason why optical communication is more effective than electrical communication is because optical communications are much faster than electrical messages. Optical requirements have also led to the creation of ultra-coherent optical carriers for transmission, which is a relatively new discovery. Fiber optic networks, which send light modified by an electrical pulse from an atomic clock, are also being utilized on a worldwide scale.

In addition to silica, the fiber processing method uses chalcogenide, fluoroziconate, and fluoroaluminate. The wavelength of sapphire is longer than the wavelengths of the other two components. The following sections discuss advancements in the material wise manufacturing of fiber fluoride glass, which is made up of different metals' fiber fluorides. These are non-oxide optical lenses with a low viscosity. Optical lenses having a high viscosity, or viscosity. Fluoride glasses have a very low optical attenuation. Silica is a mineral that may be found in almost all types of fibers. It has been chemically refined and has a minimal absorption loss. Silica has a wide transmission spectrum, but its primary benefit is that it can be doped with a variety of minerals, Aluminum oxide and germanium dioxide are two examples. Chalcogenide glass is composed of chalcogenide, a chemical that is very flexible. Phosphate glass is made from metaphosphates of various metals with high concentration doping. Despite the fact that silicon dioxide (Sio2) is the most essential component in fiber optics, it is also the most expensive. phosphorus oxychloride is often employed as the fiber's backbone (Poc13). Germanium tetrachloride may also be used to make fiber (Gecl4). The kind of material we choose determines the degree of attenuation, which is the most significant feature of fiber optic transmission [11–15]. The higher the degree of attenuation, the more distinct light wave signals that may be conveyed.

16.2 Optical Fiber Communication

In terms of technology, high-altitude systems are the focus of current fiber optic communications research (HAPs). HAPs are airships that fly between 17 and 25 km above ground level in order to maintain a laser beam free of severe atmospheric effects. In this approach, HAP is utilized as a data relation. Free-space networking, in which optical-communication links may be used for satellite-to-satellite cross linking, is another fast emerging technology. It is possible that this technology will be commercialized. The performance of physical relation connections may be enhanced by using photon counting receivers. To obtain a bandwidth of 50 mbps, which is a large quantity of data, a copper-fiber backhaul network and telecommunication switches were used. This network utilizes fiber optic cables to offer modest coverage of 450 m. In fiber optic transmission, the optical laser is the most important component. This is the part that produces light. If the laser's wave length is short, it will produce a lot of power. A laser can detect a rapid shift in wavelength. In recent years, single mode laser technology has advanced considerably. Temperature, strain, and pressure sensors may be added to optical fiber. The phase shift in light is caused by the interferometric sensor, which is a light wave transmitted via fiber. Because the total power transferred via Fiber is radial, the intensity parameter is dependent on power transmission. The telecommunications sector is highly reliant on fiber optics. It is ideal for greater bit transmission because to its low attenuation, low absorption loss, and high bandwidth properties. Fiber optic chemical and temperature sensors are integrated and utilized for environmental management and medical purposes. The fiber optics industry will continue to grow over the next decade. Fiber optics will become increasingly accessible to household applications in the next decades. These days, it's exclusively seen in commercial settings. Fiber optic infrastructure would revolutionize the telecommunications industry and change our lives forever.

The various loss mechanisms that occur inside a single mode fiber (SMF) in optical fiber communication are discussed in this article.

A variety of mechanisms cause signal attenuation in optical fibers. The optical signal is attenuated as it travels over a long stretch of fiber due to reflection, polarization, and fiber bends caused by material defects, among other factors. High-bandwidth transmission can handle massive quantities of data, and it can be made even better by lowering fiber delays, boosting data rates and lengths, and choosing the right operating wavelength for optical fiber connections. This article discusses recent developments in fiber optic networking, as well as improvements in various fiber kinds and their properties, such as attenuation or failure, and bandwidth. There is always a need for higher-capacity networks at cheaper costs in the age of connectivity. Thanks to the development of high-speed and high-density integrated circuits, recent improvements in data processing capability have exceeded previous data transmission capabilities. Massive bundles of copper cables are becoming less attractive as a data transmission medium. Because of the height, weight, bandwidth restrictions, and expense of metal conductors, scientists and engineers have been forced to explore other data processing methods. Fiber optics is one of the most significant and cost-effective communication solutions currently being explored, as illustrated in Figure 16.1.

Optical fiber networking technology has significantly developed in recent years, enabling for higher transmission capacity and longer transmission distances. In terms of height, weight, bandwidth (i.e. 1013 to 1014Hz), EMI and nuclear radiation endurance, and cost, fiber optics

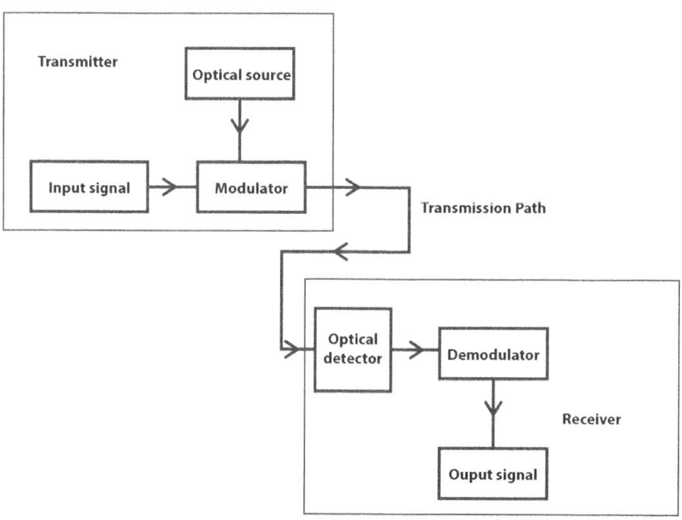

Figure 16.1 Optical communication block diagram.

may outperform metallic conductors. These benefits, however, are of little to no use in a variety of applications unless fiber optics can continue to offer efficient data processing capabilities for the duration of the system's intended lifetime. With the increasing interest in using fiber optic data links in military/defense systems, a precise prediction technique capable of dealing with the many components and assemblies used in such a device is needed. However, many attenuations occurred during information transfer through optical fiber connection, decreasing the system's output. The goal of this article is to examine the various loss mechanisms in single mode fiber and how to minimize them in optical fiber communication.

An optical fiber transmission device is similar to any other kind of communication system in terms of basic concept. An electrical signal is sent from the information source to the transmitter, which contains an electrical stage that drives an optical source to change the light wave carrier. The optical foundation for the electrical-optical conversion may be a semiconductor laser or a light emitting diode (LED). The optical carrier is demodulated using an optical detector that drives a second electrical point, with an optical fiber cable serving as the transmission medium. For optical signal processing and optical-electrical transmission, photodiodes and, in certain cases, phototransistors and photoconductors are used. As a consequence, both ends of the optical link need electrical interfacing, and signal processing is now done electronically. On the optical carrier, an analogue or digital information stream may be modified. In this technique, analogue modulation, a continuous change in the light generated by the optical source is included. Digital manipulation, on the other hand, results in distinct light intensity changes. Analogue modulation is less effective than digital modulation for optical fiber transmission equipment, but it is simpler to install. The receiver's signal-to-noise ratio must be considerably greater.

The attenuation of signals in optical fibers and metallic conductors is often represented in decibel logarithmic units. The decibel is defined as the ratio of the input transmission optical power Pi via a fiber to the output reception optical power Po from the fibers for a certain optical wavelength:

The number of decibels (dB) is equal to 10 log Pi/Po.

Addition and subtraction, on the other hand, need a numerical conversion, which may be accomplished through an optical fiber communications link.

16.3 Applications of Optical Communication

5G transition in real time Wireless networking would be transformed from a pipe dream to a reality with a fiber optic link. 5G advancements are anticipated to result in greater infrastructure capacity, increased energy consumption, decreased latency, and connectivity to a broad variety of wired devices. Despite the fact that cellular communication is the most popular form of communication, many of the problems related with 5G wireless communication need a fiber optic linked network to enable and address. With higher transmission speed and network availability, the Quality of Service (QoS) may be improved. As a consequence, the virtual world of augmented reality becomes more immersive. This article examines the optical communication paradigm, including its concepts, specifications, and modulation systems, as well as a careful comparison to emphasize the significance of optical technology in next-generation communication. Electric current and magnetic fields are produced when current flows through conductors, and EMI is a major issue in electric transmission. As a result, the current flow has changed. Our signal form, a coaxial cable and an electric current, is mainly to blame. Optical communication addresses this issue by using a light signal as a carrier and a fiber as a transferring medium. Light signals are efficiently sent via fiber, which is error-free. It is the most important feature in today's world of high-volume results. Due to the effect of an active magnetic field on electric cables used in conventional signal transmission methods, data loss is a possibility. Simple reasons such as heat dissipation and cable cutting may result in data thread failure. A dynamic set-up is often required when data is received in the midst of a transmission. The tapping method used in optical communication improves data transmission security despite high-volume data input at every stage of transmission. As a consequence, fiber virtually eliminates electromagnetic field loss, leaving just internal fields.

Road communication has grown increasingly congested as a result of the exponential development of data networks, and optical communication has become the only way out. Optical fiber networking has piqued people's attention as a cutting-edge communications technology with many benefits from the start. It has low loss, high communication capacity, electromagnetic interference tolerance, security, and other notable advantages over conventional cable transmission. As a consequence, optical fiber networking is considered as a watershed point in communications development. The United States was the first to propose the idea of an information superhighway, which had a significant effect on the industry's backbone - research and manufacturing of wide optical transmission systems. With the growth of B-ISDN to integrate broadband delivery of optical fiber networking as the primary means of connection in the twenty-first century, the telecom network will continue to thrive and exhibit vitality. Network trunk lines, power communication control systems, industrial monitoring and control, and even military applications are just a few of the applications for optical fiber. Concerns have also been expressed regarding the increasing need for knowledge in a variety of sectors related to the deployment of optical fiber in communication technology advancements. The properties of the fiber itself, as well as the fiber optic communication sector of building technologies and the environment, have an impact on a full fiber network. In this article, we examine that this study will be useful in future fiber communication research because it highlights the key characteristics and development patterns of optical fiber network technology in both conventional and mobile internet.

There are two kinds of optical modulation in optical fiber transmission theory: primary and indirect modulation. As a result, in order to communicate with a fiber optic communication cable, we must convert sound sources, via optical fiber transmission, where the sink must receive optical signal light, which is then converted into electrical signals. Optical fiber communication technology offers a larger load range and frequency spectrum than conventional materials, allowing for faster data transmission. Traditional cable and fiber optic cable cannot compare to the speed and quality of optical fiber communication transmission, which also has a high efficiency. Quartz is the current insulator material in optical fiber communication technology; It is smaller and of lower quality, but it has a high electromagnetic interference resistance. Using optical fiber as a communication link would significantly improve data transfer while simultaneously shielding it from electromagnetic interference.

An optical transmission system, on the other hand, permits the widespread use of optical technology. When the rate of integrated contact approaches the rate limit, the only option to increase the rate is to use photonic technology. For starters, optical amplification (EDFA) applications directly increase an optical signal without the need of electrical relays to convert light to electricity. This is regarded as a technological advance in telecommunications. Second, technologies like as optical time division multiplexing (OTDM) and optical soliton transmission, as well as the use of a dispersion compensating fiber have paved the way for a bright future. While commercialization of optical soliton communication is still a long way off, the experiment set a new world record for data transmission speed across a 200-km distance of 160 gigabits per second. The storage, transmission, and exchange of light signals has resurfaced as a popular study topic.

The establishment of a fiber-optic broadband integrated communications network is the third phase. Optical fiber networking applications range from trunk to LAN communications and network subscribers, thanks to the lower cost of optical fiber. Since SDH transmission infrastructure, which replaces PDH, has more complex network architecture and provides broadband networks, the importance of the delivery network and service management is growing. Fourth, new applications and technologies arise as optical integration technology improves. Gb/s systems, integrated component-based optoelectronic devices are needed. On a single chip, lasers and electric motors, as well as a light detector and preamplifier point, have all been electrically combined.

For almost 40 years, optical fiber networking technology has progressed, with delivery being the most essential component. However, as communication and computer technology converge, so will the demands Integration into a more complete connection management technology in optical networks, as well as automated discovery protection and recovery capabilities, i.e. intelligent optical network scheduling, networking, survivability, control, and other tasks, is required.. The optical fiber connection system's basic communication paradigm and fiber kinds. The numerous advancements in optical fiber networking, as well as their features, are briefly discussed after that. Figure 16.2 depicts and quickly describes a few similarities between optical communications and conventional electrical transmissions. Finally, the advantages and disadvantages of the generic optical fiber networking architecture are briefly addressed. The future of optical fiber shows how new technologies will overcome existing technological constraints.

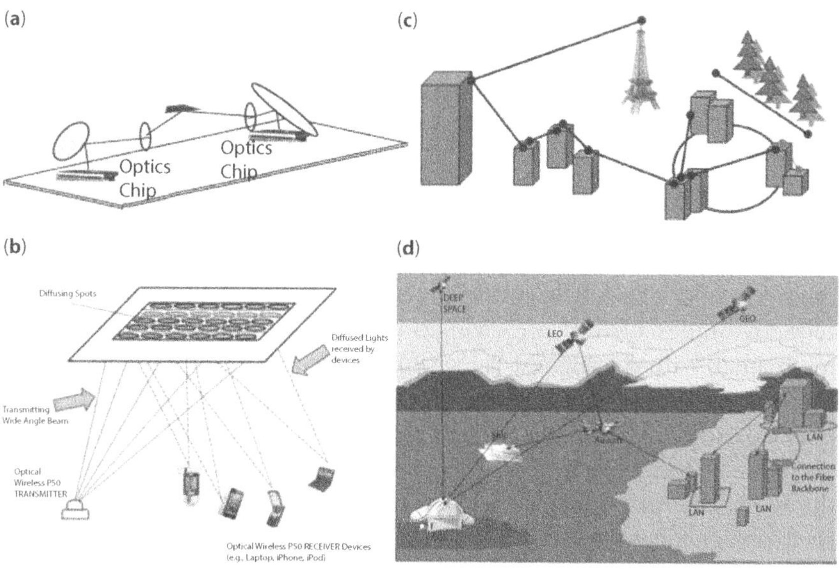

Figure 16.2 Applications of optical communication.

Fiber-optic networking is a method of sending data from one place to another using light waves sent via an optical fiber. Light is controlled and transformed into an electric carrier wave to convey data. Fiber-optic networking networks transformed telecommunications and ushered in the Information Age when they were initially deployed in the 1960s. Because of its advantages over electrical transmission, optical fibers have rapidly replaced copper wire communications in core networks in developing countries. The following are the main phases in the fiber-optic communication technique: The optical signal generation process includes using a transmitter, relaying the signal down the line, ensuring that the signal does not become too fuzzy or tiny, receiving the optical signal, and converting it to an electrical signal.

The incoming electrical signal modulates the amplitude of the optical source. Internally or externally, an electro-optic modulator (or) an acous-to-optic modulator may alter the optical carrier. Electro-optic modulators, which change the refractive index of light by altering the electrical signal input, are increasingly commonplace. In a digital optical fiber communication device, the encoded electrical signal is encoded binary pulses; it changes the light intensity from a laser diode or LED and transforms it to optical pulses. At the receiver level, a photo detector, such as an avalanche photodiode or a positive-intrinsic negative diode, converts optical signals into electrical pulses. A decoder converts the electrical pulses into the original electric signal. A typical glass fiber has a 50 mm central core glass surrounding by a cladding composed of a glass with a refractive index somewhat lower than the core. The total diameter of the fiber is between 125 and 200 mm. Cladding is needed to give the core with great mechanical strength and scratch resistance, as well as proper light direction, which means retaining light energy inside the core. There are two kinds of fibers based on their refractive index profile: The refractive index of the center in a phase index fiber remains constant throughout and abruptly changes at the core cladding boundary. At the core cladding border, meridian rays cross the fiber axis and travel in a zigzag pattern through the fiber. The refractive index of the core changes in a parabolic fashion in a graded index fiber, with the highest value of refractive index at the core's center. Skew rays, also known as helical rays, pass through it, never reaching the fiber axis and spreading in a helical (or) spiral pattern. Multimode and single mode fibers are distinguished by the number of modes flowing through the wire. The mathematical notion of mode encapsulates the essence of electromagnetic wave propagation in a waveguide. Mode refers to the electromagnetic field that exists along the light path within the fiber. In addition to TE and TM modes, optical fibers offer hybrid modes with both axial electric and magnetic fields Ez and Hz. The most common optical transmitters are semiconductors, such as light-emitting diodes and laser diodes. The difference between LEDs and laser diodes is that LEDs generate incoherent light whereas laser diodes produce coherent light. Because chromatic dispersion is reduced, the small spectrum width allows for fast data rates. Because of their low recombination time, semiconductor lasers can be modulated directly at high frequencies. VCSEL and DFB are two types of semiconductor laser transmitters used in fiber optics Distributed Feed Back. Laser diodes' light output is often freely regulated, which means the light output is controlled by a current applied directly to the semiconductor. A continuous-wave laser source may be utilized for extremely high data rates or long-distance connections, with the light modulated by an external device such as an electro-absorption modulator. This reduces laser chirp while simultaneously boosting chromatic dispersion in the fiber and directly modulated laser linewidth.

Because it uses the photoelectric effect to transform light into energy, the photo detector is an optical receiver's most important component. A semiconductor-based photodiode is often used as the image detector. Due to its circuit integration flexibility, Light detectors made of metal-semiconductor-metal (MSM) are often employed in regenerators and wavelength-division multiplexers. To produce a digital signal in the electrical domain from an incoming optical signal that is attenuated and blurred as it travels through the tube, optical-electrical converters are often used in combination with a Tran's impedance amplifier and a limiting amplifier. Further signal processing, such as clock recovery from data (CDR) using a phase-locked loop, may be done before the data is sent on.

16.4 Various Sectors of Optical Communication

In the past, fiber attenuation and interference restricted the reach of fiber-optic networking systems. To overcome these issues, opto-electronic repeaters were employed. The signal is transformed to an electrical signal and retransmitted with a greater amplitude than before. Due to the tremendous complexity of today's wavelength-division multiplexed communications, including the fact that they must be deployed only once every 20 km, these repeaters are very expensive. Optical amplification, which merely amplifies an optical signal without converting it to electrical form, is another option. It's produced by doping a stretch of fiber with erbium and filling it with light from a laser with a wavelength shorter than the transmission signal. In modern installations, amplifiers have mostly supplanted repeaters. WDM is a technique for increasing optical fiber capacity by combining several parallel channels, each with its own light wavelength. A wavelength division multiplexer is required in transmission equipment, as is a demultiplexer, which is essentially a spectrometer in receiving equipment. Arrayed waveguide gratings are extensively utilized in WDM for multiplexing and demultiplexing. WDM splits a fiber's capacity into as many as 160 channels, resulting in 1.6 terabits per second of total data speed.

The product of bandwidth and distance is critical because there is a tradeoff between the frequency of the signal and the distance it can be carried. Over a typical multi-mode fiber, a 500 MHz·km bandwidth–distance commodity may carry a 550 MHz signal for 1 km or a 1500 MHz signal for 0.8 km. NEC scientists reached a speed of 101 terabits per second by multiplexing 470 channels across a single fiber during intense growth, whereas a comparable Japanese attempt produced 108 terabits per second, but only after a laborious seven-fiber cable construction. However, this pales in comparison to backbone traffic, which is growing at a 60 percent annual rate. When choosing between optical fiber and electrical (or copper) transmission for a certain device, a variety of trade-offs must be addressed. When more capacity or longer lengths are needed than electrical wire can provide, optical fiber is often utilized. Due to its reliance on light rather than energy for transmission and the dielectric design of fiber optics, it has extremely low loss, allowing for long distances between amplifiers, the absence of ground currents and other parasitic signal and power problems common in long parallel electric conductor runs, and its inherently strong data-carrying capacitor. Thousands of electrical connections will be required to replace a single high-bandwidth fiber line. Another advantage of fibers is that, unlike other kinds of electrical transmission lines, they have very little crosstalk until they are stretched over long distances. Fiber should be placed alongside service poles, power lines, and railway tracks in places where there is a lot of electromagnetic interference (EMI). Splicing optical fibers is more complex and costly than splicing electrical conductors. Optical fibers are prone to fusing at higher pressures, resulting in catastrophic fiber core collapse and transmission component damage. In regions where lightning strikes are common, nonmetallic all-dielectric cables are often utilized.

The carrier frequency of broadcast signals is directly related to the information carrying capacity of a transmission channel. Radio waves and microwaves have frequencies of approximately 106 Hz and 1010 Hz, respectively, whereas optical carrier frequencies are in the range of 1013 to 1015 Hz. As a result, optical fiber networking infrastructure has a greater transmission capacity than traditional communication networks, as well as a greater data rate or bit rate per second. Furthermore, the transmission rate or information carrying capacity of optical fibers improves the wavelength division multiplexing process considerably. Near-lossless communication may be accomplished by using ultra-low loss fibers with erbium doped silica fibers as optical amplifiers. Advanced optical fiber communications networks utilize fibers with a propagation loss of 0.004 dB/km. Furthermore, suitable optical amplification for a limited distance at specific places may be achieved by using erbium doped silica fibers in the propagation path. As a result, the distance between repeaters has grown to more than 100 km. Because the signal is enhanced in the optical domain, the interference caused by signal augmentation is virtually non-existent. Silica, an electrical insulator, is used to make optical fibers. As a result, they are unaffected by gravitational waves or high-current lightning. It is not safe to use in potentially explosive situations. Furthermore, interference from electric cables, train power lines, and radio frequencies has little effect on optical fibers. There is no cross talk between fibers in a cable, even if there are many of them, since there is no optical contact between them.

The signal is not emitted since it is conveyed via fibers. In addition, tapping a signal from a fiber is difficult. As a result, the signal security of the optical fiber connection is guaranteed. Fiber optic cables have tiny radii and are long-lasting, thin, and light. Fiber cables are not damaged when they are bent or twisted. In terms of packaging, processing, installation, and shipping, optical fiber cables outperform copper cables while retaining the same intensity and endurance. Making the most of limited bandwidth will become increasingly essential if cellular demand continues to rise in the near future. By splitting wireless signals into separate components and transferring them, Georgia Tech's optical networking group has shown how to increase wireless power and bandwidth. Optical fiber is utilized for long-distance communications, whereas cellular is employed for the final few tens of meters. Customers connect wirelessly over much greater distances and at far higher bandwidths than before. IBM has developed a transceiver that can boost chip-to-chip bandwidth on PCB to 400 Gigabits per second (Gb/s), the greatest rate to date and a breakthrough that will enable even faster data transmission rates in homes and companies in the future. The gadget, which is made up of low-cost parts that may be mass-produced in the future, has a bi-directional data rate that is about double that of an earlier IBM transceiver. Two major improvements have resulted in this enhanced bandwidth. For starters, the new transceiver has 22 data transmission and receiving channels, compared to 14 in the previous system. Second, the modulation rate of each of the transceiver's vertical cavity surface emitting lasers has increased by 35% to 13.5 billion bits per second. To speed up commercialization, IBM utilized industry-standard 950 nanometer lasers and detectors instead of the proprietary 984-nn equipment used in the prior transceiver. Both capacity and demand for data processing have increased dramatically in recent years. Even if the present optical transmission band of 1.5 micron wavelength is adequate for the time being, scientists and engineers must begin evaluating alternative bands as soon as possible due to the anticipated massive rise in traffic. The terahertz area is mostly unexplored and undeveloped because its frequency spectrum is too wide for conventional electronics and too narrow for semiconductor lasers and detectors. However, new study to be presented at OFC/NFOEC validates what scientists have known for a long time: the terahertz spectrum has enormous promise. One of the Institute's professors will work in Berlin on the uses of the terahertz spectrum in defense, health, and materials research, as well as the role of telecommunications technology in its development. Terahertz radiation, unlike other scanning techniques, can penetrate paper, clothing, and plastics while remaining entirely safe for people. As a consequence, terahertz spectra may be utilized to detect explosives and interpret complex medical compounds in ways that traditional technology cannot.

FSO was created as a low-cost replacement for fiber-optic cable networking LEDs that operate the most advanced wavelengths are 765 nm or 840 nm in the near infrared. technology used in FSO equipment. In recent years, systems with a wavelength of 1550 nm have been developed. Manufacturers stated at first showed the wavelength of 1550 nm was better for severe weather propagation than the wavelengths of 785 nm. Those arguments were rejected after more testing and assessment. Longer wavelengths of about 10 microns have also been found to be capable of addressing FSO connection availability issues brought on by severe weather. The frenzy around FSO's magical wavelengths is detrimental to both customers and the remainder of the FSO Company, which should be establishing acceptable criteria for the capabilities of its equipment. In the circumstances that cause the greatest attenuation for FSO systems, such as coastal fog and low clouds, 10 microns offers no advantage in terms of transmission over shorter wavelengths.

For thousands of years, optical correspondence has been utilized in a number of ways. To communicate, the ancient Greeks utilized torches and a coded alphabetic system. Heliographs, or wireless sun telegraphs, were invented in the modern era and interact with their receivers through coded signals. The first wireless telecommunications transmission between two buildings was produced by Bell 213 m (700 feet) apart on June 3, 1880.

Lasers have revolutionized free space optics since their introduction in the 1960s. Military organizations were especially involved, and their growth was accelerated. In 2008, MRV Communications announced the deployment of a free-space optics-based system with a 20GB/s data capacity and a 4-km range, and high availability. The product's useful range was decreased to 350m prior to end-of-life, thus this equipment is no longer available. In 2013, MOSTCOM began mass production of a modern wireless networking infrastructure with a data capacity of 10 gigabits per second and a range of up to 2.5 km. Traditional networking applications have been opened up by recent advancements in FSO technology, ranging from short-term network bridge alternatives to a compelling and realistic solution for service providers to execute on their all-optical network promise In terms of optical technologies is a common extension of the metro optical network center that provides network edge users with cost-effective, efficient, and fast optical power. Climate variables have a significant impact on FSO-links. As a result, certain important environmental factors must be recognized before delving further into optical wireless systems. The troposphere, often known as the weather bubble, is the lowest layer of the atmosphere, rising up to 10 km above the surface of the Earth. It has a changeable refraction index that changes with height above the surface of the Earth. Despite a particular interaction during weather inversion circumstances, the refraction index typically decreases with height. Atmospheric conditions hindered laser communications over the atmosphere in two ways. First, the ambient acts as a variable attenuator between the transmitting and receiving terminals. The second use of scintillations is to operate a free-space laser connection. When it's foggy, there's more attenuation, and when it's thin, there's less. Furthermore, since the medium is air, haze has an impact on transmission. Rain has no effect on the signal since the drop size is not comparable to the wavelength of the laser.

The triple play traffic (phone, data, fax, and multimedia traffic) on today's university campuses is outpacing traditional connections. In corporate and campus networks, systems will span several buildings to provide ultra-high speeds without the cost of direct fiber optic connections. In corporate and campus networks, systems will span several buildings to provide ultra-high speeds without the cost of direct fiber optic connections. Cryptosystems today can only provide cryptographic security within the constraints of ordinary processing power, and quantum computers, for example, may make electronic money obsolete. Quantum cryptography, which is based on physical principles, offers a whole new way of encrypting data and guarantees full secrecy. Quantum cryptography techniques are often employed in conjunction with fiber optic technology. When fiber optic deployment is too expensive or impractical, FSO connections provide a flexible alternative.

Signals from a camera or a group of cameras must be sent to a transmitting truck, which is connected to a central office by satellite uplink, allowing for news reporting from far-flung locations and conflict zones. The cameras and the vehicle will be able to communicate in high-quality via an FSO link. Even the most discerning customers would be satisfied with FSO connections. FSO has shown to be a feasible high-bandwidth wireless alternative to fiber optic cable. FSO has a number of benefits over fiber, the most prominent of which are its speed of installation and substantial cost savings. Because it is a weather airport, the availability of any FSO device may be assessed as a function of distance. The unexpected and difficult-to-predict attenuation of laser power in the air is a drawback of FSO over fiber. These availability curves show how far The FSO systems in a certain geographic region should be connected. Carriers and Internet service providers are another potential consumer of FSO systems, especially for last-mile urban access applications. Figure 16.3 shows that if FSO devices are to be used in telecommunication applications, they must satisfy more stringent availability criteria. Carriers-class services are usually considered to be accessible 99.999 percent of the time. According to a research of connection budgets and visibility-limiting weather circumstances, FSO connections should be less than 140m to provide carrier-class availability. In locations like Phoenix and Las Vegas, however, the 99.999 percent distance limit is significantly increased. The budget for this computation is 53 decibels. This idea is realized in the best FSO device now available, which has a 10 W transmitter and a 1 new image counting detector. The 100 Db connection margin will simply raise the 99.999 percent link gap to 286 m in this FSO system. If the FSO connection is backed up by a lower data rate radio frequency (RF) link, expanding the high availability spectrum will be easier. With this hybrid FSO/RF technology, the 99 percent connection range would be increased, giving carriers access to a considerably wider metro/access sector. It's crucial to remember that when the connection spectrum expands, the average bandwidth decreases. The first map of FSO availabilities contoured over region is required to demonstrate the geographical dependency of FSO results. Similar to how ITU/Crane maps are used to predict microwave performance, this map is the first stage in generating an attenuation map for anticipating FSO results.

The triple play traffic (phone, data, fax, and multimedia traffic) on today's university campuses is outpacing traditional connections.

In corporate and campus networks, FSO systems will span several buildings, allowing for ultra-high speeds without the expense of direct fiber optic connections. Back-end technologies like FSO, which allow for much greater throughput, are becoming more essential as the number of bandwidth-intensive mobile phone networks grows. Quantum cryptography, which is based on physical principles, offers a whole new way of encrypting data and guarantees full secrecy. Quantum cryptography techniques are often employed in conjunction with fiber optic technology. When fiber optic deployment is too expensive or impractical, FSO connections provide a flexible alternative. In order to broadcast live events such as sports and ceremonies, or news reporting from remote regions and war zones, signals from the camera or a number of cameras must be sent to the transmitting truck, which is connected to a central office via satellite uplink. The cameras and the vehicle will be able to communicate in high-quality via an FSO link. Even the most discerning customers would be satisfied with FSO connections. FSO has shown to be a feasible high-bandwidth wireless alternative to fiber optic cable. FSO has a number of benefits over fiber, the most prominent of which are its speed of installation and substantial cost savings. Because it is a weather airport, the availability of any FSO device as a function of distance may be determined. The unexpected and difficult-to-predict attenuation of laser power in the air is a drawback of FSO over fiber. The unexpected and difficult-to-predict attenuation of laser power in the air is a drawback of FSO over fiber. Carriers and Internet service providers are another important prospective user of FSO systems, especially for last-mile urban access applications. FSO devices must satisfy much higher availability criteria if they are to be used in communications applications. Carriers-class services are often considered as 99.999 percent available. To guarantee carrier-class availability, FSO connections should be less than 140m, according to a study of connection budgets and visibility-limiting weather conditions. The 99.999 percent distance restriction is considerably raised in places like Phoenix and Las Vegas. This calculation has a budget of 53 dB. The best FSO device currently available realizes this concept, using a 10 W transmitter and a 1 nW image counting detector. The 100 Db connection margin will simply raise the 99.999 percent link gap to 286 m in this FSO system. If the FSO connection is backed up by a lower data rate radio frequency (RF) link, expanding the high availability spectrum will be easier. With this hybrid FSO/RF technology, the 99.999 percent connection range would be increased, giving carriers access to a considerably wider metro/access sector. It's crucial to remember that when the connection spectrum expands, the average bandwidth decreases. The first map of FSO availabilities contoured over

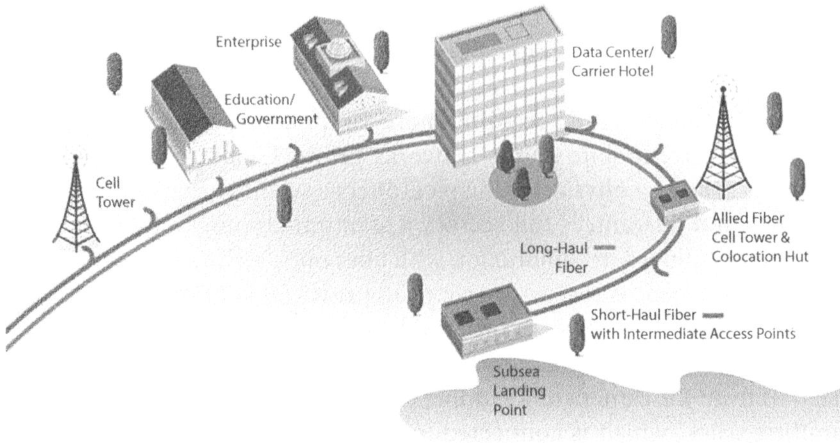

Figure 16.3 Various sectors in optical communication.

region is required to demonstrate the geographical dependency of FSO results. Similar to how ITU/Crane maps are used to predict microwave performance, this map is the first stage in generating an attenuation map for anticipating FSO results.

Researchers from a variety of fields, including electronics, communications, photonics, and signal processing, have contributed their knowledge to meet the growing demand for higher capacity, lower energy consumption, and lower cost in the field of optical communication system design, which is a vast and rapidly changing research domain. The continuous development of fast interconnects and optical transceivers for data-center servers have enabled this advancement. Recent advances in optical connectivity have increased the framework's dynamism and resilience while also expanding its boundaries. Various sophisticated technologies are now being developed to determine the number of wavelengths transmitted per fiber, as well as the data transfer of specific wavelength channels Several current and emerging optical technologies have previously been developed to serve a variety of changing applications in a flexible, power-efficient, and cost-effective manner, including new and future on-demand and high-demand data rate applications. They can create and rebuild networks on demand, and they have network capabilities, performance, and flexibility. The telecoms industry may offer services up to Gbps and more for the user end due to the ability to give larger wavelength range and wavelengths per fiber in optics. By 2020 and beyond, mobile traffic is projected to grow. Operator networks will need more bandwidth and lower latency as LTE

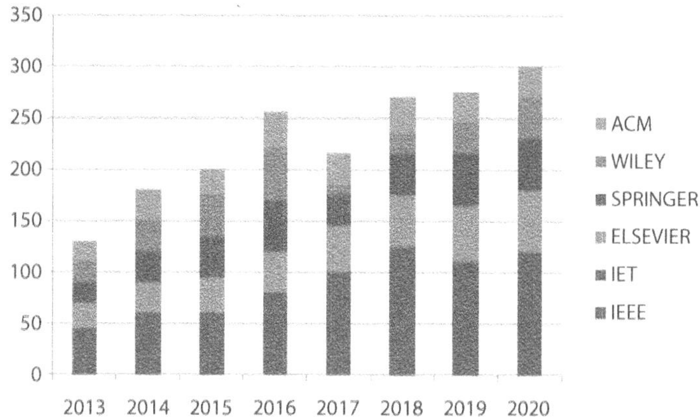

Figure 16.4 Papers published in optical communication network.

evolves to LTE-Advanced with VoLTE and 5G, since close collaboration is needed for optimal spectrum use. Operators have the most difficulty fulfilling these requirements in metropolitan and dense urban regions, as shown in Figure 16.4, where space is limited and contemporary alternative optical network technologies with high spectrum performance at reduced prices are required.

All customer service providers across the globe are increasing their bandwidth capacity at any moment to satisfy the inexorably rising demand for phone calls, on-demand streaming video, and cloud-based services over two copper using DSL applications. CSP is able to offer these services efficiently because to minimal attenuation and ultra-bandwidth capacity at the back-haul network through convergence switch. Through a single centralized billing and administrative structure, CSPs in urban integration must keep their TCO, or total cost of ownership, low for all services supplied to end-users. A provision enables all permitted facilities to be used at the same time, provided that no other services are harmed. In order to provide multimedia services throughout convergent network architecture, with a high output capacity. To make a significant contribution to the digital environment, CSPs must collaborate and efficiently upgrade current and future technologies into a single optical packet network. Optical networking infrastructure is being improved in a variety of ways. Diverse service providers are implementing various solutions based on the trade-offs that exist between different technologies and their requirements. Because of interconnections between sub-networks using various technologies, the network has grown more diverse than before. Maintaining and protecting the network is a difficult task. However, before a technology can be used successfully in the telecommunications industry, it must first overcome its own set of difficulties. An optical system's numerical aperture (NA) is a dimensionless number that specifies the range of angles from which it may receive or emit light. When a beam passes from one substance to another with no optical power at the point of contact, it is said to be travelling invisibly. NA has the property of being constant, and its description includes the index of refraction. The precise meaning of the term varies depending on the field of optics. In microscopy, numerical aperture is used to define the acceptance cone of a target (and therefore its light collecting power and resolution), while in fiber optics, numerical aperture is used to describe the cone of light admitted through or leaving the fiber. The numerical aperture is the greatest angle at which light impinging on the fiber is fully internally reflected and may be distributed efficiently through the fiber.

If the bandwidth is increased to more than Gb/s while building next-generation technology, there may be substantial difficulties keeping the same splitting ratio and coverage due to the deployment of TDM PON. Reduce the difficulty of transmission in the upstream, as well as the many maintenance costs incurred as a result of the complexity of transmission, is a major issue in the case of WDM PON. Wavelength continuity limitations are one of the main problems that all-optical networks face. This may happen if the communication of a light route is interrupted due to a lack of wavelength on one or more of the way's links. All connections with adequate bandwidth, however, are still functioning. Other problems may arise as a result of transiently designing the sensitivity of optical amplifiers. Rapid shifts and toggling up and down in Level of Control throughout the transmission line of fiber are caused by erroneous spikes in the optical amplifier. The next major challenge is to keep track of and better manage the total processing power needed to differentiate signals in different cores or modes, as well as to use For an all-optical system, optical amplifiers and ROADMs are used to provide optimum coverage while operating on all signals in the fiber. One of the major challenges of SDM systems is ensuring minimal crosstalk across long optically networked transmission lengths with integrated amplifiers, transponders, switches, splices, SDM fibers, and NEs such as spectral and spatial cross-connects. In recent years, hollow core fiber has been suggested. It's challenging to process fiber with a diameter of a few nanometers and a length of 100 km while reducing failure windows. The various published papers in the domain of optical communication networks is shown in Table 16.1.

Table 16.1 Papers published in optical communication network.

Year	Number of papers published					
	IEEE	IET	ELSEVIER	SPRINGER	WILEY	ACM
2013	20	25	25	20	20	20
2014	30	30	30	30	30	30
2015	40	20	35	40	40	25
2016	50	30	40	50	50	35
2017	60	40	45	30	10	30
2018	75	50	50	40	20	35
2019	85	25	55	50	30	30
2020	90	30	60	50	40	30

16.5 Conclusion

The highest carrier frequencies and, as a consequence, the fastest data rates are provided via optical networking networks. One of the most important technological developments in contemporary history is the creation of wireless communications. Due to its numerous benefits, researchers are paying careful attention to free space optical (FSO) connections, which is a developing field of study. This kind of wireless optical networking technology sends data from one stage to the next using an extremely narrow beam. Customers and telecoms providers will profit from this LOS (line of sight) technology. It provides high data speeds of several gigabits per second, is immune to radio frequency interferences, does not require licensure, due to the use of an extremely narrow beam angle, it offers a highly secure communication link and is less costly, quicker, and simpler to install than fiber optic implementation. Wireless gadgets and technology have advanced far more quickly than anybody could have predicted thirty years ago, and they will continue to play an important part in contemporary life for some time to come.

References

1. Noshada, M. and Rostami, A., FWM minimization in WDM optical communication systems using the asymmetrical dispersion managed fibers. *Int. J. Light Electron. Opt.*, 123, 9, 758–760, 2012.
2. Wang, X. and Kitayama, K., Analysis of beat noise in coherent and incoherent time-spreading OCDMA. *IEEE/OSA J. Light. Technol.*, 22, 10, 2226–2235, 2004.
3. Shake, T.H., Confident performance of encoded optical CDMA. *IEEE/OSA J. Light. Technol.*, 23, 1652–1663, 2005.
4. Sharma, P. *et al.*, A review of the development in the field of fiber optic communication systems. *Int. J. Emerg. Technol. Adv. Eng.*, 3, 5, 113–119, 2013.
5. Gude, V.G., Nirmalakhandan, N., Deng, S., Maganti, A., Low temperature desalination using solar collectors augmented by thermal energy storage. *Appl. Energy*, 91, 1, 466–474, 2012.
6. Fidler, F., Knapek, M., Horwath, J., Leeb, W.R., Optical communications for high-altitude platforms. *IEEE J. Sel. Top. Quantum Electron.*, 16, 5, 168–175, September/October 2010.
7. Fuqaha, A.A., Guizani, M., Mohammadi, M., Aledhari, M., Ayyash, M., Internet of Things: A survey on enabling technologies, protocols, and applications. *IEEE Commun. Surv. Tutor.*, 17, 4, 2347–2376, 2015.
8. Schulz, P., Matthé, M., Klessig, H., Simsek, M., Fettweis, G., Ansari, J., Ashraf, S.A., Almeroth, B., Voigt, J., Riedel, I., Puschmann, A., Thiel, A.M., Müller, M., Elste, T., Windisch, M., Latency critical IoT applications in 5G: Perspective on the design of radio interface and network architecture. *IEEE Commun. Mag.*, 35, 158–165, Feb. 2017.
9. Palattella, M.R., Dohler, M., Grieco, A., Rizzo, G., Torsner, J., Engel, T., Ladid, L., Internet of things in the 5G era: Enablers, architecture, and business models. *IEEE J. Sel. Areas Commun.*, 34, 3, 510–527, March 2016.
10. Hassan, W.A., Jo, H.-S., Rahman, T.A., The feasibility of coexistence between 5G and existing services in the IMT-2020 candidate bands in Malaysia. *IEEE Access*, 99, 99, 115–125, 2017.
11. Buraczynski, J.J. and Li, T.K., Tunnel lighting systems. *Tunnel Safety and Security*, vol. 56, pp. 553–556, 2010.
12. U.S. Department of Energy, Lighting market characterization. *Solid-State Lighting Program*, 2012.
13. Huang, B.J., Wu, M.S., Hsu, P.C., Chen, J.W., Chen, K.Y., Development of high-performance solar LED lighting system. *Energy Convers. Manage.*, 51, 8, 1669–1675, 2010.
14. Kandilli, C., Ulgen, K., Hepbasli, A., Exergetic assessment of transmission concentrated solar energy systems via optical fibres for building applications. *Energy Build.*, 40, 8, 1505–1512, 2008.
15. Wang, F. and Long, W., Application status and development prospect of solar lighting vessel technology. *Build. Sci.*, 24, 109–113, 2008.

17

Photonic Communication Systems and Networks

Naitik S.T.[1]*, J.V. Gorabal[2], Shailesh Shetty[3], Srinivas P.M.[3] and Girish S.[4]

[1]ISE & CoE-Cyber Security, Sahyadri College of Engineering and Management, Karnataka, India
[2]CSE, ATMECE, Sahyadri College of Engineering and Management, Karnataka, India
[3]CSE & CoE-Cyber Security, Sahyadri College of Engineering and Management, Sahyadri College of Engineering and Management, Karnataka, India
[4]CSE, Sahyadri College of Engineering & Management, Sahyadri College of Engineering and Management, Karnataka, India

Abstract

Photon is the particle of light, which is extensively used in modern digital communication systems as a signal carrier. In the previous era, electromagnetic waves were used for communication systems development. But due to the high cost for infrastructure setup and maintenance, it lags now its usage for digital communication systems. The availability of light as an indigenous source is the major merit in driving the communication systems towards digitally reliable and accessible to all. The photons are used in the manufacturing of PIC (Photonic Integrated Circuits) are optically semiconductor active devices used in most optical communication systems. Indium phosphide III V wafer substrate is used as a raw material for Integrated PIC. PICs are used as optical transceivers for data center optical networks. Photonic communication is used in a variety of applications like detection, amplification, information processing, metrology, spectroscopy, holography, biophotonics, etc. As we increase devices connected to the same Wi-Fi, the data transfer rate would decrease. LiFi is one solution to this problem because radio waves have a small spectrum available for data transfer. Data rate faster than 10 megabits per second can be produced by LiFi, which is very fast as compared to broadband connections. LiFi proposes the use of increased bandwidth with better speeds using visible lights as compared to electromagnetic waves used in Wi-Fi. As an idea, we can say LiFi can offer a bandwidth of 300 THz as compared to 300 GHz in RF communication. Photons of light are used in LiFi as data carries which cannot be seen with the

**Corresponding author*: naithikenator@gmail.com

Chandra Singh, Rathishchandra R Gatti, K.V.S.S.S.S. Sairam and Ashish Singh (eds.) Modeling and Optimization of Optical Communication Networks, (303–330) © 2023 Scrivener Publishing LLC

naked eye. So, one of the benefits can include the replacement of other technologies in environments where more security is required and the typical framework is not trusted easily. We're interested in data transfer via LiFi and all elements of data transmission by light through various materials, which will give us a general idea of where and how we may employ LiFi for data transmission in the future. Although there is still a long way to go before this technology becomes commercially viable, it has a lot of promise in the realm of wireless internet. However, it poses few challenges which could be resolved if do careful design and proper implementation.

Keywords: Communication, photon, LiFi, system, networks, LiFi technology

17.1 Introduction

Wi-Fi technology has many advantages, but it also has certain drawbacks. Wi-Fi is a wireless technology that allows many individuals to connect to other computers on a network or the internet without having to physically lay wires. It has enormous power, and it is causing a revolution in LAN networks all over the world. Those who were unable to access the internet via Wi-Fi no longer need to be concerned. Because Wi-Fi provides you with a direct internet connection via a method of communication that does not require the usage of a physical wire or cable. Wi-Fi's portability allows individuals to connect their PDAs, laptops, and other devices to a network without having to deal with wires. Wi-Fi has many advantages, especially today that everyone's main priority is the internet, yet Wi-Fi networks have limitations. The Wi-Fi network's drawback is that it only gives a connection in a small region. Its radio connection is limited to 20 to 25 meters. You won't be able to connect to the internet or a local wireless network if you're more than 25 meters away. The Wife antenna broadcasts signals in specified places all over the world, but your connection may weaken as you travel further. Wi-Fi connections have a grading system for connection strength, which ranges from good to excellent, but if you are outside of your designated locations, the connection will be bad.

In most cases, Wi-Fi networks enable two types of technology: Both infrastructure mode and ad hoc mode are available. You can connect to the internet using Ad hoc mode without the use of a third- party access point or router. As a result, the majority of users choose Ad hoc mode to infrastructure mode. When using Ad hoc mode in a Wi-Fi network, several complications arise. When Wi-Fi devices are configured in Ad hoc mode, ostensible security against network intruders is lost. Ad hoc device setup never prevents SSID access, whereas infrastructure mode does. Existing network attackers will not require much effort in ad hoc mode. When you have a signal difficulty in Ad hoc mode, you should know that infrastructure mode provides full power signals. In a Wi-Fi network with a bandwidth of 11Mbps, the 802.11g standard requires ad hoc mode. When a user configures a Wi-Fi network in infrastructure mode, the data transmission rate can reach 54 Mbps; however, when the setup is in ad hoc mode, the data transfer rate is only 11 Mbps. When compared to infrastructure mode, ad hoc mode is extremely slow.

Various security difficulties may lead to Wi-Fi limitations because while setting up a Wi-Fi network is fairly simple, maintaining security requires a lot of effort because there are no encryption mechanisms organized on the Wi-Fi network's access point. When hackers attack a Wi-Fi network, they could steal your personal information and slow down your network traffic. The slowness with which videos and audio were transferred drove them to their breaking point. When a large number of people use the same network to access the internet, the pace of data transfer slows down. When a huge number of devices are connected to a wireless network, it is called a mesh network, this problem becomes worse and attempting to download multiple huge files at once, and you'll never be able to do so because sharing bandwidth among all devices diminishes network speed and renders it unusable. Overall, there are some limits in Wi-Fi networks, however, even with those restrictions; Wi-Fi has outstanding connectivity features that help to overcome them. As a result, in the next few days, everyone will desire a Wi-Fi network to have speedy access to the internet from everywhere, and they will want to use Wi-Fi hotspots to make their businesses more profitable and successful, despite the limitations.

17.2 History of LiFi

LiFi technology is largely credited to Professor Harald Haas of the University of Edinburgh's Chair of Mobile Communications. When he campaigned for this technology in a 2011 TED Global talk and helped start a company to sell it, he coined the term LiFi (light fidelity). LiFi may be used in traffic control systems that rely on automobile headlights, as well as chemical production plants where radio frequency is too hazardous and could create antenna sparks. Formerly called pureVLC, PureLiFi develops and sells LiFi-enabled products for use with current LED lighting founded the company.

The world's first commercial LiFi technology was introduced in September 2013 by the firm. BeamCaster, a LiFi wireless local network, was announced by Stins Coman, a Russian company, in April 2014. One gigabit per second (GB/s) of data is sent by their current module, but they expect to increase this to 5 GB/s soon. Sisoft (a Mexican business) set a new record in 2014 when it was capable of light spectrum LED bulbs that can transmit data at speeds of up to 10GB/s.

A new wireless communication technology made its debut with the Li-1st - the first LiFi device to hit the market. As of February 2015, the Li-Flame gadget claimed to be the first LiFi device capable of mobile wireless connection. Lucile, a French lighting manufacturer, partnered with PureLiFi to produce the world's first industrialized LiFi system, which has since been installed in several places, including Microsoft's Paris headquarters.

The LiFi-XC system was announced in October 2017. If you're in the market for a new laptop, tablet, or smart appliance, this is a plug-and-play option for USB devices. It was announced in June that PureLiFi was launching a channel program for IT resellers interested in adding LiFi to their portfolio, as well as LiFi starting kits for university researchers.

LiFi passed an industrial environment test conducted by a BMW facility in Munich in June 2018.

BMW project manager Gerhard Kleinpeter anticipates that LiFi transceivers will be miniaturized, allowing LiFi to be used more efficiently in manufacturing processes. Kyle Academy, a Scottish secondary school, began testing LiFi in August 2018. Using a USB link between their laptop computers and a device that converts the fast on-off current from the ceiling LEDs into data, students may obtain data.

Oledcomm, a French company, will exhibit at the 2019 Paris Air Show in June, demonstrated their LiFi technology. In the future, Oledcomm plans to work with Air France to test LiFi aboard a plane while it's in the air.

17.3 LiFi Standards

Unlike Wi-Fi, which relies on radio frequency waves, LiFi employs ultraviolet, infrared, and visible light to transmit data, which has far higher bandwidth.

A large chunk of VLC's code is derived from the IEEE 802 workgroup's communications technology. For its part, the IEEE 802.15.7 standard has fallen behind because it ignores recent technological advancements in optical wireless communications, such as the introduction of O-OFDM modulation schemes that have been optimized for data rates, multiple-access, and energy efficiency. The adoption of O-OFDM has several consequences. The IEEE 802.15.7 standard does, however, specify the physical (PHY) and media access control (MAC) layers. High-speed data transmission is possible with this standard for audio, video, and multimedia applications. Optical transmission's mobility, compatibility with artificial illumination in infrastructures, and the likelihood of interference from ambient lighting are all taken into account. The MAC layer, like the TCP/IP protocol, enables other layers to utilize the link. Three PHY layers are defined in the standard, each with a distinct rate:

- The PHY 1 was designed for outdoor use and operates at speeds ranging from 11.67 kbps to 267.6 kbps.
- Data rates of 1.25 to 96 Mbps can be reached using the PHY 2 layer.
- With a modulation approach termed color shift keying, the PHY 3 is employed for a variety of emissions sources (CSK). PHY III can deliver data at speeds ranging from 12 to 96 megabits per second.

PHY I and PHY II recognize two modulation formats: on-off keying (OOK) and variable pulse position modulation (PPM). Data is sent with a clock. It has a DC component and expresses logic zero with an OOK symbol "01" and logic one with an OOK symbol "10" in the PHY I and PHY II layers. DC component avoids light extinction in the case of a lengthy series of logic zero's.

From January 7–10, 2014, the Prototype of the VLC smartphone was shown at the Consumer Electronics Show in Las Vegas. The phone employs SunPartner's Wysips CONNECT technology, which converts light waves into usable energy, allowing it to receive and decode signals without draining the phone's battery. Small screens, such as watches and smartphones, can be solar powered by adding a clear thin layer of crystal glass. During an average day, smartphones might get 15% more battery life. In 2015, the first smartphones with this technology should be available. Here you'll be able to watch VLC videos and even use your smartphone's camera.

A VLC system for store shoppers has been created by Signify Lighting (previously Philips Lighting). They must first download an app to their smartphone, which then connects to the store's LEDs. This is determined by their current location and what they are looking at; the LEDs can detect their location in the shop and provide them with relevant coupons and information.

17.4 Related Work

An evaluation of the IEEE 802.11 protocol in a Medium Voltage network management system was conducted out utilizing a model for simulating Wi-Fi chain architectures [1] in this paper. The simulation findings provide some practical recommendations for using the Under situations where the IEEE 802.11 protocol is not typically suited, such as when broadcasts are sent at regular intervals of time We end with a visual representation of the whole communication system. The performance of a Wi-Fi communication system that will be used to administer MV networks automatically has been investigated. The behavior of the IEEE 802.11 protocol has been researched for this purpose, and its extremely low reliability has been acknowledged, using an appropriate simulation model.

The radio propagation model's required parameters were explained in [2], are not readily available in an anonymous context, computing where Wi-Fi access stations are (AP). When there isn't any knowledge about the radio propagation model. Range-based AP localization using received signal strength (RSS) is investigated in the following section. To do this, we use a multilateration technique to approximate the exponential relationship between RSS and distance in a linear way. Our simulation findings show that estimating an AP location from four or more RSS readings taken from various sites is double. In an anonymous context, we developed a Wi-Fi AP localization algorithm. The path loss exponent and transmission power are not required to be known a priori by the algorithm, hence there is no need for offline training in the AP domain. The sole requirement is that RSS be measured in at least four different sites. The results of the simulations suggest that RSS may be used to estimate position without knowing anything about radio propagation characteristics.

The effect of radiofrequency radiation (RFR) from a 2.4 GHz laptop antenna on human sperm was investigated in this *in-vitro* pilot study in [3]. Ten samples of sperm collected from donors aged 20 to 30 were exposed to the RFR source when it was inactive mode. Following the exposure, sperm concentration, motility, and morphological grading were assessed in both, the same donors' exposed and unexposed samples. The Mann-Whitney U-test was used to examine the significance of the results of these semen parameters at the 0.05 level of significance revealed that RFR exposure had a significant influence on the semen parameters studied. The impact of 2.4 GHz RFR on human ejaculated sperm was investigated in *in-vitro* pilot research. RFR radiated from a laptop antenna in active mode at 2.4 GHz was found to have a substantial impact on sperm concentration, motility, and morphological grading of the sperm. Because this was a pilot study, the findings can be used as a starting point for future research, especially since wireless communication is widely used around the world and among the reproductive population. This investigation also raises awareness of the potential for RFR at 2.4 GHz to influence the results of semen analysis if it is performed near the RFR source.

In [4], the study continues on the biological impacts of blood count (hematology) and biochemistry. As for the wireless signal generator, it was decided to use 2.4 GHz Wi-Fi. Thirty white albino mouse control samples are exposed to this frequency through an antenna. The antenna is set at a distance of 1 meter to represent the quantity of radiation energy received by a human in a regular condition when receiving a Wi-Fi signal. Mice were exposed to radiation for 6 months and were subjected to it for 8 hours every day. The control and exposed samples were taken to the Veterinary Laboratory every two weeks for six months to undergo a blood test process for the biological test. The average packed cell volume (PCV), hemoglobin, red blood cell, and white blood cell counts were all checked in the blood, as well as blood biochemistry. The RF signal intensity received by the samples was monitored using a spectrum analyzer. The goal of this study is to see how biological features affect people's exposure to Wi-Fi radiation. For this investigation, 30 male mice were kept in cages at a distance of 1.0m from the RF generator's antenna. There was also a cage with 30 male mice that were positioned far away from the radiation region and was essentially radiation-free. Pathology Lab, Blood Count, and Biochemistry Lab were the tests utilized to classify the results. Anatomical studies indicated that the deterioration occurs inside the cells themselves of several organs as a result of the examinations. According to the investigations, there were some negative biological impacts. Electromagnetic radiation alters the organs in exposed samples, according to histopathology tests. Some tissue begins to degrade after the fourth test. The Packed Cell Volume (PCV), Hemoglobin (Hb), and Red Blood Cell (RBC) all fell outside of the normal range when compared to the control's results. According to the findings of this study, users who are exposed to Wi-Fi electromagnetic radiation may experience adverse effects on their biological systems.

Several known methods for WIFI-based indoor positioning systems [5] employ the received signal strength indicator (RSSI) of all access points in range to determine a mobile device's present location. External interferences, on the other hand, have a significant impact on the accuracy of those systems, which is affected by both changes in the environment, such as short and long-term. Through the use of distance as a factor and the amount of time it takes for a radio signal to go that far, time-of-flight (TOF) techniques avoid these issues. In addition, RSSI-based fingerprinting methods do not necessitate a lengthy calibration procedure or a huge database. We examine and contrast many ways in the literature in terms of communication flow and hardware components, time measurement method, and positioning accuracy in this paper. We also provide a unique method based on NULL-ACK sequences, commercially accessible hardware, and the CPU's nanosecond-resolution timestamp clock. As a result, no affiliation with access points is necessary, and no changes to client or infrastructure WIFI components are required. They have tested the system in a variety of situations. The findings show that the accuracy of TOF-based techniques is influenced by both the hardware employed and the features of the environment. They discovered that temporal variations induced by the interrupt service routine's variable delays, precision distance forecasts according to one measurement are unachievable due to multipath effects and other factors. We strive to reduce these impacts by using the use of a high number of NULL-ACKs to achieve consistent ranking results. We examine and contrast numerous filters and statistical estimations inside our system settings. With the average of numerous measurements and the usage of a band-pass filter, we strive to reduce these impacts by using a large number of NULL-ACK-sequences as a means of producing reliable range results in our system settings; we compare a range of filters and statistical calculations. It's possible to get a range accuracy of fewer than 1.33 meters by employing band-pass filters and average measurements as long as the conditions are suitable.

ToF-based WIFI-positioning systems are examined in this research; such systems can determine a mobile device's location on the fly without the need for a lengthy offline process. Furthermore, because of the relationship between distance and propagation delay, External interferences are less likely with them than with RSSI-based approaches. There is no need to associate with the access point if NULL-ACK communication is employed. They have demonstrated that the precision of such systems is hampered by the lack of precise and appropriate hardware timers. To circumvent this issue, they created their RTOF approach based on TSC and a huge number of NULL-ACK sequences exchanged. Unfortunately, time fluctuations induced by variable TSC and ISR delays degrade the accuracy of their technique. They studied several filters and statistical estimators to lessen these oscillations. The research demonstrates that the best range results are obtained by utilizing as well as a band-pass filter and RTT0 average, but precision is dependent on both the hardware and the environment. For all measured distances, a mean absolute range inaccuracy of 1.33 meters in the best-case scenario is achieved by the researchers. When compared to other RTF-based strategies that have been published, this is a good result. A range-finding technique that relies on distances that are never shorter than 6.27 meters does not provide the precision needed under indoor settings. As a result, we conclude that WIFI location based on TOF is feasible, on readily available devices is not currently possible.

More research and development in the area of more precise WIFI clocks are required. As a result of IEEE 802.11v, there are now new time measurement functionalities, the new IEEE 802.11 standard, which replaces the existing IEEE 802.11 standard, provides a fantastic chance for a more precise Positioning system using TOF and WIFI. As a result, more research on ranging accuracy using this new WIFI standard appears promising.

The energy consumption of a wireless sensor node with a Wi-Fi transceiver is investigated in this research [6], its application in smart city initiatives, among other things. Detailed explanations of energy budgeting methods are presented as well as an energy-use-justification calculus. Following that, using TCP/UDP protocols for transmission and different circumstances, practical measurement of consumption is performed. Lastly, a common battery module is used for the final comparison to determine the battery's lifespan. A low-power wireless sensor node's energy consumption was analyzed in this paper under various operational scenarios to test the practicality of deploying a Wi-Fi hotspot in a large metropolis. Models are used in part of this research that has already been published that looked at different WSN communication methods. The greatest consumption states have been selected based on the theoretical analysis and the proposed technique. A transceiver's status is enabled when it is transmitting a radio signal, which has the highest energy cost appears to be incompatible at first glance. In actuality, the sleep state has the greatest impact, as it is the active phase for more than 90% of the time. When using an OTMT technique, minimizing the sleep current will result in a large reduction in overall power demand, even more than lowering the current in the active state. According to the findings, The use of a Wi-Fi node with a medium range and low power can be put to with the correct energy budgeting strategies, underscoring the fact that Wi-Fi cannot be ruled out for WSN deployments. Without using more current, it is feasible to mediate between the two parties an acceptable number of sensing repeats and network broadcasts. It is feasible to keep the node running for a few days without having to recharge the battery if the energy storage is adequate. Furthermore, because of the high data transmission ratio, enormous amounts of data can be sent without considerably increasing the overall consumption. It is easier to communicate with non- technical consumers because it is a frequently used protocol that is well-known, allowing for the creation of a DIY sensor kit for usage in a range of urban settings. IP-based devices also promote contact with Smartphones and other mobile devices make cloud computing nodes possible. Even though Wi-Fi technologies are not now ideal for all sensing tasks, they can be used in scenarios where a for data mining and other comparable circumstances, a specific variable must be recorded. It may be possible to reduce the amount of energy consumed using cutting-edge low-power electronics, allowing for, With a lower battery recharge rate, nodes will last longer, potentially in comparison to estimate the lifetime of a ZigBee node. Aside from that, by reducing the power, it may be able to energize the node using ambient sources, lowering the device's interaction with the user and increasing its autonomy This paper looked into the low-power wireless sensor node's energy consumption in several operational conditions to see if it was feasible to install a city's Wi-Fi hotspot. A component of this study is based on models that have already been published that looked at different WSN communication methods.

Because of its inherent large bandwidth, ease of deployment, and lack of licensing requirements, optical wireless communications (OWC) systems [7] are appealing for providing broadband services. Atmospheric transmission originated with the creation of the laser LiFi, also known as optical Wi-Fi, which would be the best optimum solution over Wi-Fi technology. LiFi technology was first described, and then the performance of both LiFi and Wi-Fi technologies in wireless communication was compared. We're using LiFi and Wi-Fi to implement OWC and comparing the two types of communication. The ultrahigh capacity wireless optical communication systems have been discussed, with LiFi coverage services covering a larger area than Wi-Fi approaches. Using light media, LiFi gives covering service areas indoors (LED or LD) or outside (Bulb) with a bandwidth of 5GHz, compared to Wi-Fi's 700 MHz. If you use optical wireless communication, you can get ultrafast processing data rates by using a multi-optical channel (LiFi). They also have employed OOK and PPM formats for modulation and coding because the LiFi system relies on digital techniques and coding formats to provide ultrahigh security. In connection to optical wireless communication and infrared communication, the key concepts in visible light communication (VLC) have been discussed. The channel parameters of the VLC system were discussed in comparison to infrared communication, and the VLC transmitter and receiver were detailed, as well as the fundamental LED or Bulb features.

Visible light communication (VLC) in [8] offers a large unlicensed bandwidth, can communicate in RF-sensitive situations, is energy-efficient, and allows wireless access networks to enhance their capacity through spatial reuse. In contrast, Wi-Fi offers a greater range of coverage than VLC and is less likely to be blocked due to VLC's light of sight (LOS) requirement. To make use of both Wi-Fi and VLC, we propose and construct two heterogeneous systems with Internet connections. The first is a hybrid Wi-Fi-VLC system that employs the unidirectional VLC channel for the downlink and the Wi-Fi back-channel for the uplink. When used with full-duplex VLC transmission, this asymmetric approach solves optical uplink issues. In the second technique, we use the Linux operating system's bonding capability to combine Wi-Fi and VLC simultaneously, which is demonstrated to further improve the robustness and increase throughput. In terms of throughput and web page loading time, the hybrid system outperforms standard Wi-Fi in congested locations, and the aggregated system performs even better when time and distance between the access point and the user device that is blocked are taken into account. In this study, we compare and contrast Wi-Fi and VLC are part of two heterogeneous systems. Our goal is to demonstrate that these two communication bands can coexist in the same space. In a busy wireless environment, the hybrid VLC could function substantially Within a certain distance between the transmitter and the receiver, this method outperforms the Wi-Fi system. VLC should be investigated further as a complementary method. Wi-Fi infrastructures, on the one hand, are common and well accepted by most users; When it comes to long-distance data transfer, though, Wi-Fi may be superior to VLC in the presence of obstructions. As a result, we believe that Wi-Fi-VLC aggregation merits additional investigation to maximize the aggregated bandwidth and reduce network latency. On the hybrid VLC system, they intended to employ aggregation in the future. A strategy that merges the symmetric Wi-Fi alone link and the asymmetric hybrid VLC link should be investigated to tackle the issues of optical uplink in our implemented aggregated system. The problems surrounding the spatial reuse of VLC links is another area of future study to look into. This involves the utilization of multiple VLC front-ends. VLC is a promising and evolving wireless technology that contributes significantly to next-generation heterogeneous wireless networks, based on the benefits and results discussed in this paper.

Is it possible to use Wi-Fi signals for sensing? in [9], Wi-Fi signals can now be reused for both communication and sensing thanks to the expanding PHY layer capabilities. Sensing via Wi-Fi would allow for remote sensing without the use of worn sensors, simultaneous perception and data transfer without the need for additional communication infrastructure, and privacy-preserving contactless sensing. Due to the widespread use of Wi-Fi devices and the widespread deployment of Wi-Fi networks, When completely integrated, Wi-Fi-based sensor networks may become one of the world's largest wireless sensor networks. Wireless and sensorless sensing, on the other hand, is a new notion that is more complicated than just combining Wi-Fi and radar. Wireless and sensorless sensing aim to resolve the tension between Wi-Fi's limits and the growing demand for environmental awareness in daily life, researches solutions using frequency and spatial diversity, and discovers applications in wireless communications and mobile computing that were previously inconceivable. We believe that technological advancements will increase the granularity and sensitivity of wireless sensing, allowing for a wider range of new applications. This article is simply intended to provide an overview of the wireless and sensorless sensing concepts. Wi-Fi-based sensorless sensing can be considered one of the largest wireless sensor networks, if Wi-Fi is considered a side sensor in the world, passively monitoring office buildings, shopping malls, other public locations, and people's residences human activity. Every person in the physical world has been awarded a unique identity in the digital world by living within such a network. So, after closing the doors, close the drapes and check for wiretaps beneath the table the next time you want a peaceful meeting, and don't forget to turn off the Wi-Fi.

Indoor location systems based on Wi-Fi, Bluetooth, and UWB have recently been presented in [10]. The solution to achieving a low-cost and precise positioning system remains open due to the limitations and complexity of the indoor environment. This article describes a Wi-Fi-based positioning strategy that can increase localization performance in ToA/AoA by removing the bottleneck. Unlike existing methods, our suggested technique eliminates the requirement for a large signal bandwidth and a large number of antennas by transmitting several predetermined messages while retaining high accuracy. The complete system structure is presented by comparing localization performance in 20/40 MHz bandwidth Wi-Fi APs with varied quantities of messages. Simulation findings show that our Wi-Fi-based positioning strategy can achieve 1 m precision in commercial Wi-Fi devices without any hardware changes, which is far superior to traditional solutions from academia and industry in terms of cost and system complexity. In terms of cost and complexity, we show how different techniques and their applications are used in indoor localization. The traditional ToA/AoA methodologies for overcoming bandwidth constraints and multipath numbers are introduced and addressed. We show that our suggested method outperforms the super-resolution method in terms of ToA resolution when signal bandwidth is constrained and the calculation burden is low. Furthermore, our suggested AoA technique can eliminate the demand for numerous antennas by utilizing multiple message help. We show the mechanism's capabilities by using it in a Wi-Fi- based indoor positioning system that can handle localization with just one Wi-Fi AP. When there are two or more nearby Wi-Fi APs, AoA cooperative positioning can improve position accuracy even further. Finally, simulations revealed that in a practical Wi-Fi AP situation, our proposed technique gives higher performance without modifying hardware parameters.

The IEEE 802.11 Working Group has launched a new study group, known as IEEE 802.11ax, to figure out how to enhance spectrum efficiency, especially in extremely crowded conditions, Overlapped Basic Service Set scenarios are another name for this type of scenario (OBSS). In this article, we'll look at some of the most prevalent issues in [11] and explain how their impacts are magnified in dense installations, especially in co-channel settings, in this study. Following that, we use a simulation-based study to highlight our findings and offer conclusions. Some of the study's important findings include: In co-channel deployments, the impacts of link suppression and stalemate may be amplified, resulting in significant throughput degradation. Furthermore, this means that increasing the number of access points (APs) in a given area may not always result in improved outcomes, hence AP placement in OBSS settings must be carefully regulated. The findings highlight the necessity for appropriate To avoid the above effects, load balancing and channel selection strategies is where AP placement is impossible to control owing to uncontrolled situations. The workgroup IEEE 802.11, just started working under the 11ax group's umbrella on a dense Wi-Fi deployment. There is an increase in the number of stations serviced and the number of stations served when the Basic Service Set (OBSS) deployments overlap per unit of time and the number of APs. This research described some of the common wireless LAN phenomena that are known to decrease performance, as well as how these behaviors could magnify in Deployments of the OBSS. A simulation-based investigation was also carried out to find insights that could be relevant when building a solution for feeding issues. As a result of extensive research, it has been determined that AP distance affects total saturation throughput (DL There is link suppression, interference amplification, STA broadcasts that cause nearby access points to temporarily stop, and hidden/exposed nodes that cause wasteful medium occupancy are just a few of the effects over an extended period were identified in an OBSS setup with co-channel, resulting in lower link utilization. It was also revealed that simply it's possible that raising the quantity of APs in a certain location won't help, be enough to enhance performance in such installations. To lessen the above-mentioned negative consequences, the placement of advanced placement students must be closely monitored. While this is feasible in planned deployments, it is problematic in unforeseen deployments. In unexpected contexts, our findings highlight the necessity for a smart load balance and OBSS channel selection methods that are self-contained, where there is little coordination between adjacent APs. To verify the simulations results, in that paper, it would also be interesting to analytically study the OBSS problem (e.g. using the model of Bianchi). These are intriguing topics for future research in this area, and they should aid in addressing some of the network densification concerns.

A LiFi is a new wireless technology [12] that provides network connectivity. LiFi is the German physicist Herald Haas's Light-fidelity and LiFi. It sends data by using an LED light bulb that changes intensity quicker than the human eye can keep up with. Infrared remote controls work on the same principle, but the device is far more powerful. The Haas describes his innovation, which he calls DELIGHT; data speeds of more than 10 megabits per second may be created, which is quicker than your normal broadband connection. In various fields of everyday life, LED is used. It can use the ability to light data from one to the next. A massive LiFi use can resolve a certain bottleneck in Wi-Fi technology data transmission. Lastly, The scientists also looked at the future possibilities of using visible light as a data transfer and networking carrier for this new technology. LiFi is the newest and fastest-growing technology that is responsible for a variety of previously created and developing technologies. LiFi has become quite appealing, not least because it can give another viable radio- based wireless solution. The growing numbers and devices are accessing the wireless Internet. The airwaves are gradually getting tighter and harder to get a consistent, high-speed signal. Every bulb can use this wonderful technology in the future, a Wi-Fi hotspot of some kind, to make the future more brilliant. Every bulb can be used to transfer wireless data via a Wi-Fi hotspot, which means to make the future cleaner, greener, safer and brighter. The LiFi concept attracts great interest at present and very effective wireless alternatives. Wireless Internet access is available to an increasing number of people and their many devices.

The WIFI is available to the public in business and industries via WIFI technology. It functions using radio waves in the spectrum. These waves harm sick people in areas that are sensitive to signals. Therefore, in environments like hospitals, scan centers, airlines, etc., it cannot be used. LiFi is a technology that was created to work in these types of locations to overcome these limitations. When compared to Wi-Fi, this paper discusses LiFi technology, its operating principles, problems, and applications. Both technologies have their unique peculiarities in the comparison study [13]. In comparison to Wi-Fi, this paper examines the fundamentals of LiFi technology, including its operating principles, problems, and applications.

The comparative study provided both technologies with features and limitations. This study found that LiFi efficiently improves data transmission over Wi-Fi. The Internet of Things (IoT) devices for actuating devices and communicating effectively are also linked. It is observed that This observation is to be enlarged in the future with IoT devices for the application of intelligent traffic control.

In our modern lives, communication plays a very important role. Everywhere with very fast internet, people always need precise information. People favor a wireless network for fast communication rather than a wired network. Some new wireless technology has been developed in recent years by researchers. LiFi is a new wireless technology in [14] with high-speed connectivity, providing better efficiency, bandwidth, availability, and safety. This paper will focus on finding and detailed study of this awesome new LiFi technology with other wireless technology. LiFi is the next technology that is growing fastest. The world is now moving into a digital environment. This Internet data can be transmitted using LiFi Technology through light in a room for laptops, tablets, and smartphones. LiFi technology increases the transfer speed and can be used in numerous prohibited locations. When practically employed with this technology, every bulb can transmit high-speed wireless data using something like a Wi-Fi hotspot. We can therefore move into a cleaner, safer and brighter future. The omnipresent trend was the download of mobile Internet information via Wi-Fi. Many commercial and public organizations (e.g., libraries, cafés, and restaurants) participate in this program and commercial agencies are already wide-ranging Wi-Fi to provide their customers with alternative Internet access and to mitigate the mobile load. In addition, intelligent cities start installing Wi-Fi infrastructure [15], based e.g. upon a sensor network or Internet of Things, for current and future municipal services. A simple model is available for distributing Wi-Fi hotspots in urban areas. The hotspots are modeled on a uniform angle distribution and exponential distance distribution that is cut to the urban limits. The features of this model are compared with the actual distributions in detail. Furthermore, this model's applicability, as well as its limits, are demonstrated. The findings suggest that the model can be employed in situations where precise spatial hotspot collocations, such as the download potential, are required, coverage or signal strength are not required. This study examined the characteristics of Wi-Fi hotspot distribution in cities. It is also possible to build a basic model that may be used to generate Wi-Fi space distributions in any city. We have detailed investigated and compared the characteristics of the hotspot sites generated with the original data. Spatial patterns of actual hotspot sites were not reproduced in every detail, which led to errors of accuracy with applications such as handovers, interferences, bandwidth shared, or networks, considering a collocation or clustering of hotspots. Higher fitting accuracy with e.g. Gamma distribution cannot resolve the problem because the distance and angle of the hotspots are independent. This problem cannot be overcome. However, we found that high accuracy can be achieved through the single model for applications that do not require an accurate spatial arrangement of the hotspots but which employ other characteristics such as distance from the closest hotspots. For example, the model could accurately replicate the download potential, coverage, or signal strength in a town. Therefore, the model can be used for these applications to generate hotspot distributions to assess current, hypothesis, or future scenarios in which there is not a real distribution. This can help develop scalability mechanisms that rely on Wi-Fi and evaluate their performance for a variety of applications.

Users and operators will be more satisfied with Network deployments in the future (5G and beyond), as well as present Wi-Fi advancements like Densification of networks, traffic dumping, and Internet of Things compatibility. If you have a mix of new and old Wi-Fi devices, as well as Wi-Fi networks that are heterogeneous, cannot cohabit well, Wireless access that is both pervasive and painless may be the objective. Based on recent research trends and IEEE 802.11 advancements, they have identified many future Wi-Fi use cases in this study [16]. Consider two crucial elements when analyzing the functionality required in these use cases: coexistence of vintage and modern Wi-Fi devices, as well as inter-network interference. Our findings suggest that, despite several flaws and unresolved concerns, Wi-Fi will undoubtedly be an important part of future network deployments. Future network deployments will face coexistence challenges as Wi-Fi networks evolve and become more widely used. According to this paper, we identified future Wi-Fi use cases, analyzed their requirements (based on the most current and forthcoming 802.11 amendments, as well as existing Wi-Fi solutions linked to innovative network paradigms), and assessed the performance of the most modern devices from two perspectives: coexistence between Wi-Fi devices and network devices. Guidelines for ensuring interoperability have been identified, as well as potential areas for study. Fortunately, the 802.11 standard, which is still in development, is aimed at addressing some of these concerns. 802.11ax, for example, promises to increase spatial reuse. As new networking technologies (such as automaticity and software-defined networking) are adopted, more solutions will be found.

The most recent development in the realm of distant communication is the LIFI [17]. Many people nowadays, whether they are connected or not, use the internet to complete their work. The remote system's data transfer rate slows as the number of consumers increases. WIFI provides a speed of roughly 150mbps.11n, but it is not yet able to meet the needs of the client, As a result, the LIFI is being introduced. Harald Hass, a German researcher, claims that LIFI uses visible light to transmit data at a faster rate (10 megabits per second). So we're dissecting the LIFI/WIFI in this situation. It's comparable to the thought band that powers infrared remote controls, It is, however, far more effective. Haas believes that his D-LIGHT device can provide data rates quicker than our present broadband connection. In this article, we'll look at and break down the speed of LIFI and WIFI, as well as address stuck issues that arise as the number of clients grows. With the advancement of technology and the increasing usage of internet offers, there are numerous prospects for the use of the lithium-ion era to soon become a reality. The concept of LiFi is catching on so quickly because it is so simple to use that it is piquing people's interest. The use of LiFi technology provides a fantastic chance to replace or offer an alternative to radio-based Wi-Fi technologies. As the number of people and their access to the internet grows on such a big scale, having an internet connection via wireless will soon become insufficient, as usage grows while bandwidth remains constant. In this recording paper, we conclude that the opportunities are enormous and should be pursued. Furthermore, this technology is in the manufacturing system to enable each bulb to function as a Wi-Fi hotspot for wireless data transmission.

WPA2, all modern secured Wi-Fi networks that use this protocol to keep their data safe have been found to have serious flaws. An attacker within the range of a victim can take advantage of these flaws, issues by In [18], they used key reinstallation assaults (KRACKs). To put it another way, attackers can use this innovative attack approach to decode previously thought-to-be-safe data. Credit card numbers, passwords, chat conversations, emails, photographs, and other personal information can all be stolen using this method. The attack is effective against all modern WPA2-protected Wi-Fi hotspots It's also feasible to inject and alter data depending on the handshake mechanism. A solution is provided to offer capturing and analyzing EAPOL packets to prevent nonce reuse for a secure handshake during Pairwise Transient Key (PTK) reinstallation, which occurs in the event of an attack. Our changes prevent the attacker from accessing the victim system via a rogue AP generated by KRACK, alerting the client to unusual activities and halting the attack. It is demonstrated how to identify and prevent the KRACK assault. The 4-way handshake mechanism is used to authenticate clients utilizing WPA2 Wi-Fi APs. By watching the transmission of message3 twice during authentication, we were able to detect nonce reuse. As a result, for KRACK attacks we developed a solution that involves disconnecting the victim from the rogue AP and delivering a message to the victim informing them that the rogue AP has been detected. The authentication mechanism of WPA2 Wi-Fi networks is examined in this technique. Our solution protects the user from being a victim of a KRACK attack. The detection of KRACK attacks in Windows and IOS platforms should be investigated as a future feature that will make Wi-Fi provides a secure networking environment in which to test and receive services.

Photonic Integrated Circuits enable us to fulfill the expanding need for internet communication systems, which are developing at a pace of roughly 40% each year. The rise in online video traffic is primarily responsible for this growth. Mobile access has expedited this expansion, due to the availability of video clients on all smartphones and tablets, it is now easier to view video through a network connection from anywhere at any time. This research [19] compares the performance of multiple material platforms for photonic integrated circuits. The article examines the most important applications for rising internet traffic demand and optoelectronic transport networks, cloud computing, and high-performance computer systems have all entered the terabit age. The novel technologies and signal processing in photonic integrated circuit-based coherent optical transport networks are explained. The characteristics and performance of the three main photonic integrated circuit material platforms, as well as their future technology nodes, are examined. Infrastructure-based networks will be unreachable in the case of infrastructure failures (such as earthquakes and tsunamis) or highly inhabited regions (e.g., concert and conference hall). Because location-based communication networks are encouraged, this helps offload computing, mobile edges computing, and mobile crowdsourcing. The local communication system is built using Wi-Fi Direct (WFD) in this study [20]. We describe a generic solution for bidirectional inter-group communication that is not covered by WFD standards, as well as an intra-group communication solution based on WFD's native features. According to testing findings, the highest intra-group communication throughput is 31.7 Mbps and the highest inter-group communication throughput is 8.26 Mbps. A local communication system based on WFD technology is being developed and tested to address the risks of infrastructure failures as well as the requirements of performing cooperative activities in mobile edge computing. We offer not just intra-group communication based on WFD's inherent qualities, but also a generic inter-group communication solution supported by WFD's current implementation. The greatest intra-group communication throughput might reach 31.7 Mbps, whereas the maximum intergroup communication throughput is 8.26 Mbps, according to the findings of the experiments.

In this study [21], we'll Light-Fidelity (LiFi) is defined as a 5th Generation (5G) technology. Cellular networks based on LiFi have been created, with peak transmission speeds of 8 Gbps shown from a single light source. We'll debunk some prevalent technology myths and demonstrate how they might be dispelled, it can impact a range of current and future enterprises. The future of LiFi is also discussed, as well as what new uses it. We show in this work that in wireless communications, there has been a definite trend towards higher and higher frequencies. Because RF spectrum in lower frequency bands is restricted, this is the case, as well as the decade-long exponential growth in wireless data traffic. This expansion will continue. As a result, future wireless communication systems will virtually probably need to use frequencies other than those available in the RF band.

LiFi is an example of this new paradigm change in wireless communications as we migrate from millimeter wave to nm-wave, which utilizes light. Data throughput from a single LED has risen from a few Mbps in 2002 to 8 Gbps in 2016 according to several studies. Physical layer technologies for LiFi have improved over the previous 15 years, and data speeds from a single LED have increased from a few Mbps in 2002 to 8 Gbps in 2016. Multiuser access, interference avoidance, and mobility aid are all LiFi networking strategies that have all seen an upsurge in the study over the previous five years, and LiFi products that enable wireless networking using light have entered the market. As a result, LiFi is now a reality, and this technology is set to revolutionize the way we communicate will be around for quite some time.

Basic information is communicated through radio communication technologies such as Bluetooth, the global positioning system (GPS), and Wi-Fi. Even though radio communication technologies are common, there is a need to transport data wirelessly and more efficiently. Overcrowding and interference from other radio applications are key components of radio communication technology. The use of visible light for data transfer has been demonstrated in a study over a wireless medium. As a result, instead of using ordinary radio waves as a communication medium, we now use visible light, which is referred to as Light Fidelity (LiFi). The LiFi technologies have piqued the interest of the research community. There is a necessity to support technology because of modernity. Many leading companies are working on light fidelity (LiFi) initiatives, and numerous surveys and research studies have been undertaken. Data is collected from a variety of surroundings using visible light, and it is used to improve services in a variety of industries by making intelligent decisions. The information gathered is analyzed and processed. The terms LIFI (Light Fidelity), wireless communication, and Internet of Things (IoT) are used in this work [22] (Internet of things).

Light Fidelity (LiFi) is a Visible Light Communications (VLC) technology in [23] that allows data to be exchanged via illumination by using a LED is an acronym for Light Emitting Diode. With this technological advancement, light signals are received by a photodetector, and the data is converted into a stream-able substance by a signal processing unit. Other uses include It is possible to install a system that monitors the health of a patient, to have all benefited from LiFi technology implementation and development, according to this survey paper. Only a little amount of study has been done on LiFi communication networks and implementation. Despite this, there are several opportunities to put it into practice. There are numerous applications in the Data systems that are still necessary for sensitive electromagnetic interference environments such as airplanes and power plants, where Wireless Fidelity (Wi-Fi) has proven unsuccessful yet is fast and networked. In businesses that require secure networks, LiFi has been used and where signal availability and transmission radius must be tracked, such as banking and intelligence gathering in the military and government special operations. In densely crowded areas such as campuses, malls, and conference centers, it can also provide high data capacity. Integration of the LiFi network with both the mobile and optical fiber networks is still unattended scopes. There have been no recommendations produced for the establishment of LiFi networks with efficiency in a variety of settings, including hospitals, clinics, and banks.

LiFi stands for light-based networked bidirectional wireless communication. It uses visible light and the infrared spectrum to connect stationary and mobile devices at exceptionally fast data speeds. When all of these spectral resources are combined, it will be possible to create a new spectral resource, the total radio frequency (RF) spectrum is 2600 times larger. This paper discusses why LiFi is such an important technology, especially in the context of Cellular communications of the 6th generation (6G). Interference mitigation and hybrid LiFi/Wi-Fi network topologies are among the issues covered, as well as other important networking technologies. Furthermore, we examine the implications and solutions for load balancing, as well as how LiFi may be seamlessly incorporated into current To develop heterogeneous networks, wireless networks are used that span the optical and radio domains. Finally, we provide the outcomes of a real-world hybrid LiFi/Wi-Fi network configuration performed on a software-defined networking testbed. Furthermore, results from a LiFi implementation in a school classroom are shown, demonstrating that offloading traffic to the LiFi improves Wi-Fi network performance dramatically. This work in [24] has demonstrated that future cellular systems can be built using Light communication in open space. It has been emphasized in this context that, For free-space light communications to achieve this, The emphasis in wireless networks must change from point-to-point data rate increases in VLC to data density maximization. LiFi has been demonstrated to improve Wi-Fi networks by offloading data traffic. End customers, or our mobile devices, will be able to access data rates that are currently only available via fiber-optic transmission.

By integrating diverse technical areas, network convergence will help keep up with the rapid growth of mobile devices and their increasing demand for Internet services. LiFi and Wi-Fi hybrid networks (HLWNets) are intriguing methods for indoor wireless communications [25]. High-speed data transport is provided via LiFi, while Wi-Fi provides extensive coverage. As a survey-style introduction to HLWNets, we begin with a system design framework that covers network topologies, cell deployments, various access, and modulation techniques, lighting demands, and backhaul. HLWNets outperform stand-alone networks is then demonstrated using key performance metrics and recent developments. Furthermore, major research subjects such as user behavior modeling, interference control, load balancing, and handover are discussed in detail to highlight the particular challenges that HLWNets face. Furthermore, Indoor location and physical layer security are used as examples to highlight the possibilities of HLWNets in application fields. Finally, future research difficulties and opportunities are addressed. LiFi has evolved as a viable solution for indoor wireless communications in recent years, owing to the impending RF spectrum bottleneck. Meanwhile, Wi-Fi's pervasiveness in daily life continues. The existence of both Wi-Fi and LiFi is gaining traction, thanks to commercial LiFi devices from firms like pure LiFi and Signify. The structure of HLWNets is formed by the easy administration of LiFi and Wi-Fi by a central control unit in the same local region. HLWNets can provide higher network performance by the high data rate of LiFi with the ubiquitous coverage of Wi-Fi than a single wireless technology. Implementing HLWNets in practical contexts and enhancing network performance are currently being researched.

Using the visible light spectrum and existing lighting infrastructure, such as light-emitting diodes (LEDs), A new wireless networking technology called Light Fidelity (LiFi) was introduced rival (LEDs).

Communication between two or more points can be done over an extremely high-speed bidirectional channel. As a result, LiFi has a restricted coverage area and its optical gain is largely reliant on the receiver orientation, regarding the transmitter, it is prone to frequent service interruptions. The use of wireless fidelity is another example of (Wi-Fi), which is utilized in combination with LiFi to construct a hybrid system that provides reliable coverage. Many issues confront the hybrid LiFi/Wi-Fi system, including smooth integration with Wi-Fi, mobility support, changeover management, resource sharing, and load balancing, to name a few. The existing literature has dealt with one or more aspects of the problems that LiFi systems face. There are just a few free open-source technologies that can address these issues comprehensively and in a scalable way. In [26], based on the network simulator 3, we created an open-source simulation framework (ns-3) for that aim that implements key components of the LiFi wireless network. There are just a few free open-source technologies that can address these issues comprehensively and in a scalable way. Network Simulator 3 was used as the basis for our simulation system (ns-3) that incorporates important LiFi wireless network components for this purpose. There are two layers to our ns-3 LiFi architecture: a fully working access point (AP) and a medium access (MAC), as well as a user device mobility model and LiFi and Wi-Fi integration with handover functionality. As a result of these simulations, the hybrid LiFi-Wi-Fi system is shown to have mobility and handover capabilities as well as in packet latency, throughput, and PDR as well as fairness to consumers. Researchers now have access to the framework's source code.

17.5 Methodology

Figure 17.1 is a block diagram of the LiFi system. The lamp driver receives steady power from the power supply. The lamp driver is connected to the internet. With fiber optics connections, the switch and LED lamp are connected to the lamp driver. A communication source is provided via an LED bulb. The data is converted into light by a microchip in an LED bulb. A light beam from an LED lamp is used to deliver high-speed data to a photodetector. The receiver detects changes in light beam intensity and turns the information into an electrical signal. These data are transformed and sent to the technical equipment.

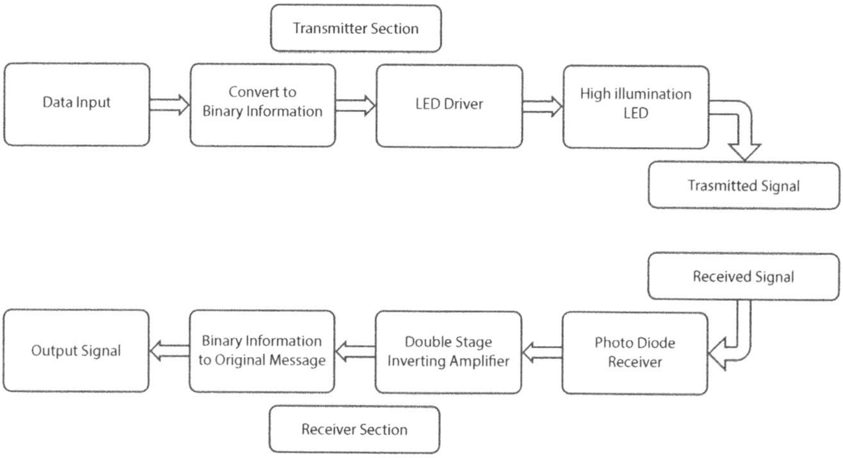

Figure 17.1 The operation of the LiFi module.

17.6 Proposed Model

Figure 17.2 demonstrates how to construct the two circuits using hardware components. A wire should be connected to digital port 10 and the LED's plus side. The minus side of the LED should be connected to the Arduino's ground connector with a cable. To prevent the LED from overheating, use a 220-ohm resistor. To construct the code, use the Arduino software. This program tells the Arduino to flash the LED for a set amount of time. With this code, you can only select from a list of prepared messages due to the simplicity of the setup. However, you could have different durations match to the letters of the alphabet and transmit whole messages that way. This code instructs the Arduino to keep track of how long the LDR is lighted. It associates a message with the number of seconds and prints it on the serial monitor.

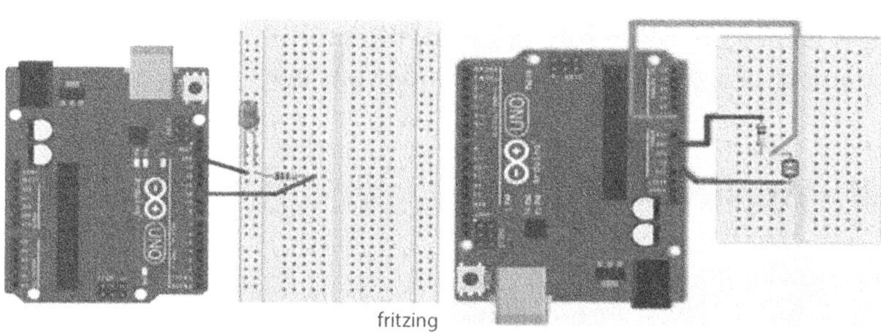

Figure 17.2 Circuit diagram for securing data communication using LiFi.

Figure 17.3 Code for sender end.

17.7 Experiment and Results

The code in Figure 17.3 tells the Arduino to blink the led for a certain amount of time. You can just choose from a list of prepared messages because this is a minimal setup, but you could have varied durations match with the letters of the alphabet and transmit whole messages that way. In that scenario, modifying the delay timings to lessen the time it takes to deliver a message is a good idea. If you're in a situation with more or less light than we are, you might need to adjust the threshold sensor value in the code. Before picking a medium value, uncomment the print sensor value command and verify the sensor's values with the led on and off.

Receiver End
The code in Figure 17.4 instructs the Arduino to keep track of how long the LDR is turned on. Make sure the serial monitor is open because it relates the number of seconds to a message and prints it.

17.8 Applications

Undersea Communication
Because radio waves absorb quickly in water, underwater radio transmissions are impossible, yet light may go great distances. As a result, LiFi

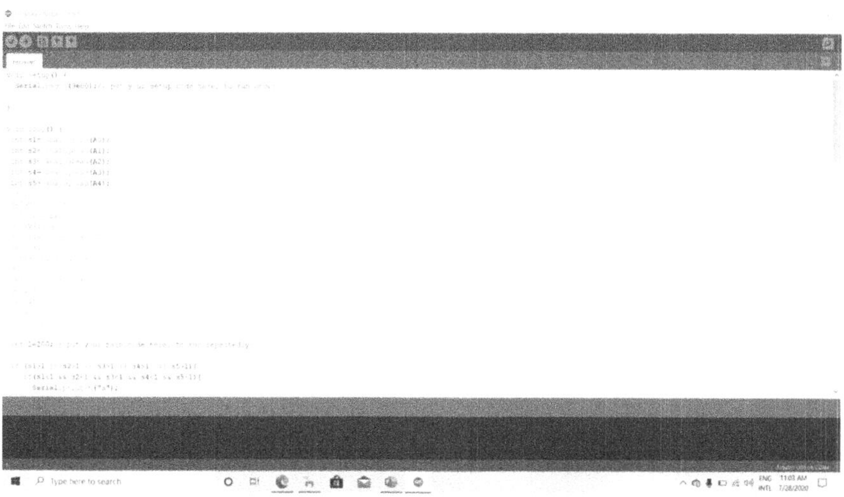

Figure 17.4 Code for receiver end.

allows communication between divers, mini-subs, and drilling rigs, among other things.

Security

The breadth of the light pool is the access zone for each channel in a meeting room environment, and It is accessible to several users. Each user can get data at a quicker pace than if they were linked to a comparable Wi-Fi channel. Bandwidth is a finite resource that must be on a one-to-one basis when using Wi-Fi. This means additional connections mean a slowdown in download speed rates become for everyone. In LiFi, however, As a result of the increased number of available access points, each pool of light provides full-channel data rates with fewer simultaneous users. Each user will benefit from speeds that are up to 1000 times quicker. In addition, a solid object cannot block the path of light, unlike radio waves. As a result, security over Wi-Fi is significantly improved, with only little precautions taken to prevent leakage from windows and other sources.

Cellular Communication

In outdoor metropolitan contexts, LiFi connected street lighting could be used to establish an internet access point network. Radio base station distances have decreased to 200-500 meters in cellular telephony. As a result, instead of installing new radio base stations in our cities, street lamps may provide both lighting and high-speed data access, day and night, seven days a week. Surprisingly, full data transfer rates are still feasible even when the lights appear to be off to the naked sight. Another benefit is cost savings, as building new radio base stations is expensive, typically entails a significant investment in terms of both installation and site lease.

Indoor Navigation
Each light may be recognized and utilized as a navigational aid in metropolitan areas (for example, using the widely known MAC codes used by data routers and PCs). Each code's identification would be tied to a specific site. The light from the nearest lamp, for example, may display a user's actual location as they walk along a corridor on a mobile device.

17.9 Conclusion

Although this technology has a long way to go before being commercially practical, it has a lot of potential in the realm of wireless internet. This concept is now being developed by a large number of researchers and companies, and it promises to tackle the problems of radio spectrum shortages, space constraints, and slow internet connection speeds. Using this technology, we can shift to communications networks that are greener, cleaner, and safer. According to LiFi's radio-frequency bandwidth limitations and other shortcomings of radio communication technologies can be overcome. As a new technology, LiFi has the potential to act as a catalyst for a range of other discoveries and technologies. Because of this, we should expect to see future LiFi apps that are extensible across multiple platforms and facets of human life.

Acknowledgment

We are thankful for CoE-Cyber Security, Sahyadri College of Engineering & Management for providing technical and infrastructure assistance. The above research work is autonomously carried at CoE-Cyber Security, Sahyadri College of Engineering & Management.

References

1. Campoccia, F., Sanseverino, E.R., Zizzo, G., Analysis of the limitations of WiFi communications managed by the IEEE 802.11 protocol in data

transmission in automated power distribution systems. *SPEED AM 2010, International Symposium on Power Electronics, Electrical Drives, Automation and Motion*, DOI: 10.1109/SPEEDAM.2010.5545051.
2. Koo, J. and Cha, H., Localizing WiFi access points using signal strength. *IEEE Commun. Lett.*, 15, 2, 187–189, February 2011, Digital Object Identifier 10.1109/LCOMM.2011.121410.101379.
3. Oni, O.M., Amuda, D.B., Gilbert, C.E., Effects of radiofrequency radiation from WiFi devices on human ejaculated semen. *IJRRAS*, *9*, 2, November 2011.
4. Ishak, N.H. *et al.*, Biological effects of WiFi electromagnetic radiation. *IEEE International Conference on Control System, Computing and Engineering*, 978-1-4577-1642-3/11/$26.00 ©2011 IEEE, 2011.
5. Schauer, L., Dorfmeister, F., Maier, M., Potentials and limitations of WIFI-positioning using time-of-flight, IEEE, DOI: 10.1109/IPIN.2013.6817861 $31.00.
6. Trasvifia-Moreno, C.A. *et al.*, WiFi sensor networks: A study of energy consumption, IEEE, 2014, 978-1- 4799-3866-7/14/$31.00.
7. Hameed, S.S. and Shukla, A.K., Design and analysis optical wireless communication based on "LiFi". *IJSETR*, 7, 94–100, 2014.
8. Shao, S., Design of a visible-light-communication enhanced WiFi system, 1, 1–10, *arXiv:1503.02367v2* cs.NI, 10 Mar 2015. https://arxiv.org/abs/1503.02367,
9. Zhou, Z., Wu, C., Yang, Z., Liu, Y., Sensorless sensing with WiFi. *Tsinghua Sci. Technol.*, 20, 1, 1–6, February 2015, ISSN 1007–0214ll01/11ll.
10. Yang, C. and Shao, H.-R., WiFi-based indoor positioning. The future Wifi-*IEEE*, 53, 150–157, © 2015, 0163-6804/15/$25.00.
11. Zhong, Z., Kulkarni, P., Cao, F., Fan, Z., Armour, S., Issues and challenges in dense WiFi networks, IEEE, ©2015, 978-1-4799-5344-8/15/$31.00.
12. Bhavya, R. and Lokesh, M.R., A survey on LiFi technology. *Int. J. Eng. Technol.*, 3, 1, 7–12, January, 2016, eISSN: 2394-627X.
13. Kuppusamy, P., Muthuraj, S., Gopinath, S., Survey and challenges of LiFi with comparison of WiFi. *IEEE International Conference on Wireless Communications, Signal Processing and Networking*, Chennai, 2016, 23-25.03.
14. Shende, A.D. and Rojatkar, D.V., LiFi technology: A new revolution in wireless technology. *IJSRSET*, 2, 6, ISSN: 2394-4099.
15. Seufert, M., Moldovan, C., Burger, V., Hoßfeld, T., Applicability and limitations of a simple WiFi hotspot model for cities. *IFIP*, 2017, 978-3-901882-98-2 _c.
16. Kosek-Szott, K., Gozdecki, J., Łoziak, K., Natkaniec, M., Prasnal, Ł, Szott, S., Wągrowski, M., Coexistence issues in future WiFi networks. *IEEE Network*, 31, 86–95, 2017.
17. Gite, B.B., Maydeo, P., Bade, S., Muluk, T., Indoor navigation using LiFi. *Int. J. Adv. Res. Comput. Commun. Eng.*, 6, 412–413, 2017, DOI 10.17148/IJARCCE.2017.6185.

18. Raiton Lobo, N.S.T., Vernekar, P.S., Shetty, V.G., Mitigation of key reinstallation attack in WPA2 WiFi networks by detection of nonce reuse. *Int. Res. J. Eng. Technol. (IRJET)*, 05, 05, 1528–1531, May-2018, e-ISSN: 2395-0056.
19. Chovan, J. and Uherek, F., Photonic integrated circuits for communication systems. *Radioengineering*, 27, 2, 357–363, June 2018, DOI: 10.13164/re.2018.0357.
20. Wang, Z., Li, F., Wang, X., Li, T., Hong, T., A WiFi-direct based local communication system, IEEE, c 2018, 978-1-5386-2542-2/18/$31.00.
21. Haas, H., LiFi is a paradigm-shifting 5G technology, ©2017, 2405-4283/ Published by Elsevier B.V., 3, 26–31, https://doi.org/10.1016/j.revip.2017.10.001. https://www.elsevier.com/locate/revip
22. Revathi, G., Sujana, G., Kavyasree, A., Srinivasan, A., Vijay Harish, S., Sona, S., LiFi based data transmission and analysis using IoT platform. *International Conference on Physics and Photonics Processes in Nano Sciences, Journal of Physics: Conference Series*, vol. 1362, 2019, 012025 IOP Publishing, doi:10.1088/1742-6596/1362/1/012025.
23. Haas, H., Yin, L., Chen, C., Videv, S., Parol, D., Poves, E., Alshaer, H., Islim, M.S., Introduction to indoor networking concepts and challenges in LiFi, *IEEE/OSA Journal of Optical Communications and Networking*, A190-A203, ©2020 Optical Society of America, 1943-0620/20/02A190-14 Journal.
24. Wu, X., Zhou, L., Safari, M., Hybrid LiFi and WiFi networks: A survey. *IEEE Commun. Surv. Tutor.*, 23, 2, 1–24, Second Quarter, 2021.
25. Ullah, S., Ur Rehman, S., Chong, P.H.J., A comprehensive open-source simulation framework for LiFi communication. *Sensors*, 21, 2485, 2021, https://doi.org/10.3390/ s21072485.
26. Singh, C., Sairam, K. V. S. S. S. S., MB, H., Global fairness model estimation implementation in logical layer by using optical network survivability techniques, in: *International Conference on Intelligent Data Communication Technologies and Internet of Things*, pp. 655–659, Springer, Cham, 2018 August.

18

RSA-Based Encryption Approach for Preserving Confidentiality Against Factorization Attacks

Raghunandan K. R.

Department of CSE, NMAM Institute of Technology, Nitte, India

Abstract

Security of the shared information is the most important aspect in today's life. Various algorithms are proposed in cryptosystem to achieve the same and RSA is one of them. AlthoughRSA is a strong encryption algorithm it can be interpreted by factorization attack. Hence, this paper proposes an enhanced RSA algorithm that focusses on the security feature of RSA by providing immunity against factorization attacks. Algorithm introduces a third variable which is deployed to the network as the public key replacing common modulus *n*. The investigational end results like differential analysis, performance analysis and statistical analysis undoubtedlyrevealed the proficiency of the proposed methodology for secure communication. Furthermorethe chapter analyses different attacks possible on the proposed system and reports the efficiencyof the proposed system.

Keywords: RSA, Fermat's factorization, differential analysis, standard deviations

18.1 Introduction

Today, when the world is focusing on recovering from the threats posed by COVID-19, cybercriminals are impersonating the popular brand's attributes, thereby misappropriating the common person through misleading information on the internet. Cryptography could be a possible solution, as data scrambled through strong cryptography is nearly indecipherable. Cryptography involves encryption of the message (plaintext) to ciphertext

Email: raghunandan@nitte.edu.in

Chandra Singh, Rathishchandra R Gatti, K.V.S.S.S.S. Sairam and Ashish Singh (eds.) Modeling and Optimization of Optical Communication Networks, (331–348) © 2023 Scrivener Publishing LLC

from the source and decryption of ciphertext at receiver's end using a Key. Although there is a possibility that the attackers can bypass the cryptography and eavesdrop on the information, cryptography makes it harder by its strong encryption algorithms [1].

Cryptography can be classified into Symmetric Key Encryption and Asymmetric Key Encryption. The Symmetric Key Encryption involves only a single key for both encryption and decryption whereas Asymmetric Key Encryption includes two keys, Private Key and the Public Key. The Private Key is known only to the receiver whereas the Public Key is disclosed to the network [2]. RSA is an Asymmetric Key Encryption approach that is significantly complex in terms of its computation level and security level.

RSA comprises of three stages: Key generation, Encryption, and Decryption. In the key generation stage, the public key and the private key are constructed using two large prime numbers. The strength of RSA depends on the fact that it is difficult to produce two enormous primes and multiply them [3].

Limitations associated with RSA are inclusive of the fact that, if the value of any of the prime factors (u, v) or the public-key exponent (α) or private-key exponent (β) is known, and then the other values can be calculated, hence eliminating the secrecy. The RSA algorithm is easily factorable through modulo n, since it is the product of two prime numbers, hence, making it easy to obtain the original message [4]. Various factorization methods can be used to obtain factors of modulo n, few of them are discussed in this paper:

- Pollard's rho [5]: It is particularly effective for composite numbers having a small prime factor. The expected period is proportional to the root of the size of the smallest divisor of the composite number being factorized and it uses a small amount of space.
- Pollard's $p - 1$ [6]: It depends on the fact that a number N with prime divisor u can be easily factored if $u - 1$ has small prime factors. This prompts the idea of safe primes,for which $u - 1$ is multiple times a Sophie Germaine prime v.
- Fermat's Method [7]: Gupta *et al.* claim that it is the (rephrase) method that can be usedto find an odd whole number n such that $n = x^2 - y^2$ where x and y are numbers. It makes some deterministic memories unpredictability of ($n^{1/4}$) number jugglingactivities.
- Trail Division [8]: It checks for each prime number u, with the end goal n such that $u \leq \sqrt{n}$, if $u|n$. Nonetheless, the

preliminary division may take more than $O(\sqrt{n})$ bitactivities, which makes it wasteful for the quantities of intrigue. This strategy can be additionally enhanced by disposing of all the even numbers.

- Euler's factorization technique [9]: It is a method for factorizing a number by writing it as a sum of two squared numbers in two different ways. The limitation is that it cannot be applied to factorize an integer with any prime factor which is of the structure $4k + 3$.

This paper concentrates on the factorization issue of the RSA and an enhancement method is introduced to overcome this problem. RSA encounters the factorization problem with modulo n, hence an efficient algorithm is developed such that the distribution of n is eliminated. In this method, a mathematical transformation is applied over n and the replacement is computed, which cannot be traced back to find the factors of n.

Section 18.2 speaks about the related work done by the researchers in the field of cryptography. Mathematical preliminary information needed to develop the proposed system, which is discussed along with the correctness of the proposed system in Section 18.3. The proposed system is presented in section 18.4 followed by the analysis of the results obtained in section 18.5.

18.2 Related Work

Numerous researchers contributed towards improvisation in the effectiveness of cryptography. In this section enhancement made in the field of RSA is dealt in brief which is essentially required for proposing a new variant on RSA.

Authors Gupta and Sharma in paper [10] propose an algorithm by combining RSA and Diffie-Hellman algorithms to obtain higher security. The time complexity of the algorithm can be revised to improve the functionality, and the size of the key can be reduced as well. Authors Patidar and Bhartiya in paper [11] proposed an algorithm that three prime numbers to compute modulus n, such that it cannot be factored easily. Keys are stored offline prior to the initialization of the process. Although the encryption and decryption process is quick, the idea of utilizing a database for storage of keys is not safe as the keys are easily accessible in case of security breach. Authors Minni *et al.* proposes a method in paper [12] which eliminates the distribution of n which is the huge number and provide alternate values X

to replace n. Disadvantage is that it requires more time for key generation than traditional RSA. Authors Jaju and Chowhan modified RSA in their paper [13] such that it incorporates three prime numbers rather than two for computing n and passes the estimation of X rather than n, where X is the value computed to replace n. The limitation is that it requires more time for encryption and decryption. M. S. Iswari in paper [14] which proposes a key generation algorithm and it maintains the security factors and complexity regardless of whether the prime numbers used are small because of complex mathematical operations like factorization and discrete logarithm calculation. Authors Panda and Chattopadhyay in their paper [15] propose an algorithm in which the calculation of public-key and private-key relies upon n, where n is the product of four small prime numbers which expands the complexity of factorizing n, and hence upgrading the security. Authors Mathur *et al.* in paper [16] present an algorithm which includes exponential powers, n prime numbers, multiple public keys, and K-NN algorithm. It provides a verification feature at both senders as well as the collector's end. The limitation is that it requires more time for the encryption and decryption process. Authors Manu and Goel in paper [17] proposed an algorithm that performs the encryption and decryption process twice using two private and public keys which is capable of resisting attacks. Utilizing 4 different keys expands the unpredictability of the algorithm, which consequently upgrades its security. Authors K R Ragunandan *et al.* in paper [18] proposes a new variant of public key cryptography using Pell's quadratic case. The use cubic power of Pell's equation in the key generation process makes it difficult to factorize the public key exponent to gain access to the private key. Authors Islam *et al.* in paper [19] proposed to build a RSA algorithm using 'n' different prime numbers. The time required to generate the key for this algorithm is higher than RSA, indicating the increase in time required to breach the security system. Authors Raghunandhan *et al.* in paper [20] propose an enhancement in RSA algorithm that uses a fake public key exponent and replaces modulus n by X to increase the factoring complexity of the public key. Authors R. K.R. *et al.* in paper [21] propose a methodology in which Public Key exponent n, is replaced by phony modulus and Mersenne prime numbers are used to build it. Time required for decryption is less and the key generation has greater complexity in this method. Authors Ragunandan *et al.* in paper [22] propose Dual RSA using Pell's equation anda fake modulus key that replaces modulus 'n', and is computed using RSA algorithm and the solutions to Pell's equation. In paper [23] to reduce the factorization attack of RSA fake publickey exponent and fake common modulus approach is used and in paper [24] using Pell's equation fake public key exponent and fake common modulus is used and proved that proposed method is secure against Factorization problem. Radhakrishna D *et al.* in paper [25, 26] used XOR free approach which efficiently reduces the time complexity. In paper [27, 28] Krishnaraj *et al.* suggested RSA2048 is strongest algorithm compared against brute force, since it uses $5.8*10^{613}$ key combinations.

18.3 Mathematical Preliminary

Complexity of any algorithm is purely depends on how strongly the algorithm was built using mathematical functions. RSA technique is built on Fermat's theorem [29, 30] and Euler's totient function [31]. The proposed algorithm is proved mathematically using the following theorems and the proofs.

Theorem 1: *If n is composite number, then n has a prime factor* $u \leq \sqrt{n}$.

Proof: If n is a composite number then it consists of at least 2 prime factors, say u, and v. Assuming $u \leq v$, then $n \geq uv \geq u^2$. Hence, $u \leq \sqrt{n}$.

Theorem 2 (Fermat's Little Theorem): *For a prime number, u and n which is an integer not divisible by u, u divides* $n^{u-1} - 1$, *that is* $n^{u-1} \equiv 1(mod\ u)$

Corollary: *Let u be a prime number and n be an integer, then* $n^u \equiv (mod\ u)$

Theorem 3 (Fermat's Little Theorem): *For a prime number u, suppose* $x \in Z$ *satisfies* $gcd(x, u) = 1$, *then* $x^{u-1} = 1\ mod\ u$.

Proof: Consider the group with Z_u^* for prime u under multiplication mod u. Note that $|Z_u^*| = u-1$, and Z_u^* has the identity element $u = 1$. For any $x, y \in Z$ that satisfies $\gcd(x, u) = 1$, then $y = x\ mod\ u \in Z_u^*$. Using Legrange's theorem we have, $x^{b-1} = y^{v-1}\ mod\ u = 1\ mod\ u$

Theorem 4: *For distinct prime numbers u and v, suppose* $n = uv$ *and* $m = (u-1)(v-1)$.

If a and b are integers such that $ab = 1\ mod\ m$, *then for all* $x \in Z_n$, $x^{ab} = x\ mod\ n$.

Proof: For integers a, and b if $ab = 1\ mod\ m$, then $ab = 1 + km$ for some $k \in Z$, and for all $x \in Z_n$ the following will hold: $x^{ab} = x^{1+km} = (x^{km}) = (x^{u-1})^{(v-1)}$. If $\gcd(x, u) = 1$ then from Fermat's Little Theorem it is evident that $xu-1 = 1\ mod\ u$, thus, $x^{ab} = (1)^{(v-1)}\ mod\ u = x\ mod\ u$. Also if $\gcd(x, u) \neq 1$, then $x = 0\ mod\ u$, and certainly $x^{ab} = x\ mod\ u$.

Similarly $x^{ab} = x\ mod\ v$ for all $x \in Z_n$. Thus, $u|x^{ab} - x$ and $v|x^{ab} - x$, and so $uv|x^{ab} - x$. That is $n|x^{ab} - x$, or, equivalently, $x^{ab} = x\ mod\ n$.

Here the proof of the correctness of the RSA algorithm is presented, that is, $C_i^{\beta}\ mod\ n = M_i$. Let M_i be the message (plaintext) sent u and v are the prime factors of modulo n, α be the public-key component and β be the private-key component.

We know that $\alpha\beta = 1$ ($mod\ \varnothing(n)$). Hence, $\alpha\beta = 1 + k\varnothing(n)$ for some integer k. Let $M_i \in \mathbf{Z}_n^*$ (integers less than and co-prime to n). Then,

$$C_i^{\beta} \equiv M_i^{\alpha\beta} (mod\ n)$$
$$\equiv M_i^{1+k\varnothing(n)} (mod\ n)$$
$$\equiv M_i \cdot (M_i^{\varnothing(n)})^k (mod\ n)$$
$$\equiv M_i\ (mod\ n) \qquad (as\ M_i^{\varnothing(n)} \equiv 1\,(mod\ n) by\ Eulers'\ Theorem\,[31]).$$

Assume that $M_i \in Z_n \setminus Z_n^*$, where $(M_i, n) > 1$. If $M_i \equiv 0\ (mod\ u)$ and $M_i \equiv 0\ (mod\ v)$, then $M_i \equiv 0\ (mod\ N)$. Then $C_i^{\beta} \equiv M_i^{\alpha\beta} \equiv M_i \equiv 0\,(mod\, n)$. On the other hand, assume, without loss of generality $M_i \equiv 0\ (mod\ u)$ and $M_i \equiv 0\ (mod\ v)$. Then $C_i^{\beta} \equiv M_i^{\alpha\beta} \equiv M_i \equiv 0\,(mod\, u)$. Hence, u divides $M_i^{\alpha\beta} - M_i$. We know that

$$M_i^{\alpha\beta} \equiv M_i^{1+k\varnothing(n)} (mod\ v)$$
$$= M_i \cdot {}^{(u-1)(v-1)} (mod\ v)$$
$$= M_i \cdot (M_i^{(v-1)})^{(u-1)} (mod\ v)$$
$$= M_i\ (mod\ v)\ \text{as}\ (M_i^{(v-1)} \equiv 1\,(mod\ v)$$

Since, v divides $M_i^{\alpha\beta} - M_i$. Hence, $uv = n$ divides $M_i^{\alpha\beta} - M_i$ i.e., $M_i^{\alpha\beta} \equiv M_i (mod\, n)$.

The relationship between public-key exponent α and private-key exponent β is $\alpha\beta$ mod $\varnothing(n) = 1$ where modules key n is the constant. To find fake modulus, α and β are kept constant and finding the ζ which satisfies the equation $\alpha * \beta$ mod $\varnothing(z_i) = 1$. Where mod $\varnothing(z_i)$ is calculated using Euler Totient function. Hence ,

$$M_i^{\alpha\beta} \equiv M_i (mod\ z_i)$$

In proposed method, fake modules key ζ is computed using the product of z_i components and common modules n, using

$$\zeta = * \quad * n \tag{18.5}$$

Now,

$$M_i^{\alpha\beta} \neq (mod\, \zeta) \tag{18.6}$$

If $M_i^{\alpha\beta} \neq (mod\, \zeta)$ decrypting with ζ is not possible in the decryption side. Since n is one of the factor of ζ and $M_i^{\alpha\beta} \equiv (mod\, n)$, n can be used for decryption.

18.4 Proposed System

This section deals with the methodology using of Enhanced RSA Cryptosystem (RSA-ζ), which has three parts: Key generation, Encryption and Decryption. Figure 18.1 displays the diagram for the proposed algorithm.

Key generation

Select two Mersenne prime numbers u, and v and compute $n = u * v$. Using Euler Totient function compute $\varnothing(n)$ using,

$$\varnothing(n) = (u - 1) * (v - 1) \tag{18.7}$$

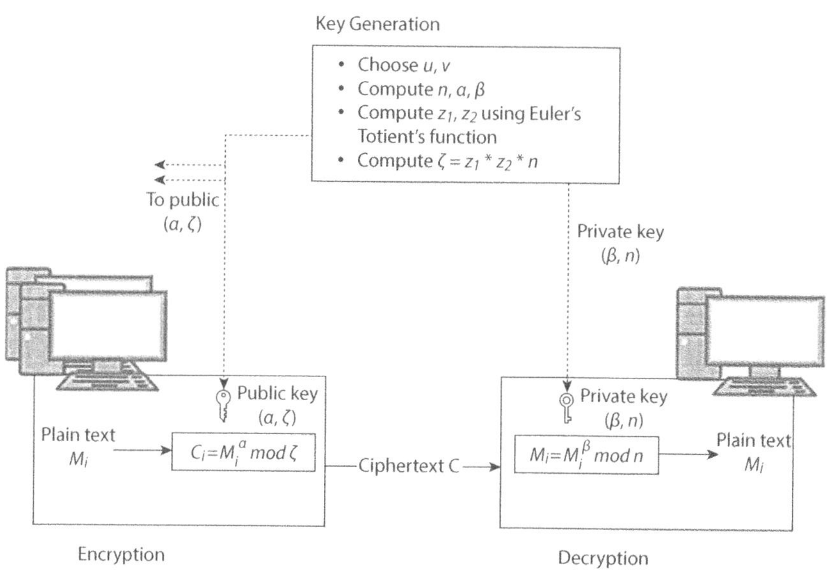

Figure 18.1 Block diagram for RSA-ζ.

Compute public key exponent α using

$$\gcd(\varnothing(n), \alpha) = 1 \tag{18.8}$$

Private-key component β can be calculated using

$$\alpha * \beta \bmod \varnothing(n) = 1 \tag{18.9}$$

Let z_i be the arbitrary integer, where i ranges from 0 *to* n, calculate z_1, z_2 using expression,

$$z_i = (* \beta)\%\varnothing(Z) = 1 \tag{18.10}$$

where $\varnothing(Z)$ is computed using Euler's Totient function Compute fake modulus ζ using,

$$\zeta = z_1 * z_2 * n \tag{18.11}$$

Share fake modulus ζ instead n as the public-key along with the public-key exponent α and keep the private-key exponent β and n as secret key.

Encryption
Let M_i be the plain text, using public key exponent α and fake modulus ζ encrypt the ciphertext, C_i using

$$C_i = M_i^{\alpha} \bmod \zeta \tag{18.12}$$

Decryption
The ciphertext is deciphered using private key exponent β and n as,

$$M_i = C_i^{\beta} \bmod n \tag{18.13}$$

Example:
Let the 2 prime numbers be $u = 283$, $v = 263$ and computed modulus $n = u * v = 74429$. Using equation (18.7) obtain $\varnothing(n) = 73884$. Select $\alpha = 41$ such that (41, 73884) = 1. Using α and equation (18.9) obtained $\beta = 41^{-1}$ *mod* 73884 = 36041.

By equation (18.10) the values for z_1, *and* z_2 are computed as, $z_1 = 14672$, and $z_2 = 11790$. Fake modulus, $\zeta = 12874942775520$ is computed using equation (18.11).

During encryption, sender selects plaintext $M = 65$. Sender generates cipher text using equation (18.12) as, $C_i = 65^{41}$ *mod* $12874942775520 = 7000164382145$.

In the receiving end, original message M_i can be obtained using equation (18.13) as, $M_i = 7000164382145^{36041}$ *mod* $74429 = 65$.

18.5 Performance Analysis

i. Differential Analysis

Differential analysis is a metric used to check the cipher resistance. Generally, when slight changes are made to the original image by an, say changes a single bit, noticing the variation in the cipher image the relation between the original and cipher image could be located [18].

The prerequisite for the image encryption scheme is that the encrypted form of the image should be significantly different from its true form. The disparity is calculated using the Number of Pixel Change RATE (NPCR) and the Unified Average Pixel change Intensity (UACI). The proposed cryptosystem will guarantee two entirely separate ciphered images of a given image, although there is only one bit of difference between them. The NPCR focuses on the total number of pixels that affect the value of differential attacks. The impact due to change in pixel is evaluated using NPCR as given in (18.14)

$$NCPR = \left(\frac{1}{W_i H_i} \sum_{i,=1}^{n,m} (i, j) \right) \times 100 \tag{18.14}$$

with $(i, j) = 1$ *if* $C_1(i, j) \neq C_2(i, j)$ *and* $D(i, j) = 0$ *if* $C_1(i, j) = C_2(i, j)$

Where W_i and H_i respectively indicate image width and height. $C_1(i, j)$ and $C_2(i, j)$ are the ciphered images before and after the one pixel change in the plain image. For the pixel at position (i, j) calculation were made *if* $C_1(i, j) \neq C_2(i, j)$, then set $D(i, j) = 1$ else set $(i, j) = 0$.

Main focus of UACI is the average difference between two paired ciphertext images. The evaluation of the impact on the pixel change on the encrypted image using UACI is as follows:

$$UACI = \left(\frac{1}{L_i} \sum_{i,=1}^{n,m} \frac{|C_1(i,j) - C_2(i,j)|}{255} \right) \times 100 \qquad (18.15)$$

Where *Li* is the length of the image with total number of pixels. $C_1(i, j)$ and $C_2(i, j)$ are the ciphered images before and after the change in one pixel of the key used.

Table 18.1 displays the findings calculated using NPCR and UACI equations. From the analysis it is found that the NPCR values exceed 99% and the UACI values exceed 33% for the proposed method, indicating that a minor change in the plaintext can affect the encryption scheme. Also the values measured are within the confidence range band specified as 98 to 99%. The NPCR numbers are comparable with the RSA scores whereas the UACI scores are considerably higher than the scores obtained using RSA. These results provide the proof that the proposed cryptosystem is immune to known differential attacks.

ii. Time Complexity

The time complexity of proposed RSA-ζ is illustrated bellow based on theoretical means using asymptotic notations is shown in Table 18.2.

To calculate modulus $n = u * v$, required time complexity is (n^2) since n uses multiplication operation on 2 integers. To find $\emptyset(n)$ using Euler totient function it time complexity of $O(n^2)$. Worst case complexity of Euclidean GCD algorithm is ( ). Hence $\gcd(\emptyset(n), \alpha) = 1$ requires the complexity $M * O(n\ )$, where M is the number of iteration of tests to meet the condition.. To compute $\alpha * \beta \bmod \emptyset(n) = 1$ it needs $M * O(2n^2)$, where M is the number of iteration of tests to meet the condition and in each iteration. Complexity to compute z_i using $(\ \alpha * \ \beta)\%\emptyset(z_i) = 1$ is $* O(n^{2\frac{1}{2}})$, where M is the number of iteration of tests to meet the condition to get *zi*.

Table 18.1 Comparison of NPCR and UACI values between plaintext and cipher of RSA and RSA-ζ.

	NPCR			UACI		
	RED	**GREEN**	**BLUE**	**RED**	**GREEN**	**BLUE**
RSA	99.37848	99.20747	99.34066	33.95049	33.80078	33.87705
RSA-ζ	99.58115	99.62158	99.62616	33.00484	33.02288	33.04785

Table 18.2 Time complexity involved in each steps in RSA-ζ.

Notation	Equation	Complexity
n	$n = u * v$	$O(n^2)$
$\emptyset(n)$	$\emptyset(n) = (u - 1) * (v - 1)$	$O(n^2)$
a	$\gcd(\emptyset(n), \alpha) = 1$	$M * O(n)$
Q	$\alpha * \beta \bmod \emptyset(n) = 1$	$M * O(2n^2)$
z_i	$(\alpha * \beta)\%\emptyset(z_i) = 1$	$M * O(n^{2\frac{1}{2}})$
Z	$\zeta = n * z_1 * z_2$	$O(n^3)$
C_i	$C_i = M_i^{\alpha} \bmod \zeta$	$O(n^2 \log_2 n)$
M_i	$M_i = C_i^{\beta} \bmod n$	$O(n^2 \log_2 n)$

Condition contains Euler totient function, hence its complexity is $\sqrt{n}$ in number of steps and in each step contain division (n^2) operation.

To find the fake modulus use the equation (18.11) and the complexity involves three multiplication operation (n^3).

Encryption uses the complexity $(n^2 \log_2 n)$ for the equation (18.12) which is used to encryption. Here modulus exponential function gives the result in $\log_2 n$ steps and (n^2) is the complexity used for multiplication function which is used as a major function used in each steps.

During decryption process, uses the complexity $(n^2 \log_2 n)$ for the equation (18.13). Here modulus exponential function gives the result in $\log_2 n$ steps and (n^2) is the complexity used for multiplication function which is used as a major function used in each steps.

The above results indicate that there is no increase of time during encryption and decryption, even after introducing the concept of fake modulus principle. During the key generation process only the value of ζ is computed which requires the time complexity of (n^3).

iii. Statistical Attacks

Histogram analysis: Histogram analysis refers to the pixel intensity distribution of the image, where each pixel has 256 occurrence values [32, 33]. Experiment is carried out and tested on Lenna image. Figure 18.2(a), (b), (c) and (d) depicts the plain-image of 'Lena' and the histogram for its RGB components respectively.

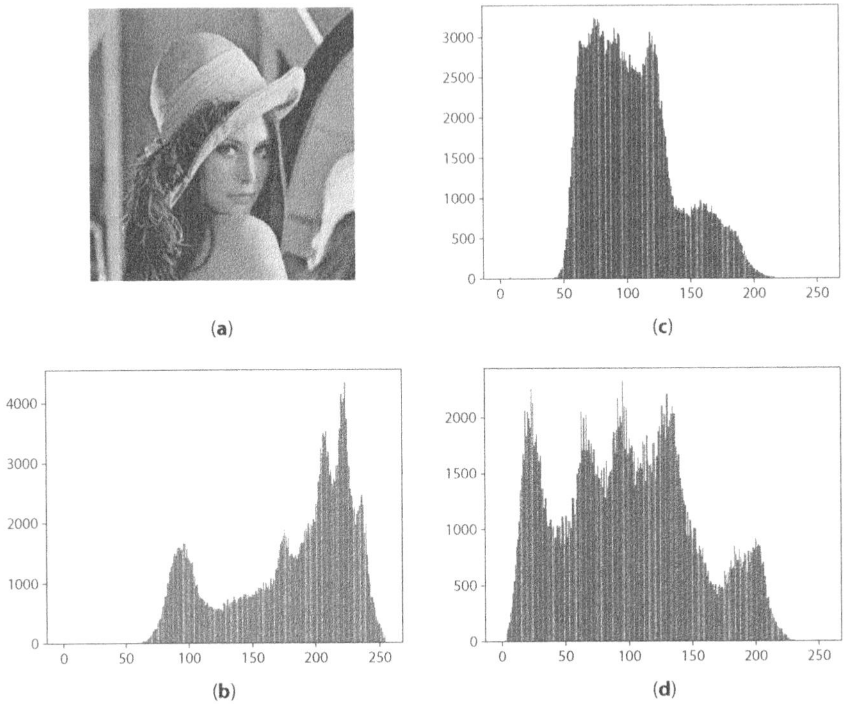

Figure 18.2 (a) Test Lena image. (b) R component. (c) B component. (d) G component.

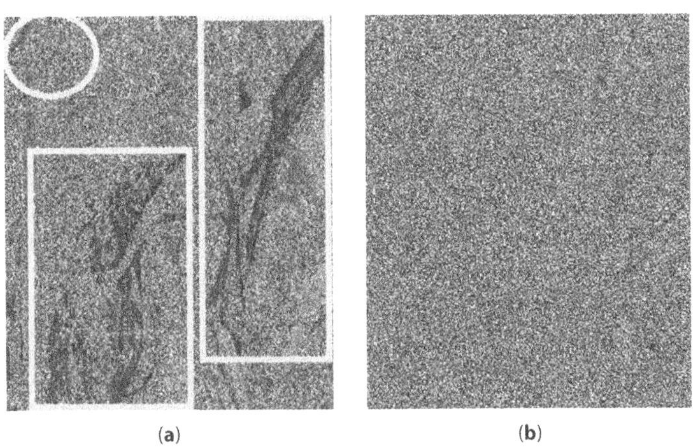

Figure 18.3 (a) Encrypted histogram using RSA. (b) Encrypted using RSA-ζ.

Figure 18.3(a), (b) indicates encrypted image of 'Lena' using RSA and RSA-ζ. From the visual observations given below, traces of image can be observed in the marked areas Standard RSA whereas in case of proposed method no trace of original image can be found.

Figure 18.4(a), 18.5(a) and 18.6(a) indicates RGB components of the encrypted image of Figure 18.2(a) using RSA and Figure 18.4(b), 18.5(b) and 18.6(b) indicates RGB components of the encrypted image of Figure 18.2(a) using RSA-ζ. The histograms obtained for RSA technique have a greater number of peaks compared to proposed method.

It is shown that the encryption using RSA-ζ is analogous to noise obtained by flat and reasonably consistent histograms. The histograms displayed above are significant enough to suggest that the new approach ensures cryptographically secure pixel distribution than the encrypted pixels obtained using RSA technique. Hence the observations indicate that information gathering is not simple through statistical attacks.

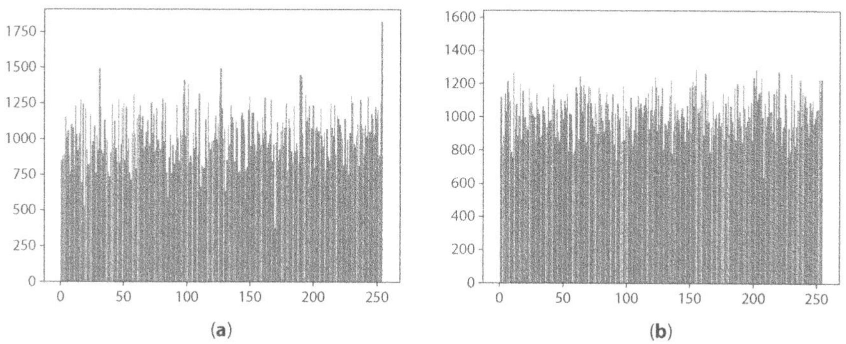

Figure 18.4 (a) R component using RSA. (b) R component using RSA-ζ.

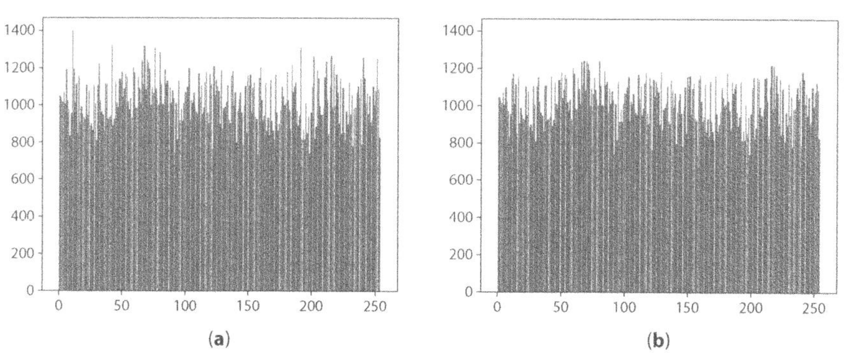

Figure 18.5 (a) G component using RSA. (b) G component using RSA-ζ.

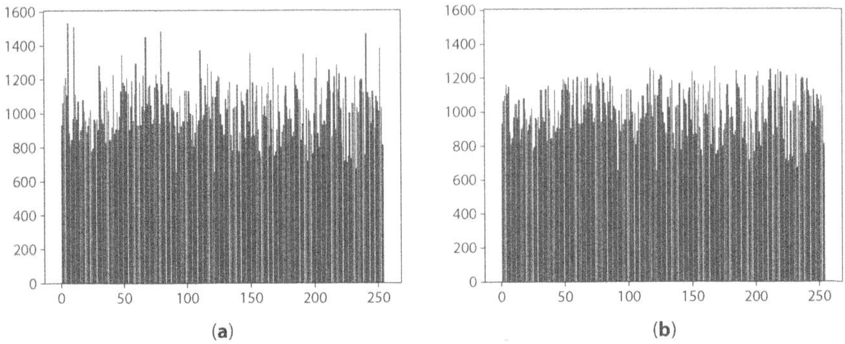

Figure 18.6 (a) B component using RSA. (b) B component using RSA-ζ.

Entropy Analysis: The entropy is defined as the degree of uncertainty in the system. Higher the entropy, higher the randomness of the image [33]. The description of entropy can be defined mathematically using (18.14).

$$Entropy_t = (PC)^1 - \sum_{i=1}^{t} -(i)\log_2 \left(p(i)\right) \tag{18.16}$$

Table 18.3 shows the results of the information entropy of RGB components of the proposed RSA-ζ algorithm, compared with the RSA algorithm. It is evident from the results that the entropy values of the proposed encrypted images (RSA-ζ) are higher(close to 8) indicating that the signal has immunity to attack, and is safe against statistical entropy attack.

iv. Mean and Standard Distribution

In the proposed method, the cipher images are uniformly distributed because of the improvement in pixel values of plain-color images. The

Table 18.3 Comparison of information entropy of images.

Methods	Entropy of RGB components		
	Red	**Green**	**Blue**
RSA	7.962641	7.96832	7.956098
RSA-ζ	7.96544	7.972283	7.961711

Table 18.4 Mean and standard deviations of RSA and RSA-ζ.

	Mean or variance(v)			Standard deviations		
	Red	**Green**	**Blue**	**Red**	**Green**	**Blue**
Original (Lenna) image	180.2237	99.05122	105.4103	49.04877	52.87752	34.05792
RSA	126.5782	125.2746	126.4324	73.6	73.21156	72.97806
RSA-ζ	129.0942	126.4333	126.9286	73.70313	73.98288	73.77808

mean values are computed for plain image and cipher images and are shown in Table 18.4 to analyze the pixels uniform distribution.

Difference between encrypted values and the mean value is shown using Standard deviation.

$$\textit{Standard Deviation} = \frac{\sqrt{\sum_{i=1}^{n} (C_i - v)^2}}{N} \tag{18.17}$$

Where C_i represents the cipher text values for the corresponding plain text value i, v is the mean and N is the number of pixels in a image.

18.6 Conclusion

In this paper, a new method, i.e. RSA-ζ, is introduced to reduce the factorization attacks on RSA. The proposed system is immune to differential attacks and very efficient as seen by the results of NPCR and UACI analysis. Through histogram analysis it is proven that more secure pixel distribution occurs in the proposed than in the traditional RSA, which implies harder access to the message by intruder. Also the entropy analysis report high immunity of the proposed method over statistical attacks. The results show the accuracy and efficiency of RSA-ζ. Concluding, the proposed system is robust and has higher immunity towards attacks than the traditional system.

References

1. Mohammed, A. and Varol, N., A review paper on cryptography, 1–6, 2019, 10.1109/ISDFS.2019.8757514.
2. Dodmane, R., Aithal, G., Shetty, S., A new approach on advanced encryption standards to improve the secrecy and speed using nonlinear output feedback mode, in: *Advances in Artificial Intelligence and Data Engineering. Advances in Intelligent Systems and Computing*, vol. 1133, N. Chiplunkar and T. Fukao (Eds.), Springer, Singapore, 2021, https://doi.org/10.1007/978-981-15-3514-7_61 2.
3. Rao, R., Ganesh, A., Surendra, S., Kallapu, B., Key generation using generalized Pell's equation in public key cryptography based on the prime fake modulus principle to image encryption and its security analysis. *Cybern. Inf. Technol.*, 20, 86–101, 2020, 10.2478/cait-2020-0030.
4. Raghunandhan, K.R., Shetty, S., Aithal, G., Rakshith, N., Enhanced RSA algorithm using fake modulus and fake public key exponent, in: *2018 International Conference on Electrical, Electronics, Communication, Computer, and Optimization Techniques (ICEECCOT)*, Msyuru, India, pp. 755–759, 2018.
5. Hegde, N., Deepthi, P., P.D., Pollard RHO algorithm for integer factorization and discrete logarithm problem. *Int. J. Comput. Appl.*, 121, 14–17, 2015, 10.5120/21639-4969.
6. Zralek, B., A deterministic version of Pollard's p-1 algorithm. *Math. Comput.*, 79, 513–533, 2007, 10.1090/S0025-5718-09-02262-5.
7. Raghunandan, K. R., Dodmane, R., Bhavya, K. Rao, N. S. K., Sahu, A. K., Chaotic-map based encryption for 3D point and 3D mesh fog data in edge computing," in *IEEE Access*, 11, 3545–3554, 2023, doi: 10.1109/ACCESS.2022.3232461.
8. Raghunandan, K.R., Shetty, R., Aithal, G., Key generation and security analysis of text cryptography using cubic power of Pell's equation, in: *2017 International Conference on Intelligent Computing, Instrumentation and Control Technologies (ICICICT)*, Kannur, pp. 1496–1500, 2017.
9. Zralek, B., A deterministic version of Pollard's p-1 algorithm. *Math. Comput.*, 79, 2007, 10.1090/S0025-5718-09-02262-5.
10. Gupta, S. and Sharma, J., A hybrid encryption algorithm based on RSA and Diffie-Hellman. *2012 IEEE International Conference on Computational Intelligence and Computing Research*, Coimbatore, pp. 1–4, 2012, doi: 10.1109/ICCIC.2012.6510190.
11. Patidar, R. and Bhartiya, R., Modified RSA cryptosystem based on offline storage and prime number. *2013 IEEE International Conference on Computational Intelligence and Computing Research*, Enathi, pp. 1–6, 2013, doi: 10.1109/ICCIC.2013.6724176.
12. Minni, R., Sultania, K., Mishra, S., Vincent, D.R., An algorithm to enhance security in RSA. *2013 Fourth International Conference on Computing, Communications and Networking Technologies (ICCCNT)*, Tiruchengode, pp. 1–4, 2013, doi: 10.1109/ICCCNT.2013.6726517.

13. Jaju, S.A. and Chowhan, S.S., A modified RSA algorithm to enhance security for digital signature. *2015 International Conference and Workshop on Computing and Communication (IEMCON)*, pp. 1–5, Vancouver, BC, 2015, doi: 10.1109/IEMCON.2015.7344493.
14. Iswari, N.M.S., Key generation algorithm design combination of RSA and ElGamal algorithm. *2016 8th International Conference on Information Technology and Electrical Engineering (ICITEE)*, pp. 1–5, Yogyakarta, 2016, doi: 10.1109/ICITEED.2016.7863255.
15. Panda, P.K. and Chattopadhyay, S., A hybrid security algorithm for RSA cryptosystem. *2017 4th International Conference on Advanced Computing and Communication Systems (ICACCS)*, pp. 1–6, Coimbatore, 2017, doi: 10.1109/ICACCS.2017.8014644.
16. Mathur, S., Gupta, D., Goar, V., Kuri, M., Analysis and design of enhanced RSA algorithm to improve the security. *2017 3rd International Conference on Computational Intelligence & Communication Technology (CICT)*, pp. 1–5, Ghaziabad, 2017, doi: 10.1109/CIACT.2017.7977330.
17. Manu, and Goel, A., Encryption algorithm using dual modulus. *2017 3rd International Conference on Computational Intelligence & Communication Technology (CICT)*, pp. 1–4, Ghaziabad, 2017, doi: 10.1109/CIACT.2017.7977331.
18. Raghunandan, K.R., Aithal, G., Shetty, S., Bhavya, K., Image encryption scheme in public key cryptography based on cubic Pells quadratic case. *Indones. J. Electr. Eng. Comput. Sci.*, 20, 1, 385–394, October 2020, DOI: 10.11591/ijeecs.v20.i1.pp385-394.
19. Islam, M.A., Islam, Md.A., Islam, N., Shabnam, B., A modified and secured RSA public key cryptosystem based on "n" prime numbers. *J. Comput. Commun.*, 6, 78–90, 2018.
20. Raghunandhan, K.R., Shetty, S., Aithal, G., Rakshith, N., Enhanced RSA algorithm using fake modulus and fake public key exponent. *2018 International Conference on Electrical, Electronics, Communication, Computer, and Optimization Techniques (ICEECCOT)*, pp. 755–759, Msyuru, India, 2018, doi: 10.1109/ICEECCOT43722.2018.9001351.
21. K.R., R., Aithal, G., Shetty, S., Comparative analysis of encryption and decryption techniques using mersenne prime numbers and phony modulus to avoid factorization attack of RSA. *2019 International Conference on Advanced Mechatronic Systems (ICAMechS)*, pp. 152–157, Kusatsu, Shiga, Japan, 2019, doi: 10.1109/ICAMechS.2019.8861599.
22. Raghunandan, K.R., Dsouza, R.R., Rakshith, N., Shetty, S., Aithal, G., Analysis of an enhanced dual RSA algorithm using pell's equation to hide public key exponent and a fake modulus to avoid factorization attack, in: *Advances in Artificial Intelligence and Data Engineering. Advances in Intelligent Systems and Computing*, vol. 1133, N. Chiplunkar and T. Fukao (Eds.), Springer, Singapore, 2021, https://doi.org/10.1007/978- 981-15-3514-7_60.

23. Raghunandan, K.R., Aithal, G., Shetty, S., Secure RSA variant system to avoid factorization attack using phony modules and phony public key exponent. *Int. J. Innov. Technol. Exploring Eng. (IJITEE)*, 8, 9, 1065–1070, July 2019, ISSN: 2278-3075.
24. Raghunandan, K.R., Dsouza, R.R., Rakshith, N., Shetty, S., Aithal, G., Analysis of an enhanced dual RSA algorithm using pell's equation to hide public key exponent and a fake modulus to avoid factorization attack, in: *Advances in Artificial Intelligence and Data Engineering. Advances in Intelligent Systems and Computing*, vol. 1133, N. Chiplunkar and T. Fukao (Eds.), Springer, Singapore, 2020, doi.org/10.1007/978-981-15-3514- 7_60.
25. Dodmane, R., Aithal, G., Shetty, S., Algorithm for clustering the moduli of RNS for the application of optimization of time complexity in standard cipher system. *Int. J. Innov. Technol. Exploring Eng. (IJITEE)*, 9, 7, 92–97, May 2020, ISSN: 2278-3075.
26. Dodmane, R., Aithal, G., Shetty, S., Time complexity reduction for the application of stream cipher system based XOR free operation. *Int. J. Recent Technol. Eng.*, 8, 3, 5402–5408, September 2019.
27. Bhat, K., Mahto, D., Yadav, D., Vantages of adaptive multidimensional playfair cipher over AES-256 and RSA-2048, *International Journal of Advanced Research in Computer Science*, 8, 5, 201–217, May – June 2017.
28. Bhat, K., Mahto, D., Yadav, D., Comparison analysis of AES-256, RSA-2048 and four dimensional playfair cipher fused with linear feedback shift register, *International Journal of Advanced Research in Computer Science*, 8, 5, May – June 2017.
29. Zhang, H. and Takagi, T., Attacks on multi-prime RSA with small prime difference, in: *Information Security and Privacy. ACISP 2013. Lecture Notes in Computer Science*, vol. 7959, C. Boyd and L. Simpson (Eds.), Springer, Berlin, Heidelberg, 2013, DOI: https://doi.org/10.1007/978-3-642-39059-3_4.
30. Maitra, S. and Sarkar, S., Revisiting wiener's attack – new weak keys in RSA, in: *Information Security. ISC 2008. Lecture Notes in Computer Science*, vol. 5222, T.C. Wu, C.L. Lei, V. Rijmen, D.T. Lee (Eds.), Springer, Berlin, Heidelberg, 2008.
31. Burton, D.M. *Elementary number theory. 6th edition*, Tata McGraw-Hill Publishing Company Limited, New Delhi, 2007.
32. Raghunandan, K.R., Nireshwalya, S.N., Sudhir, S., Bhat, M.S., Tanvi, H.M., Securing media information using hybrid transposition using fisher yates algorithm and RSA public key algorithm using pell's cubic equation, in: *Advances in Artificial Intelligence and Data Engineering. Advances in Intelligent Systems and Computing*, vol. 1133, N. Chiplunkar and T. Fukao (Eds.), Springer, Singapore, 2020, doi.org/10.1007/978-981-15-3514- 7_73.
33. Raghunandhan, K.R., Radhakrishna, D., Sudeepa, K.B., Ganesh, A., Efficient audio encryption algorithm for online applications using transposition and multiplicative non-binary system. *Int. J. Eng. Res. Technol.*, 2, 6, 472–477, 2013.

19

Sailfish Optimizer Algorithm (SFO) for Optimized Clustering in Internet of Things (IoT) Related to the Healthcare Industry

Battina Srinuvasu Kumar[1,2*], S.G. Santhi[1] and S. Narayana[3]

[1]*Department of Computer Science & Engineering, Annamalai University, Chithambaram, Tamil Nadu, India*
[2]*Department of Information Technology, Gudlavalleru Engineering College, Gudlavalleru, Andhra Pradesh, India*
[3]*Department of Computer Science & Engineering, Gudlavalleru Engineering College, Gudlavalleru, Andhra Pradesh, India*

Abstract

The COVID-19 pandemic poses a serious threat to humanity because of how deadly it is and how it spreads. Internet of Things (IoT) devices have been reimagined to play a vital part in data collection from covid-19 patients since it is challenging to acquire data from affected individuals in real time and because there are not enough medical experts. Furthermore, collecting and monitoring data offer substantial issues due to redundant data collection and insufficient data transport. The algorithms are flexible and straightforward, and they heavily use optimization techniques to adapt to diverse data collection and transmission issues. These methods work well for optimization issues without structural modifications. These algorithms work well for optimization issues without structural modifications. This work introduces the Sail Fish Optimizer (SFO), a nature-inspired optimization technique that was prompted by the sailfish group. A dangerous issue on the Internet of Things (IoT) is the monetary custom of energy. Increasing network longevity and lowering node power consumption can both be accomplished by network clustering. Many meta-heuristic-based cluster head (CH) selection strategies have been put forth to take its place. Due to early convergence from the conventional technique, the SFO in the IoT was developed to prevent this problem with locating the CH. The suggested SFO-based energy-efficient algorithm is used to find the CHs optimal state. The effects of energy usage, network durability, and throughput are simulated.

**Corresponding author*: bskgec@gmail.com

Chandra Singh, Rathishchandra R Gatti, K.V.S.S.S.S. Sairam and Ashish Singh (eds.) Modeling and Optimization of Optical Communication Networks, (349–364) © 2023 Scrivener Publishing LLC

Keywords: Sailfish optimizer, wireless sensor network, cluster head selection, energy efficiency

19.1 Introduction

The Internet of Things (IoT) is a network of sensor nodes that are powered by batteries and installed in specific locations. For example, in one study, IoT sensors were installed inside people as part of the healthcare sector. These Internet of Things sensor nodes are in charge of gathering, grouping, and delivering the data to the drain. The collection of data from a patient may be viewed as an unavailable entity with limited energy capacity, limited on-board processing and computing, as well as the hierarchical nature of IoTs. These are just a few of the challenges these networks face. There are several energy-conscious routing strategies for these problems in the literature [1–3].

IoT can be categorized into two groups based on deployment: organized and unorganized. A huge number of sensor nodes can be supported by an unstructured IoT network. Network administration is difficult in an unstructured network due to the considerable processing and fault detection barriers that a big IoT network provides. IoT nodes make up a small portion of the hierarchical network, which is easy to maintain because its use is preplanned. IoTs require energy efficiency services since the remaining energy decides how long IoT nodes may operate. Conventional routing protocols [4, 5] are unable to maintain the minimal consumption of power among the IoT nodes.

This means that nodes farther distant from the sink – like patients – will die sooner when information is transmitted straight from the sensors to the sink. Take the quickest route possible in MTE nodes where fees are determined by the quantity of electricity used for transmission. The conventional protocol from the literature was advised for this issue. The nodes can be grouped into clusters in order to improve network performance and conserve resources, with each cluster electing a cluster leader. Participating in the cluster are the remaining IoT nodes. There only has to be one cluster applied for each sensor node. Each cluster member's information is gathered by the cluster head, combined, and sent to the sink using hop-based communication. If the energy level across all nodes is the same, then the network is called as organized network in healthcare field. In comparison, a heterogeneous network is named if any percentage of the IoT nodes produces extra energy than other nodes.

A multi-objective formulation is developed in this study to enhance energy balancing between IoT nodes in the healthcare industry and path formations between the cluster head (CH) IoT nodes and gateway [25, 28]. Objectives including proximity, connectivity costs, residual energy, and coverage are taken into account for node clustering and CH collecting. To determine the best data transfer path from CH to gateway, the evolutionary approach SFO is employed [6]. Advantages of SFO involve the abundance of hardware, software and efficiency options to avoid local optimum conditions and rapid convergence. The resulting results are compared to different approaches: salp swarm algorithm [7], slime mould algorithm [8], Harris hawk optimizer [9, 10], sine cosine algorithm [11–13], gravitational search algorithm [14], grey wolf optimizer [15–18], whale optimization algorithm [19–21] and Multi-verse optimizer [22, 23].

IoT is crucial in the healthcare industry to reduce energy consumption during data transport. There are various methods for resolving this problem, each of which focuses primarily on one energy metric while paying little attention to other relevant parameters like service effectiveness, reach, connectivity, etc. Some of the protocols involve finding the best routes while also dealing with the difficulties of route selection. As a result, improving clustering, CH selection, and routing in a wireless network of sensors both a challenge and an inspiration.

The suggested strategy employs a multi-objective optimization technique and the Sail Fish Optimizer (SFO) algorithm to provide a solution to the issue of lengthening the lifetime of IoT.

19.2 Related Works

A clustering heuristic based on the Ant Lion optimization (ALO) algorithm of Ant Lion proposed by Yogarajan *et al.* [24]. The fitness function is modeled on the ALO algorithm for CH selection to improve the lifespan of sensor networks. An ALO algorithm is used to obtain the optimum data collection at sink. A discrete ALO algorithm is often used to compute the possible probability for the gateway to meet the allocated CH nodes.

The solution to the energy problems in IoTs around gateway due to heavy traffic conditions was given by Aziz *et al.* [26]. The authors used an energy consumption algorithm for Grey Wolf Optimization (GWO). The GWO is used or routing and clustering fitness functions. The exercise feature is dependent on two goals, i.e. total distance crossing and hop distance reduction. The clustered fitness function assigns the full load for CH-to-gateway distances, and is an ideal solution for clustering within the population.

Sheik *et al.* [27] suggested an evolutionary task plan strategy based upon the MOEA, i.e. the Multi-Objective Evolutions Algorithm, to identify

Pareto's optimum solutions, maximizing resources, temperatures and efficiency. It uses problem-centered encoding solutions to evaluate the initial population of space and genetic solutions that work together to provide fast turnaround solutions.

Esmaeili *et al.* [29] suggested a clustering approach based on a Genetic Algorithm (GA) to increase network longevity. The GA-centered self-organization network clustering approach is used to automatically improve IoT clusters. Through neighboring nodes, anticipated energy use, distance from a gateway, and leftover energy considerations, a complex network structure is created. Each cluster in which a node achieves this balance is referred to as a CH for the duration of that cluster's optimal configuration. Additional clusters are constructed in areas where the sensor field is far from the gateway to maintain optimal energy levels and extend longevity. Elhabyan *et al.* [30] have created a two-tier directed clustering and protocol for routing Particle Swarm Optimization (PSO). The problem is expressed by the formulation of linear programming (LP). In the clustering stage, the collection of energy-efficient CHs is chosen. The particle encoding and fitness function is used to identify a feasible routing tree connecting chosen CHs to gateway.

Using an RFID reader and sensor, Abuelkhail *et al.* [31] created a smart node network that closely monitors channel congestion and further aids in minimizing network interferences. It is deployed by identifying the CH between nearby IoT nodes, and following the selection of the CH, a cluster is formed. In line with this theory, the current study applied a meta-heuristic optimization model to CH creation and clustering in the healthcare industry.

19.3 Proposed Method

SFO comprises grouping IoT nodes, choosing a CH, and routing for the best data transmission in the healthcare sector, where it is crucial to optimize. Near the sink, the congestion of traffic is the main problem. Network configuration, energy modelling, node clustering, and route recognition were the four steps that were created. The creation and deployment of the sensor network are described by the networking procedure. The energy model analyses energy use and transmits nodes. How the nodes are clustered and network clusters are made is demonstrated by the clustering and subsequent selection of CH. During the route recognition stage, the SFO algorithm is used to determine the best route for data transmission.

19.4 System Model

In this section, the study considers a two dimensional network model with IoT nodes considering the assumptions given below:

In this section, the analysis takes into account a 2D network model taking into account the following assumptions:

- All the IoT nodes are fixed and it is placed in human body for data acquisition from covid-19 patients.
- The 2D network has a gateway in which the nodes are intended for data collection and transmission.
- The nodes of same communication capacity with same initial energy are deployed.
- Arbitrarily deployed IoT nodes with their x and y coordinates are often situated over the patients' body.
- Distance between the sensors fit in the walls is determined by Euclidean distance from two neighboring IoT nodes.

19.5 Energy Model

The analysis uses a radio model of the first order to simulate the transmission and reception of signals by an IoT node in the healthcare industry.

Homogeneous Internet of Things nodes are thinking about an energy dissipation module for extensive radio transmission. The goal of this model is to calculate the nodes' transmission rate. Energy is calculated as follows when sending an n-bit data packet over a distance of d:

$$E_{tx}(d) = n_{tx} r_{tx} \phi_{amp} d^{\alpha} + nr\phi_{cir}$$

where,

ϕ_{cir} is the Energy dissipation at transmitter nodes and

ϕ_{amp} is the distance of transmitter nodes from the receiving nodes.

α is the path loss component.

The energy consumption at receiver occurs at the rate r_{rx} that entirely depends on the following expression:

$$E_{rx} = n_{rx} r_{rx} \phi_{cir}$$

For a hop based distance estimation, the intermediate node *i* consumes an energy E_i for relaying at an optimal distance *d* and it is defined as below:

$$E_i = E_{tx} + E_{rx} = 2\left(n_{tx} r_{tx} \varphi_{amp} d^{\alpha} + n_{rx} r_{rx} \varphi_{cir}\right)$$

19.6 Cluster Formation Using SFO

Node clustering is conducted out using a different methodology than node clustering. In the IoT healthcare industry, the uneven approach to clustering provides numerous advantages, including a longer life cycle, scalability, and load balancing. It reduces intracluster traffic in the gateway zone and prevents overloading of nodes close to the sink node. Additionally, the generated cluster will be rebuilt with the suggested solution where a node in the network is failing due to energy loss or death. In healthcare IoTs, the gateway transmits the beacon alert across the network in irregular clusters based on the location of the node. After receiving the message from the gateway, healthcare IoTs capture their position data using angle-based positioning algorithms. Gateway also determines the upper network area evaluating the use of SNs before and after clustering. The selection method for clustering and CH is as follows:

CH IoT Node Selection

To collect CH, a probabilistic mechanism is employed. Based on the multiple objectives of proxy, connectivity costs, residual energy, and coverage, the cluster chooses the best CH node. Network nodes need resources for operations including data storage, transport, and receipt. CH nodes will require more resources than other nodes, such as transmissions, data receiving from many IoT nodes, and data aggregation of acquired data. Additionally, these nodes require more resources to complete these duties. In this case, an efficient CH selection approach is necessary.

The multi-objective functions are used for the selection of CHs as below and it includes the following parameters

- Neighboring nodes proximity Euclidean distance (O_D)
- Communicating cost with neighbor node (O_C)
- Residual node energy (O_R)
- Node coverage (O_N)

$$MOO = W_1 O_D + W_2 O_C + W_3 O_R + W_4 O_N$$

$$W_1 + W_2 + W_3 + W_4 = 1$$

where,

$$O_D = \frac{1}{N_T} \sum_{i=1}^{N} N_{prox}(N_i)$$

Where

N_{prox} is the Neighboring nodes proximity
N_T is the number of nodes

$$O_D = \frac{1}{N_T} \sum_{i=1}^{N} C_{com}(N_i)$$

Where

C_{com} is the communication cost with the neighboring node.

$$O_R = E_T - \left(E_{Tx} + E_{Rx} + E_{col} + E_{agg}\right)$$

Where

E_{tx} is the transmission energy
E_{rx} is the receiver energy
E_{col} is the data collection energy
E_{agg} is the aggregation energy

$$O_N = \frac{1}{N_T} \sum_{i=1}^{N} N_C(N_i)$$

Where

N_C is the node coverage area

SFO is regarded as a population-based meta-heuristic algorithm. The position of the sailfish in the search area is referred to as a vector for problems; the candidate approach is often referred to as sailfish. The population creation in the solution area takes place at random. The search output of sailfish will occur in 1D space depending on the vector location.

Algorithm 1: CH selection

Input: IoT nodes
Output: Best nodes for CH election
Start
For $i = 1: N_T$
    While CH selection do
        For each node i
            Estimate O_D
                If min is O_D
                    Select the node i
                Else
                    Move to next node $i+1$
                End
            Estimate O_C
                If min is O_C
                    Select the node i
                Else
                    Move to next node $i+1$
                End
            Estimate O_R
                If min is O_R
                    Select the node i
                Else
                    Move to next node $i+1$
                End
            Estimate O_N
                If min is O_N
                    Select the node i
                Else
                    Move to next node $i+1$
                End
        End
    Choose Sensor node based on maximum $MOO = W_1O_D + W_2O_C + W_3O_R + W_4O_N$
End

Algorithm 2: SFO Routing

Input: Possible routing paths to gateway
Output: Optimal paths
Initialize CH and gateway locations.
Initialize parameters of SFO

```
Validate CH fitness
While iter = 1: N
        For each CH
              Find throughput
              If high is the network throughput
                     Select the CH healthcare IoT node, which is selected by
                     Algorithm 1 as the CH
              Else
                     Reject the CH healthcare IoT node, which is selected by
                     Algorithm 1 as the CH
              End
              If good is the link quality
                     Select the CH healthcare IoT node, which is selected by
                     Algorithm 1 as the CH
              Else
                     Reject the CH healthcare IoT node, which is selected by
                     Algorithm 1 as the CH
              End
              If high is the node energy
                     Select the CH healthcare IoT node, which is selected by
                     Algorithm 1 as the CH
              Else
                     Reject the CH healthcare IoT node, which is selected by
                     Algorithm 1 as the CH
              End
        End For
        If optimal route is found
              Update path as optimal and forward the data
        Else
              Find other optimal routes
        End
End
```

19.7 Results and Discussion

The simulation is conducted using NS2.34 simulation and the nodes are designed with batter module. The parameters considered for simulation is given in Table 19.1 and the proposed method is compared with four other algorithms that includes: Grey Wolf Optimizer, Whale Optimization Algorithm and Multi-Verse Optimizer. The simulation setup is given in

Table 19.1 Simulation parameters.

Parameter	Value
Size of the cluster	1000 × 1000 m^2
Traffic Type	CBR
Propagation Space	Free Space
Protocol Type	802.11
Data rate (MBPS)	2.4
Propagation limit	-100 dB
Packet size (bit)	5000
Message size (bit)	1000

Figure 19.1, where it shows the development of cluster head using Algorithm 1 and formation of cluster and routes in Figure 19.2.

At first, a response was sent by the high energy nodes. The healthcare IoT nodes set the cluster head according to the highest energy obtained from the four nodes and set the most likely cluster head with maximum energy. The identification of the cluster head is then assigned and set as the target node for the packet transmission, which establishes the forming of the cluster. This approach uses a route failure model to predict the whole scenario and the path is generated by the algorithm chain formation

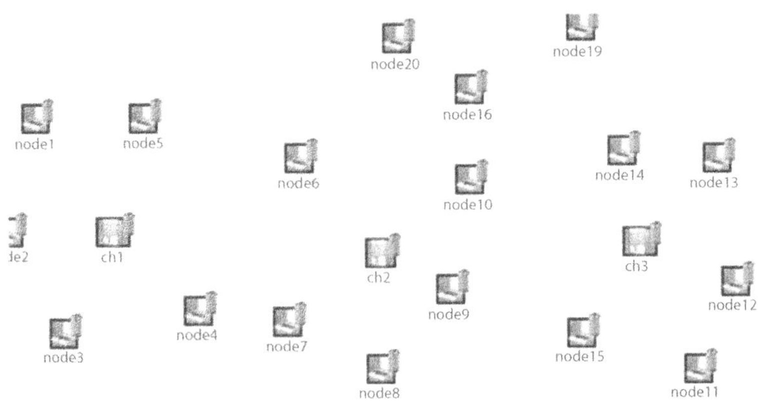

Figure 19.1 Network setup.

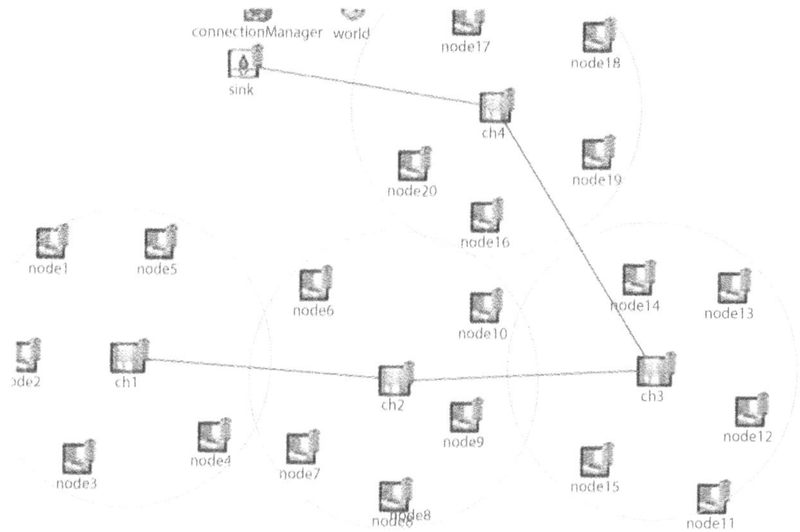

Figure 19.2 Path between CH and gateway.

method but is close to adding hops by the flood system. Table 19.1 shows the simulation parameters for simulating the proposed process.

Figure 19.3 shows the results of latency, where the proposed Sail Fish Optimization obtains reduced latency than Multi-Verse Optimizer, Whale Optimization Algorithm, Grey Wolf Optimizer and Gravitational Search Algorithm. The results of latency tends to get increased with increasing number of nodes, which is due to increased number of packets being transmitted via the healthcare IoT nodes.

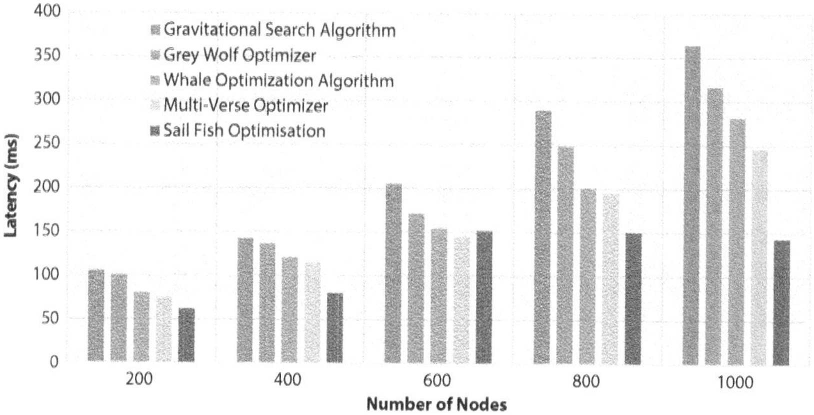

Figure 19.3 Latency.

Figure 19.4 shows the results of network lifetime, where the proposed Sail Fish Optimization obtains increased network lifetime than Multi-Verse Optimizer, Whale Optimization Algorithm, Grey Wolf Optimizer and Gravitational Search Algorithm. The results of lifetime tends to get reduced with increasing number of nodes, which is due to increased number of packets being transmitted via the healthcare IoT nodes.

Figure 19.5 shows the results of network throughput, where the proposed Sail Fish Optimization obtains increased throughput than Multi-Verse Optimizer, Whale Optimization Algorithm, Grey Wolf Optimizer and Gravitational Search Algorithm. The results of throughput tends to get reduced with increasing number of nodes, which is due to increased number of packets being transmitted via the healthcare IoT nodes.

Figure 19.6 shows the results of residual energy, where the proposed Sail Fish Optimization obtains increased residual energy than Multi-Verse Optimizer, Whale Optimization Algorithm, Grey Wolf Optimizer and Gravitational Search Algorithm. The results of residual energy tends to get reduced with increasing number of nodes, which is due to increased number of packets being transmitted via the healthcare IoT nodes.

At first, a response was sent by the high energy nodes. The healthcare IoT nodes set the CH according to the highest energy obtained from the four nodes and set the most likely cluster head with maximum energy. The identification of the cluster head is then assigned and set as the target node for the packet transmission, which establishes the forming of the cluster. This approach uses a route failure model to predict the whole scenario and the path is generated by the algorithm chain formation method but is close to adding hops by the flooding system.

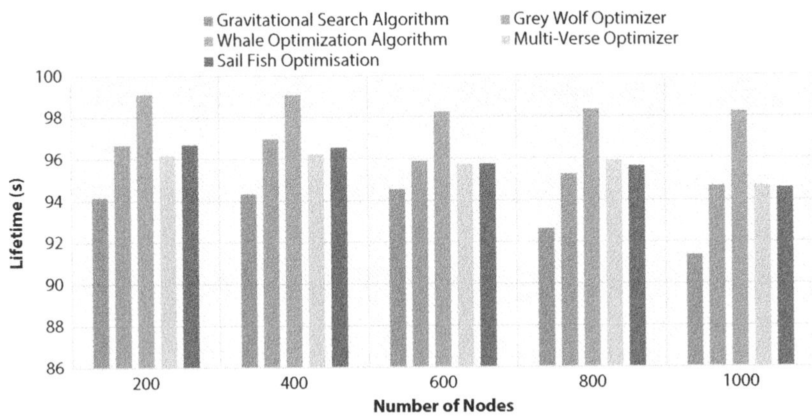

Figure 19.4 Network lifetime.

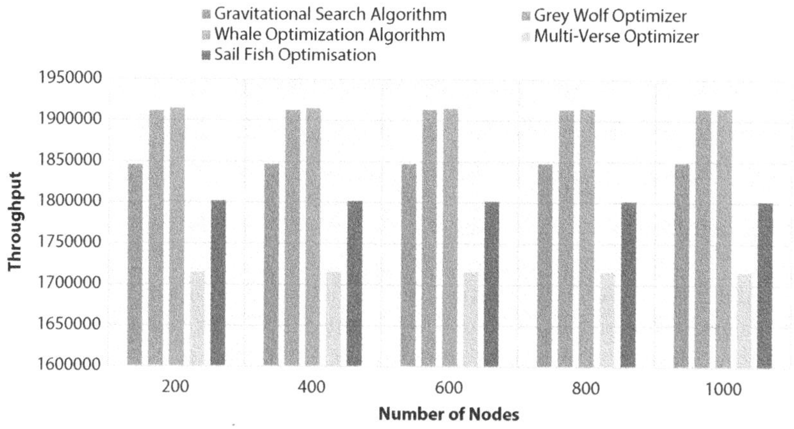

Figure 19.5 Network throughput.

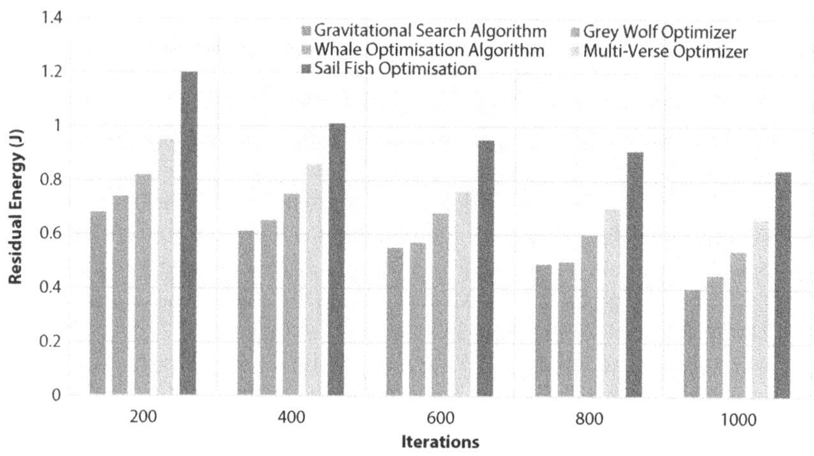

Figure 19.6 Residual energy.

19.8 Conclusions

In this study, we employed SFO, which creates cluster head formation in the Internet of Things before creating stable routing paths. Increasing network longevity and lowering node power consumption can both be accomplished by network clustering. The SFO was suggested in IoT to prevent problems with finding the CH by taking into account many objectives and avoiding the issue with early convergence from the conventional approach. The CHs optimal scenario is found using the proposed

SFO-based energy-efficient algorithm. It is demonstrated through simulations that SFO adheres to better rates of network throughput, less energy usage, and longer network lifetimes. Healthcare can handle the issue of IoT node congestion in the future.

References

1. Abdul-Qawy, A.S.H., Almurisi, N.M.S., Tadisetty, S., Classification of energy saving techniques for IoT-based heterogeneous wireless nodes. *Proc. Comput. Sci.*, 171, 2590–2599, 2020.
2. Jayakumar, H., Raha, A., Kim, Y., Sutar, S., Lee, W.S., Raghunathan, V., Energy-efficient system design for IoT devices, in: *2016 21st Asia and South Pacific Design Automation Conference (ASP-DAC)*, pp. 298–301, IEEE, 2016, January
3. Singh, U. and Chana, I., Enhancing energy efficiency in IoT (Internet of Thing) based application, in: *International Conference on Inventive Computation Technologies*, pp. 161–173, Springer, Cham, 2019, August.
4. Ganesan, M. and Sivakumar, N., An energy efficient IoT based healthcare system based on clustering technique, in: *2019 3rd International conference on Electronics, Communication and Aerospace Technology (ICECA)*, pp. 1395–1399, IEEE, 2019, June.
5. Ghosh, A., Raha, A., Mukherjee, A., Energy-efficient IoT-health monitoring system using approximate computing. *Internet Things*, 9, 100166, 2020.
6. Saba, T., Haseeb, K., Ahmed, I., Rehman, A., Secure and energy-efficient framework using Internet of Medical Things for e-healthcare. *J. Infect. Public Health*, *13*, 10, 1567–1575, 2020.
7. Mirjalili, S., Gandomi, A.H., Zahra, S., Saremi, S., Salp swarm algorithm: A bio-inspired optimizer for engineering design problems. *Adv. Eng. Software*, 114, 163–191, 2017.
8. Li, S., Chen, H., Wang, M., Heidari, A.A., Mirjalili, S., Slime mould algorithm: A new method for stochastic optimization. *Future Gener. Comput. Syst.*, 2, 13, 2020.
9. Heidari, A.A., Mirjalili, S., Faris, H., Aljarah, I., Mafarja, M., Chen, H., Harris hawks optimization: Algorithm and applications. *Future Gener. Comput. Syst.*, 97, 849–872, 2019.
10. Kamboj, V.K., Nandi, A., Bhadoria, A., Sehgal, S., An intensify Harris Hawks optimizer for numerical and engineering optimization problems. *Appl. Soft Comput.*, 89, 106018, 2020.
11. Nenavath, H., Kumar, R., Das, S., A synergy of the sine-cosine algorithm and particle swarm optimizer for improved global optimization and object tracking. *Swarm Evol. Comput.*, 43, 1–30, 2018.

12. Mirjalili, S., SCA: A sine cosine algorithm for solving optimization problems. *Knowl. Based Syst.*, 96, 120–133, 2016.
13. Gupta, S., Deep, K., Mirjalili, S., Hoon, J., A modified sine cosine algorithm with novel transition parameter and mutation operator for global optimization. *Expert Syst. Appl.*, 2020, 113395, 2020.
14. Al-betar, M.A., Awadallah, M.A., Faris, H., Aljarah, I., Hammouri, A.I., Natural selection methods for Grey Wolf Optimizer. *Expert Syst. Appl.*, 113, 481–498, 2018.
15. Sivaram, M., Yuvaraj, D., Mohammed, A.S., Manikandan, V., Porkodi, V., Yuvaraj, N., Improved enhanced Dbtma with contention-aware admission control to improve the network performance in manets. *CMC-Comput. Mater. Contin.*, 60, 2, 435–454, 2019.
16. Yuvaraj, N., Karthikeyan, T., Praghash, K., An improved task allocation scheme in serverless computing using gray wolf optimization (GWO) based reinforcement learning (RIL) approach. *Wireless Pers. Commun.*, 117, 3, 2403–2421, 2021.
17. Miao, Z., Yuan, X., Zhou, F., Qiu, X., Song, Y., Chen, K., Grey wolf optimizer with an enhanced hierarchy and its application to the wireless sensor network coverage optimization problem. *Appl. Soft Comput. J.*, 96, 2020, 106602, 2020.
18. Mirjalili, S., Mohammad, S., Lewis, A., Grey wolf optimizer. *Adv. Eng. Software*, 69, 46–61, 2014.
19. Kousik, N., Natarajan, Y., Raja, R.A., Kallam, S., Patan, R., Gandomi, A.H., Improved salient object detection using hybrid convolution recurrent neural network. *Expert Syst. Appl.*, 166, 114064, 2021.
20. Jadhav, A.N. and Gomathi, N., WGC: Hybridization of exponential grey wolf optimizer with whale optimization for data clustering. *Alexandria Eng. J.*, 57, 3, 1569–1584, 2018.
21. Mirjalili, S., Mirjalili, S.M., Hatamlou, A., Multi-verse optimizer: A nature-inspired algorithm for global optimization. *Neural Comput. Appl.*, 27, 2, 495–513, 2016.
22. Gowrishankar, J., Kumar, P.S., Narmadha, T., Yuvaraj, N., A trust based protocol for manets in IoT environment. *Int. J. Adv. Sci. Technol.*, 29, 7, 2770–2775, 20202020.
23. Golzari, S., Zardehsavar, M.N., Mousavi, A., Saybani, M.R., Khalili, A., Shamshirband, S., KGSA: A gravitational search algorithm for multimodal optimization based on k-means niching technique and a novel elitism strategy. *Open Math.*, 16, 1, 1582–1606, 2018.
24. Yogarajan, G. and Revathi, T., Improved cluster based data gathering using ant lion optimization in wireless sensor networks. *Wireless Pers. Commun.*, 98, 3, 2711–2731, 2018.
25. VeerappanKousik, N.G., Natarajan, Y., Suresh, K., Patan, R., Gandomi, A.H., Improving power and resource management in heterogeneous downlink OFDMA networks. *Information*, 11, 4, 203, 2020.

26. Aziz, A., Osamy, W., Khedr, A.M., El-Sawy, A.A., Singh, K., Grey wolf based compressive sensing scheme for data gathering in IoT based heterogeneous WSNs. *Wireless Netw.*, 26, 3395–3418, 2020.
27. Sheikh, H.F., Ahmad, I., Fan, D., An evolutionary technique for performance energy-temperature optimized scheduling of parallel tasks on multi-core processors. *IEEE Trans. Parallel Distrib. Syst.*, 27, 3, 668–681, 2016.
28. Yuvaraj, N., Nandhini, A.S., Vivekanandan, P., A survey on energy efficient routing protocols for MANET. *Int. J. Adv. Eng. Technol.*, 6, 1, 370, 2013.
29. Esmaeili, M., Jamali, S., Fard, H.S., Energy-aware clustering in the Internet of Things by using the genetic algorithm. *J. Telecommun. Electron. Comput. Eng. (JTEC)*, 12, 2, 29–37, 2020.
30. Elhabyan, R. and Yagoub, M., Two-tier particle swarm optimization protocol for clustering and routing in wireless sensor network. *J. Network Comput. Appl.*, 52, 116–128, 2015.
31. Abuelkhail, A., Baroudi, U., Raad, M., Sheltami, T., Internet of Things for healthcare monitoring applications based on RFID clustering scheme. *Wireless Netw.*, 27, 1, 747–763, 2021.

20

Li-Fi Technology and Its Applications

Sumiksha Shetty[1]*, Smitha A.B.[1] and Roshan Rai[2]

[1]Department of Electronics & Communication Engineering, Sahyadri College of Engineering & Management, Adyar, Mangalore, Karnataka, India
[2]Department of Civil Engineering, NMAM Institute of Technology, Nitte, Karnataka, India

Abstract
Communication technology, explicitly remote Communication, remains one of the quickest developing advances in olden times. Some years ago, we progress from substantial huge appliances employed conspicuously for speech Communication, headed for tiny acute appliances that can ensure various abilities which encompass getting to the Cyberspace and audio-visual streaming. Data rate is considered to be the main difference between today's mobile technologies. As we equipped to rise the data percentage, remote gadgets will actually want to accomplish more capacities. In this era of communication technology, Li-Fi is another and proficient method of remote communication. Li-Fi utilizes LED to communicate information. The data transmission of information is done remotely. The existing remote system that links us to the Cyberspace turns out to be moderate when numerous gadgets are associated. Likewise with the expansion in the numerous gadgets, which utilizes the Cyberspace, the accessibility of fixed transmission capacity makes it significantly harder to appreciate high data transmission rates and to link a protected system.

***Keywords*:** Li-Fi, LED, LASER, Visible Light Communication (VLC)

20.1 Introduction

Important research efforts have been undertaken in the past ten years to utilize a large portion of the electromagnetic spectrum. Li-Fi is an internet communication system which does not practice cables, but rather practices

**Corresponding author*: Sumikshashetty7@gmail.com

Chandra Singh, Rathishchandra R Gatti, K.V.S.S.S.S. Sairam and Ashish Singh (eds.) Modeling and Optimization of Optical Communication Networks, (365–380) © 2023 Scrivener Publishing LLC

the light from a diverse LED to communicate data. Li-Fi presents nearly zero capacity limitations [1]. The visible light spectrum is much greater than the radio frequency spectrum. It is a communication structure that utilizes light as a transporter as a substitute of RF waves. With the use of light as a medium to convey information, an electronic device optically senses the switching of light at incredibly high speeds and converts them to electronic pulses. The use of light to transmit data provides many benefits, such as wider bandwidth range, better protection and greater transmission speeds. Visible Light Communication (VLC) is a very important factor in Li-Fi. VLC is a rising innovation that plans to empower high speed internet access fundamentally for indoor conditions and deals with the principle of intensity modulation of existing lighting devices which are usually LEDs. It uses rapid pulses of light, which cannot be detected by the human eye, to transmit information wirelessly. Though the LEDs are used to transmit data, they still need to be used for illumination and should not provide any hindrance to the user. If continuous 0 bits are sent then the LED will remain OFF for a long time, thus causing hindrance to the user, we use visible light communication to solve this specific problem. The duty cycle of the pulses is adjusted so that the ON time is greater than the OFF time. The point of focus of the project is multiuser support. The techniques used to realize this is CDMA and Walsh codes. CDMA is a technique which allows multiple transmitters to transmit their signals in a single channel. So, multiple users can share a single frequency band. Every CDMA user has a different code encoded to distinguish the different users and these codes are called Walsh codes. Walsh codes are uniquely decodable codes which are used to recover a signal when only a part of it is known (Sahitya).

The only minor portions of the electromagnetic spectrum obtainable for data transmission are Radio waves. Li-Fi has a lot more extensive range for transmission of information contrasted with regular strategies for remote correspondences that are done on radio waves. The essential thought behind this innovation is that the data could be moved over LED by changing light powers quicker than the normal humanoid eyes which cannot identify (a review paper on Li-Fi technology). Li-Fi, Light Fidelity is another remote communication innovation which empowers a remote data broadcast through LED. It depends on an exceptional capacity of robust state lighting frameworks to make a binary code that is 1s as well as 0s with a LED gleaming that is imperceptible for natural humanoid eyes. Data could be gotten by electronic gadgets with photodiode inside space of light perceivability. This implies that wherever LEDs are utilized, illumination of bulbs can bring the light as well as remote connection simultaneously. With expanding interest for remote data, absence of radio range and problems with dangerous electromagnetic contamination, Light Fidelity seems as another greener, better and less expensive option in contrast to Wi-Fi. The conglomerate confidences it is feasible to undertake in excess of 10 Gbps, suppositionally approving a fine coating to be taken on an average of 30 seconds. Li-Fi has the benefit of having the option to be utilized in complex areas, for example, in airplane without producing interference. The development of Light Fidelity is to flatten the deficiency of the existing revolution. A the same instant everyone in general should apprehend that as of this day and age Wi-Fi is the utmost employed revolution to link frequent appliances to the net. As possibility appears by, the employment of web-based appliances is lengthened.

20.2 Technology Portrayal

The abbreviation of word Li-Fi is light fidelity, which implies communicating remote computerized information via elucidation of light. Li-Fi word was chosen to address the ophthalmic rendition of Wi-Fi innovation. Rather than utilizing the RF which is used for modulating the advanced information and communication taking via radio wire, the information is regulated in the sunlit recurrence range as well as remitted via LED sunlit. At each point when we turn ON the light, 1 is sent, and 0 would be sent to turn OFF the light. The light appears to be as consistently on the grounds that varies somewhere in the range of 0 and 1s is done rapidly for the humanoid eye in order to be recognize. Rather than utilizing single information stream, it can be feasible to utilize equal transmission or utilize a variety of LEDs to send a large number of information streams and in this manner, we can accomplish extremely high information levels.

Figure 20.1 sums up the instrument of innovation. The advanced streams information from the Web which is given to lamp driver, this is dependable to alter the evidence in a kind of sunlit sign. The evidence is directed via LED lamp, which is taking place at transmitter side. The photosensor is used in the receiver side which is used to distinguish the changes in a light emitted by the LED which then converts photons to electrical sign. The recipient gadget will get the information which is been processed, signals is been strengthened and changed over back for its inventive arrangement.

20.2.1 Li-Fi Modulation Methods

By and large, the average balance procedures utilized for RF correspondences which is for single-carrier regulation could be utilized aimed at

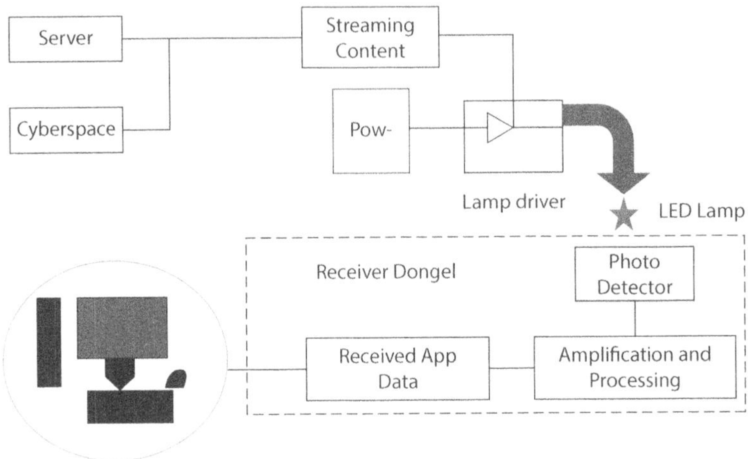

Figure 20.1 Li-Fi structure.

Li-Fi frameworks, from the time when both the waves sunlit as well as the Radio Frequency waves is the electromagnetic waves. Nonetheless, because of the sunlit sign highlights, Li-Fi fit for utilizing some exceptional modulation procedures.

Instances of Single carrier modulation plans comprise for off keying also, pulse position modulation, which is to be considered as well as thought about in remote infrared light correspondence frameworks [2]. OOK is considered as basic modulation scheme, simple to execute and gives a worthy presentation. The PPM is considered to have force proficiency, yet it is less otherworldly proficient. Another modulation termed optical spatial adjustment is projected in [3] and end up being both force as well as transmission capacity effective for interior optical remote interchanges.

A strategy for sending the computerized information with distinctive carrier frequencies is called MCM. The fundamental benefit of MCM strategy is to adapt to unembellished conduit conditions, such that fading and, attenuation and also impedance. The MCM is robust against the fading brought about by the transmission over additional than each way in turn. It is additionally insusceptible against inter-symbol interferences, narrow band co-channel impedance and less delicate Single carrier Modulation to interference which is caused due to the impulse noise. Furthermore, MCM is more fit than SCM to adapt the diminution in high-recurrence correspondences, with the end goal that in the apparent light band. Also, MCM has complex spectral productivity than which the double sideband modulation plans, and also comparatively less vulnerable to time harmonization errors. Comparing the mcm and SCM, MCM is been more efficient regarding the bandwidth and it also has high rapidity for optical communication. But there are a few drawbacks of MCM plans.

They are considered to be sensitive to recurrence synchronization issues and also, they are considered to be low energy efficient from the SCM plans. The most widely recognized strategy in MCM utilized for Li-Fi interchanges is orthogonal recurrence division multiplexing [4, 5]. Making the utilization of this plan, equal information streams can be communicated all the while through an assortment of orthogonal subcarriers. An OFDM technique is executed by a reverse discrete Fourier transform block which is followed by computerized to-simple converter which is also known as DAC. Accordingly, the OFDM produced signal is intricate and bipolar. In any case, the light power can't be negative; the Li-Fi sign ought to be unipolar. Additionally, there are a few necessities which include the intensity modulation along with the direct detection in order to meet with the accessible LEDs. Hence, a few alterations to the traditional OFDM procedures are needed to be utilized in Li-Fi system. Instances of these regulation strategies incorporate DC one-sided optical OFDM which is also known has DCO-OFDM, lopsidedly cut optical OFDM (ACO-OFDM) and unevenly cut DC one-sided optical OFDM (ADO-OFDM). An itemized examination amongst the three procedures in optical frameworks is introduced [6].

20.3 Distinctive Modulation of Li-Fi

In contrast to radio transmitters, the Li-Fi transmitters are having the illumination work adjacent to the correspondence work. Luminaires outfitted with the multi color LEDs which can give different conceivable outcomes of other modulation methods. The CSK Modulation which is also know has color shift keying communicates information via variety of yield sunlit tone [7, 8]. It consists of two benefits for the CSK contrasted along with IM plans. To begin with, it ensures that the brightness of light won't vary, and the medical problems of sunlit glinting are restricted. 2nd, explains about the current which is driven from LED is practically steady, subsequently the brightening force is consistent.MM is a particular sunlit modulation technique, where it is dependent on CSK method. It is equipped for giving greater oomph proficiency along with the capacity, which regulates the shading excellence. To fulfil the color coordinating and darkening necessities in the color space, and to expand correspondence limit, for symmetrical also non-symmetrical ophthalmic channels [9, 10].

20.4 Antiquity of Improvements and Li-Fi Innovation

An overall idea of the "Li-Fi" is to concentrate on apparent light correspondence, which is also known, has VLC from the time 1880s, yet the word "Li-Fi was devised by Prof Harald Hass during his talk at TED Global in the month of August 2011. Prof Harald Hass, the Chairman of the Mobile Communications at the University of Edinburgh also the prime supporter of Pure Li-Fi [11, 12]. Afterward, prof also published a paper to portray his innovation. Prof clarified in subtleties the contrasts amongst the VLC term, Li-Fi term. The term VLC was imagined as a highlight objective information correspondence method, fundamentally as a link substitution. Then again, Li-Fi addresses a total remote systems administration framework, which incorporates bi-directional multiuser correspondence, which is the multipoint-to-point and highlight multipoint communication. Besides, Li-Fi comprises numerous passages (APs), empowers full client versatility and develops another layer inside the current heterogeneous remote organizations. In the month of October 2011, following the 2 months of Harald's TED talk, a few organizations along with some businesses established the "Li-Fi Consortium" [13, 14]. Li-Fi is anything but a non-benefit association given to progress and present the ocular remote innovation, which consortium's individuals are the main gathering of global innovation organizations and also research organizations in ocular correspondence innovation. In the year 2012, the confederation illustrated the guide for various kinds of ocular correspondence, for example, gigabit-class correspondence and also the Li-Fi cloud. In the year 2013, PureVLC, was coined as Pure Li-Fi, exhibited information rate to 1.67 Gbps on the solitary color LED, which is predictable to accomplish 2 Gbps for every one of greens, red also blue channels by the end of the year. In the year 2014, the BeamCaster unit created by the Russian organization "Stins Coman" if information rate is 1.25 Gbps, and it tends to be supported to 5 Gbps. The center of the organization is a switch that is fit for sending information with a sunlit bar with the range of 8 m, and also the information is conveyed to the eight gadgets. Simultaneously, an immense upgrading in Li-Fi information rate is accomplished by a Mexican programming advancement organization Sisoft, which had the option to move information at rate of 10 Gbps utilizing LED lights [15]. In the year 2015, the Li-Fi focus at Edinburgh University thought of an answer for significant distance optical comminutions by making the recipients more delicate for frail signs. Around the same time, Prof. Hass conveys another TED talk, through which he exhibited sending Internet information utilizing LED sunlit and also the stellar cells The advancement in Li-Fi innovation is developing every now and then. The principal improvements in innovation were featured, as a few mechanical gatherings work to upgrade this innovation and advance Li-Fi items [16, 17]. This is a direct result of the interesting benefits of Li-Fi versus regular remote radio organizations.

20.5 Li-Fi Technology and Its Advantages

Li-Fi is predominantly exhilarating when we gossip about the forthcoming of IoT, augmenting further consideration to the debates for IoT products on "Wi-Fi as well as Bluetooth" and "Wi-Fi as well as Cellular". IoT devices mainly depend on connectivity that is nothing but "internet" part of the "internet of things". The recent development in Li-Fi shows that it has greater advantage than Wi-Fi. Li-Fi guarantees highest connectivity despite of using fewer power, dropping software as well as hardware costs and complication while creating. Lots of possibilities open up when we think of the other part of IoT i.e. the "things" now connectable via Li-Fi [19]. Illuminations can link with IoT wireless for sending information and distantly regulate devices across industries. Thus in future Li-Fi can be integrated with IoT in order to provide Li-Fi enabled IoT applications that are mentioned below.

20.5.1 Free Spectrum

In view of the extent of inaccessible applications and also the clients, which origins enormous importance for the information communication, this creates the spectrum scarcity issue. The greater part of the assortment in the broadcasting is completely utilized, also it is nearly inflexible to get data transfer capacity. Furthermore, there should be a permit to get to the greater part of the groups in the Radio Frequency assortment. Nonetheless, the apparent sunlit band is free-range band, which is ordinarily used to see things. By making use of the band in inaccessible correspondence will moderate the range deficiency concern.

20.5.2 Efficiency

Ordinary RF correspondence frameworks burn-through a great deal of energy and a large portion of energy which is utilized to nonchalant base stations which is also known has BSs or the APs. Notwithstanding, Li-Fi correspondence frameworks depend on LED lights which burn-through less energy and give correspondence notwithstanding the enlightenment work. Along these lines, Li-Fi is supplementary energy-productive.

20.5.3 Accessibility

Ordinary RF correspondence frameworks burn-through a great deal of energy and a large portion of energy is utilized to nonchalant base stations (BSs). Notwithstanding, Li-Fi correspondence frameworks depend on LED lights which burn-through less energy and give correspondence notwithstanding the enlightenment work. Along these lines, Li-Fi is more energy-productive.

20.5.4 Complexity

Light fidelity is a basic innovation in examination with the wireless innovation, which depends on the direct modulation along with the direct demodulation, considering in the transmission side the light source, and also in receiver side photodetector. In any case, any radio innovation framework, Wi-Fi for instance, requires an unpredictable RF circuit which is used in adjusting the data and a receiving antenna to communicate the information. Additionally, the wireless collector is extra intricate because it involves simultaneous demodulation circuit and the recipient receiving wire.

20.5.5 Security

In contrast to the radio waves, the light waves can't enter through the dividers. Subsequently, the information will be safely bound around the source of light, it is not possible to access by the interloper for any kind of terrible aims.

20.5.6 Safety

An enormous amount of studies in order to attempt the analyze of impacts on an electromagnetic waves in wireless groups to the humanoid, considerably lot of the investigations presumed that the wireless signs, like Wi-Fi along with portable signs, is contrarily influence generally speaking human body wellbeing, particularly in kids. Exceptional normal openness to the radio portable signs will change the innovations along with conduct. Likewise, it adds to the advancement in sleep deprivation, which vintages battles for resting around evening time, and which could be the start of bigger medical conditions in the upcoming years. Also, it is tentatively demonstrated by the demonstration of a wireless radiations which influences a cell development and also upset ordinary fetal improvement. Moreover, a few specialists tracked down that 4 G remote radiations decrease the mind usefulness [20]. Furthermore, a few potential risks incorporate fertility issues [21], and cardiovascular pressure [22]. Be that as it may, the electromagnetic rushes of the sunlit sign are innately protected from the previously mentioned concerns. The lone conceivable medical problem may happen only when extreme focus light is coordinated to the natural eye. Along these lines, by depending on Li-Fi innovation, particularly in indoor conditions, the medical problems resulted because of remote correspondences can be relieved.

20.5.7 No Fading

Probably the greatest issue for signaling in wireless rang could be channel blurring brought about by multipath proliferation. The signals which are been reflected is anti-phased with the communicated sign, and consequently they will also drop one another and also origin signal blurring. Accordingly, multipath blurring isn't an concern for the sunlit signals.

20.5.8 Cost-Effective

Other than each of the past benefits, the expense of the Li-Fi framework is significantly lesser than the Wi-Fi framework or any of radio practically identical framework. This is on the grounds that the LED lights and the necessary circuit parts in the Li-Fi framework are less expensive than the necessary segments in the radio framework. Indeed, no requirement for distinctive LEDs in the Li-Fi framework. One among the LED could be utilized for information communication, yet for all intents and purposes brilliant outcomes is accomplished by utilizing COTS of LED gadgets.

20.6 Confines of Li-Fi Innovation

Considering the innovation, other than knowing its benefits, understanding its impediments to more readily recommend potential uses of this innovation. In this part, a short portrayal of certain difficulties of Li-Fi innovation contrasted with radio innovation will be introduced. Some potential answers are present for the limits.

20.6.1 Obstructions

The sunlit sign is supplementary delicate to squares has well as hindrances than the wireless sign. In internal circumstances, in contrast to radio sign and also light signals can't go through dividers. This is inclusion drawback, yet security advantage simultaneously. In outside correspondence conditions, the quality corruption of information communication is seriously influenced by utilizing light waves rather than wireless waves.

20.6.2 High Path Forfeiture

The way forfeiture of any correspondence framework is relative which the square of a working recurrence is. The recurrence of noticeable sunlit waves is a lot superior to the recurrence of wireless electromagnetic waves. The most extreme Radio frequency will be in scope of gigahertz, while the sunlit source wave would be in the scope of terahertz. Along these lines, commonly the light sign is exposed to greater attenuation while considering the radio signs. Accordingly, the Li-Fi innovation is hard to practice for significant distances. A potential answer for the problems would build the quantity of a LED lights to expand the inclusion region.

20.6.3 Uplink Problems

The vast majority of the introduced trials and showings of this innovation are accomplished for simply the downlink heading. Because of useful reasons, the converse correspondence course, from remote gadgets to either an AP or also the BS isn't possible. Despite the fact that there are numerous distributions that advance or survey this innovation, a large portion of these investigations overlook this problem from thought. The uplink either bearing in Li-Fi framework, to utilize infrared waves or utilizing greater-recurrence wireless waves for two potential options.

20.6.4 NLOS Problems

Li-Fi framework in a smaller amount is consistent, which will have no line of-sight amongst sending cause and the receivers. Thusly, a quantity of the transmitters ought to been expanded to build the chance of LOS among the transmitters and the collectors.

20.7 Application of Li-Fi Technology

The particular highlights of Li-Fi innovation, which is helpful in genuine applications in this segment, a short depiction of roughly recommended uses will be presented.

20.7.1 Spaces wherein Exploiting of RF would be Controlled

Since the sunlit sign is protected also accessible all over, which is very well be utilized to give remote correspondence in places where the utilization of the RF signals is restricted because of the likely perils. Instances of such places incorporate emergency clinics, planes and delicate plants.

20.7.1.1 Hospitals

For the most part utilizing the Wi-Fi organization or the versatile organization in clinics is confined, particularly next to clinical observing gadgets. This is on the grounds that the radio electromagnetic transmissions may meddle with clinical gadgets and cause issues. By fusing the Li-Fi innovation in such places, remote interchanges and getting to the Internet will be simple and safe for all individuals inside [23]. Besides, we could approach the Internet even in activity rooms, since they are furnished with lights. Accordingly, specialists and specialists might be consulted online based during basic tasks, or perhaps activities can be communicated online for instructive and clinical.

20.7.1.2 Airplanes

Utilizing cellphones or else Wi-Fi to get to the Internet is prohibited on planes, because of the uncertainties of trepidation of impedance with delicate plane devices. In only couple of years prior, a few aircrafts offered restricted admittance to the Internet utilizing Wi-Fi with exceptionally high charges. In any case, Li-Fi can use each seat's lamp to furnish admittance to the Internet with an exceptionally highest data rate, and with no worries of obstruction.

20.7.1.3 Sensitive Floras

The RF interchanges are restricted in susceptible plants, for example, power floras, petrochemicals floras and atomic floras, as a result of the ignition chances in these conditions. Be that as it may, since the light is protected in these spots, the Li-Fi framework can give simple inclusion and information transmission in such conditions. Subsequently, checking the floras or identifying deficiencies should be possible distantly.

20.7.2 Traffic Flow Management

On the off chance that the vehicles' headlights as well as backlight illuminations are substituted by LED lights, the Li-Fi system can be set up between vehicles. Vehicles can impart one another to lessen odds of fender benders. Likewise, the lanterns can be furnished with photographic camera to screen the streets and distinguish any clog or crisis cases, and afterward send the data straightforwardly through Li-Fi innovation to the Traffic flow management office to make a rapidly conceivable move.

20.7.3 Submerged Applications

For submerged distantly worked automobiles, radio interchanges, for example, Wi-Fi bombs totally. Such automobiles utilize long channels for communications. By utilizing powerful lights with Li-Fi innovation, we can dispose of these channels and subsequently the automobiles can interchange effectively to investigate bigger regions, and send the information remotely.

20.7.4 Outdoor Permission to the Cyberspace

These days, the solitary conceivable alternative for getting to the Internet outside is done with the cell organization. Wi-Fi is restricted for indoor circumstances, and can't be utilized outside in most real-world circumstances. However, the Li-Fi innovation will utilize the real-world lamps to empower getting to the Internet without any problem. In this way, in the event that somebody is strolling in the city, sitting at the sea shore, or playing in the recreation center, he can have simple admittance to the cyberspace if the lamps are in the vicinity of him.

It is observed that the accomplished data rate of Li-Fi innovation in outside circumstances is simply somewhat inferior to in inside circumstances. The caused date rate for various spaces between client gadgets and the AP is displayed. Not withstanding, the deliberated outside circumstances has most good circumstances, which is no downpour or haze. In any case, if such situations occur, the presentation outside Li-Fi structure will corrupt essentially.

20.7.5 Educational Tenacities

Because Li-Fi offers excessive data rate as well as it has an exceptionally widespread transmission capacity, it might substitute the Wi-Fi frameworks instructive foundations to serve countless understudies, educators, also added to this, organization team with great data rate. Hence, exploring and getting to this data would be quick as simple. Moreover, it will be very basic and quick to download or transfer addresses, or any superior quality video web-based action.

20.7.6 Amalgamation of Wi-Fi vs. Li-Fi

Contemplating the benefits and detriments of Li-Fi techniques, nearly proposed situations that connect the Wi-Fi v/s Li-Fi techniques are presented in several ways. The joined framework has numerous new fascinating highlights. These highlights comprise sanctuary upgrade, highest data rate, further all-encompassing insertion locale as well as upgraded indoor stationing.

The preliminary three high point commencing from the Li-Fi invention, however the reportage attribute emanates from the Wi-Fi innovation. Fixed as well as semi fixed clients would have inaccessible statistics association done through Li-Fi invention, but further versatile clients would depend on Wi-Fi system. These tactics of connotation will moderate the blockage of RF system and will loose Wi-Fi structure ability to indulge budding forthcoming petition progress. The twofold procedures for amalgamating the Wi-Fi with Li-Fi. First approximate is hybrid methodology and succeeding one is the aggregated methodology. The hybrid methodology utilizes Wi-Fi intended for uplink bearing whereas Li-Fi intended for downlink bearing. The aggregated methodology utilizes Wi-Fi as well as Li-Fi equally. Further, single direction Li-Fi linkage is misused in hybrid methodology to help conventional Wi-Fi downlink. In the aggregated methodology, collectively bi-directional Li-Fi as well as Wi-Fi linkages stays completely used. Obviously utilizing the mixture of Wi-Fi as well as Li-Fi stretches highest data rate other than utilizing just Wi-Fi innovation. Here the outcomes illustrates that the aggregated methodology is far superior to the hybrid methodology [18].

20.7.7 Optical Attocell

In the remote cordless systems, it has revealed that lessening the hexagonal cell dimensions would multiply network ghastly productivity.

By diminishing cell dimensions, the exposure would improve as well as the recurrence, reuse would increment, also thusly, data rate also incremented. Thusly, little cells, for example microcells or picocells or femtocells have started to introduce. This little cell idea can be handily stretched out to the Li-Fi innovation. The attocell is treated as heightened wireless capacity. This attocell doesn't meddle with the RF cell systems, since it works in the optical range. Due to the properties of light waves, optical attocells permit substantially extensively thick data transmission reprocess. In this manner, by conveying attocells, we acquire two advantages: working on indoor inclusion and improving the limit of the RF remote webs. Notwithstanding, because of the great way forfeiture of light as a wave, the inclusion of each and every attocell is extremely restricted, and simultaneously, dividers keep the framework from encountering co-channel obstruction among places. This implies we need enormous wireless capacity to refuge a specific region. Luckily, the necessary framework as of now exists, due to the enclosed light prerequisites.

The arrangement of numerous attocells in a chamber gives widespread data exposure and simultaneously gives unvarying enlightenment. Nonetheless, the arrangement of the ophthalmic APs could influence the framework execution. Further, the illumination plan in a chamber influences the attocells link. 20*20 m dimension is considered for four dissimilar arrangements of ophthalmic APs in a chamber. It incorporate hexagonal, square, homogenous Poisson point cell model.

For normal as well as unchanging lighting, the hexagonal as well as square prototypes are deterministic prototypes, which suit the lighting. In any case, in utmost genuine circumstances, the lighting won't be unchanging, because of irregular places of celling illuminators, work area lights and surprisingly LED canopies. Thusly, some irregular prototypes are presented to give further exact execution results. The Poisson point cell prototype adopts the quantity of APs tracks the Poisson appropriation also the APs are physically autonomous of one another. The outcome of this is that, two APs could be discretionary near to one another in PPP prototype, which is not practical. In this way, HCPP prototype is presented, which consumes an extra boundary that controls the base detachment between every two APs.

20.7.8 Multiple User Permission

Li-Fi innovation can give various clients with coinciding network permission. It has revealed that because of utilizing ophthalmic space division multiple access method, accomplished system quantity is multiple times additional than the ordinary time division multiple access strategies. Yet, space division multiple access method is a complicated to enterprise and tedious. Orthogonal Frequency Division Multiplexing access method gives a simple answer for multiple user access, where clients are assisted and isolated by various symmetrical subcarriers.

Since Li-Fi frameworks don't have quick blurring similar to RF frameworks, and the indoor optical remote channel has comparable qualities of a low-pass filter, Orthogonal Frequency Division Multiplexing access needs proper client booking strategies to keep up with decency in the designation of assets. To build the data rate of a user where the received power is same as threshold, (NOMA) Non-Orthogonal Multiple Access method was presented in the case of RF correspondence structures. By utilizing this procedure and using the telecom description of LEDs, the exhibition of a Li-Fi framework can be upgraded meaningfully. Non-Orthogonal Multiple Access method can aid an expanded number of clients by means of Non-Orthogonal resource allocation (RA).

References

1. Badeel, R., Subramaniam, S.K., Hanapi, Z.M., Muhammed, A. A review on LiFi network research: Open issues, applications and future directions. *Appl. Sci.*, 11, 11118, 2021. https://doi.org/10.3390/app112311118
2. Wakchaure, S.L., Pawar, S.D., Thitme, V.V., Shinde, B.B., Overview of Li-Fi technology. *Int. Res. J. Eng. Technol.*, 4, 9, 422–426, 2017.
3. Mesleh, R., Elgala, H., Haas, H., Optical spatial modulation. *J. Opt. Commun. Netw.*, 3, 3, 234–244, 2011.
4. Ayyash, M., Elgala, H., Khreishah, A., Jungnickel, V., Le, T., Shao, S., Rahaim, M., Schulz, D., Hilt, J., Freund, R., Existence of WiFi and LiFi toward 5G: Concepts, opportunities, and challenges. *IEEE Commun. Mag.*, 54, 2, 64–71, 2016.
5. Haas, H., Yin, L., Wang, Y., Chen, C., What is LiFi? *J. Light. Technol.*, 34, 6, 1533–1544, 2015.
6. Chatterjee, S., Agarwal, S., Nath, A., Scope and challenges light fidelity (LiFi) technology in wireless data communication. *Int. J. Innov. Res. Adv. Eng. (IJIRAE)*, 2, 6, 1–9, 2015.
7. Shao, S., Khreishah, A., Rahaim, M.B., Elgala, H., Ayyash, M., Little, T.D.C., Wu, J., An indoor hybrid WiFi-VLC internet access system, in: *2014 IEEE 11th International Conference on Mobile Ad Hoc and Sensor Systems*, IEEE, pp. 569–574, 2014.
8. Alfattani, S., Review of LiFi technology and its future applications. *J. Opt. Commun.*, 42, 1, 121–132, 2021.

9. Butala, P.M., Chau, J.C., Little, T.D.C., Metameric modulation diffuse visible light communications with constant ambient lighting, in: *2012 International Workshop on Optical Wireless Communications (IWOW)*, IEEE, pp. 1–3, 2012.
10. Komine, T., Haruyama, S., Nakagawa, M., Performance luation of narrow-band OFDM on integrated system of power line communication visible light wireless communication, in: *2006 1st International Symposium on eless Pervasive Computing*, IEEE, p. 6, 2006.
11. Yin, L., Wu, X., Haas, H., On the performance of non-orthogonal multiple access in visible light communication, in: *2015 IEEE 26th Annual International Symposium on Personal, Indoor, and Mobile Radio Communications (PIMRC)*, IEEE, pp. 4–1359, 2015.
12. Saito, Y., Kishiyama, Y., Benjebbour, A., Nakamura, T., Anxin, Higuchi, K., Non-orthogonal multiple access (NOMA) for cellular future radio access, in: *2013 IEEE 77th Vehicular Technology Conference (VTC Spring)*, IEEE, 2013.
13. Dimitrov, S. and Haas, H. *Principles of LED light communications towards networked Li-Fi.* Cambridge University Press, Cambridge CB2 8BS, United Kingdom, 207 pp., 2015.
14. Polshetwar Poonam, V. and Siddiqui, S., Li-Fi technology. *Int. J. Comput. Sci. Inf. Technol. (IJCSIT)*, 5, 6, 1–8032, 2014.
15. Asadzadeh, K., Farid, A.A., Hranilovic, S., Spectrally factorized optical OFDM, in: *2011 12th Canadian Workshop on Information Theory*, pp. 102–105.E, 2011.
16. Chatterjee, S., Agarwal, S., Nath, A., Scope and challenges light fidelity (LiFi) technology in wireless data communication. *Int. J. Innov. Res. Adv. Eng. (IJIRAE)*, 2, 6, 1–9, 2015.
17. Sarkar, A., Agarwal, S., Nath, A., Li-Fi technology: Data transmission through visible light. *Int. J. Adv. Res. Comput. Sci. Manage. Stud.*, 3, 6, 1–12, 2015.
18. Sharma, R.R. and Sanganal, A., Li-Fi technology: Transmission of data though light. *Int. J. Comput. Technol. Appl.*, 5, 150, 2014.
19. Vinay Kumar, S., Sudhakar, K., Sudha Rani, L., Emerging technology Li-Fi over Wi-Fi, *International Journal of Inventive Engineering and Sciences (IJIES)*, 2, 3, February 2014, ISSN: 2319–9598.
20. Gupta, S.U., Research on Li-Fi technology & comparison of Li-Fi/Wi-Fi. *Int. J. Adv. Res. Comput. Sci. Software Eng.*, 5, 6, 429–33, 2015.
21. Nivrutti, D.V. and Nimbalkar, R.R., Light-fidelity: A reconnaissance of future technology. *Int. J. Adv. Res. Comput. Sci. Software Eng.*, 3, 11, 753–756, 2013.
22. Khandal, D. and Jain, S., Li-fi (light fidelity): The future technology in wireless communication. *Int. J. Inf. Comput. Technol.*, 4, 16, 1687–1694, 2014.
23. Raj, H., Mitra, C., Shankar, G., Kumar, C., Raj, H., LiFi wireless communication. *Int. J. Innov. Res. Phys.*, 2, 2, 15–18, 2021.

21

Smart Emergency Assistance Using Optics

Chandra Singh[1]*, Sachin C. N. Shetty[1], Manjunatha Badiger[1] and Nischitha[2]

[1]Sahyadri College of Engineering & Management, Mangalore, Karnataka, India
[2]Mangalore Institute of Technology and Engineering, Moodbidri, Karnataka, India

Abstract
Many people meet unfortunate events like road accidents. The number of people killed per day in road accidents has been increasing every year. Delay in getting immediate medical assistance at times of emergency costs many lives. As the number of accidents is increasing, there has to be some device that can help in providing medical assistance immediately. This gives rise to the need for a system which can assist people in such unprecedented emergencies. We therefore intend to provide a solution by building an emergency assistance system that assists people in getting ambulance services in need. Smart emergency assistance system is a system inside a vehicle, based on a microcontroller platform that checks the heart beat rate and notifies concerned people, detects the accident and notifies concerned people. The basic idea of this system is to call the ambulance whenever a person is in some emergency. Accidents are detected with the help of impact sensors. The pulse rate is monitored using a pulse rate sensor. This sensor is also used to understand the abnormalities in heart beat rate. With the help of GSM, a message is sent to the ambulance service. The location of the accident is determined using GPS. All the information is stored in EEPROM for further reference.

Keywords: Emergency, GSM, SMS, GPS

21.1 Introduction

There is an increase in the number of road accidents happening. The delay in the ambulance in reaching the accident location raises the victim's chances of dying. Many of the fatal accidents happen due to the abnormal

**Corresponding author*: chandrasingh146@gmail.com

Chandra Singh, Rathishchandra R Gatti, K.V.S.S.S.S. Sairam and Ashish Singh (eds.) Modeling and Optimization of Optical Communication Networks, (381–396) © 2023 Scrivener Publishing LLC

pulse rate while driving that lead to loss of control of the vehicle. We have introduced a system to limit the loss of life by safeguarding the person in case of abnormal heart rate as well as assisting the person in getting ambulance help in need while notifying the concerned people.

In our system, we use an impact sensor to detect the accident. If an accident is detected, then SMS is sent to the ambulance and an emergency contact which is the contact of the concerned member of the victim. If the person does not need ambulance's help after an accident, then the person can press the switch. On pressing the switch, SMS notification is sent to the ambulance that person does not need the ambulance. A pulse sensor is included in the vehicle to check if the person's heart beat is normal. If it is abnormal, it implies some health issue. An emergency contact is notified instantly. Buzzer is used to alert surrounding vehicles when an accident occurs.

Whenever a SMS is sent to the ambulance, SMS is also sent to an emergency contact. The SMS contains the location of the person who needs an ambulance. GSM is used for the communication between rescue team and vehicle when a person needs an ambulance. GPS assists the ambulance to reach the right place by sending the live location of the person. Separate switches are used for both the cases.

21.2 Literature Survey

Nimisha Chaturvedi *et al.* [1] have proposed a solution to make emergency facilities available during road accidents. When an automobile is in an accident, a sensor on the vehicle detects the collision, sends a signal to the microcontroller. The alert message is transmitted by GSM to a police control center or rescue squad, along with the position, which is determined via GPS. An alert message containing the accident location will be sent to the relative's victim. When accidents are not severe, the driver can stop sending SMS to the ambulance with the help of a switch in the system. As a result of this, the ambulance's time gets saved. With the aid of sensors and microcontrollers, the proposed solution project can accurately identify the accident. Sumit Chavan *et al.* [2] have proposed a system that will be more beneficial to sick drivers who are suffering from high blood pressure and are at risk of heart attacks. They've also integrated an Alcohol sensor for added security and to combat the causes of drunk driving incidents. In case of any abnormal blood pressure, heartbeat, or drug level, a message will be shared to the doctor with the help of GSM.

Pragati Dhake *et al.* [3] have suggested a system that deals with the many features neededto make a smart vehicle. Human safety, bus tracking, no authorization to unauthorized users, alcohol detection, and over speed detection are among the features provided. In the smart bus, a PIR sensor is used. It could identify any live person within its range, and if it did, the bus would come to a complete stop and the accident would be avoided.

K. Hari Babu *et al.* [4] have proposed a solution for detecting and reporting accidents at greater speeds. They suggest a method that makes use of a GPS receiver's capacity to track a vehicle's speed, detect accidents based on that speed, and report accident locations to an alert service center. The vehicle's speed is measured every second and compared with the prior speed. It will be assumed that an accident has occurred if the speed falls below the prescribed speed due to an unnatural slowdown. The device will use GPS to determine the location of the accident. And it is shared along with the time and the speed through GSM. By pressing the manual button, the car occupant will be able to manually communicate the accident scenario in addition to the automated detection system.

Sivaraman Karthikeyan *et al.* [5] have proposed a system to call the ambulance. Components like an Arduino microcontroller, a SIM800A Quad Band GSM/GPRS module, a Neo 6M-0-00-1 U-Blox GSM module, and a vibration sensor are used to summon an ambulance in the event of an emergency or accident. This entire setup was deployed inside the vehicle as a black box to ensure the appliance's reliability. After the accident, an ambulance was called to the scene, and the location was sent to emergency contacts. This system notifies the victim's emergency contacts of the accident, reducing legal issues.

Prabakar *et al.* [6] proposed an improved accident detection system that would show the status of victims in the accident location. Smart biomedical sensors and microcontroller-based mobile technologies are used to build and deploy the system. When an accident occurs, the system provides the accident site as well as several parameters such as temperature, heartbeat, and coma stage. The physiological parameters of the injured person are communicated to the emergency contact via SMS. In the SMS, the victim's physiological characteristics, such as body temperature, were sent. As a result, the suggested approach aimed to reduce the number of people killed in car accidents. When the accident is not severe, the victim can hit the button to avoid getting ambulance help Gowshika *et al.* [7] have proposed an Arduino-based vehicle accident alert system using GPS, GSM, and Accelerometer. The accelerometer is used to detect the sudden change of the axis. The coordinates of the accident site are obtained using GPS. GSM module is used to send messages to all the concerned members. After the accident, SMS is sent to the ambulance and police station. Google Maps is used to get accident's exact location of an accident. The system sends SMS to the nearest ambulance to the accident location. A switch is used to avoid sending messages when accidents are not severe. The number to whom a message is to be sent is saved in EEPROM well in advance.

C. Prabha *et al.* [8] proposed a useful system for accident detection using both vibration sensor and Micro Electro Mechanical System (MEMS) or accelerometer. They developed a system that detect, notifies traffic accidents by sending messages in the form of SMS to a specific mobile phone number. In the field of embedded systems, GSM GPS signaling and tracking algorithms are developed and implemented using LPC2148MCU. The microcontroller transmits an alarm message to the police control center or a rescue squad via GSM modem, including the position. A switch is used to stop the sending of a message when accidents are less severe. The EEPROM establishes a connection to permanently store the previously saved mobile phone number.

S. Parameswaran *et al.* [9] have proposed a system to provide immediate medical help during road accidents. The vibration sensor receives the signal and sends it to the microcontroller when the car crashes. The microcontroller sends a message to the police or rescue squad via GSM. After receiving the message, the police can track the location via GPS. Then make the necessary measurements after adapting to the position. The driver can turn off the alarm message using a switch when the accident is not severe. The proposed system detects accidents with the help of vibration sensors. EEPROM interconnection is used for the permanent storage of mobile phone numbers.

Niranjan Kumar Mandal *et al.* [10] have proposed an Arduino-based device that can detect an accident and send the location of the rider to a predefined number. Along with that, there will be one band in the rider's hand which will note the rider's pulse rate and upload it to a website. They have used two Arduino and one ESP8266 in the proposed system. Knock sensor, hall-effect sensor, and pressure sensors are used with the first Arduino. The bike makes a jerk when the accident takes place; thus, the knock sensor detects the jerking of the bike. Then the pressure sensor sends a high signal to the hall-effect sensor. Another device is used to measure the heart rate of the rider. A heart rate sensor is used to determine the driver's heart rate. The data will be uploaded to the web using Wi-Fi module ESP8266. The data from the heart rate sensor is always sent to the ESP8266. So ESP8266 will upload heart rate continuously. Anyone can check a rider's heart rate by checking the web address where it is uploaded. A panic button is used to indicate that the rider is not in serious condition. A message will be displayed that the user is not in serious condition on the web.

Tushar Rahman Khan *et al.* [11] have developed a real-time vehicle safety system that can reduce the number of accidents with emergency braking systems, which can also be used to alleviate traffic congestion. The STM32 chip is used in conjunction with various sensors such as infrared sensors and probe sensors. The sensor determines whether an impediment is in front of the car and communicates the value to the STM32 microprocessor.

The activation of the braking system is done by the STM32 chip. When the traffic is heavy, the GSM module is connected to the STM32-IC, which searches for an alternate path. To check the module performance Arduino is used instead of STM32. They found that the object detection and braking system has an accuracy rate of 100% for STM32 devices and 60% for Arduino devices. STM32 was proven to be better than Arduino. The car's motor is powered by a motor driver shield (L293D).

Sahil Haria *et al.* [12] presented the concept of preventing and detecting automobile crashes at the same time. In this paper, an obstacle will be identified using ultrasonic sensors, and the automobile will come to a halt as soon as the obstacle is recognized, preventing an accident. Nonetheless, if an accident happens, the wounded should seek emergency medical attention as quickly as possible. A vibrator sensor will be used to detect the crash, which will activate the circuit and send a message to the nearest hospital, police station, fire department, and the victim's emergency number. In this paper, a novel concept of collision prevention is proposed where, on the mountain, if it is difficult to see the corner on the roadway, the smart bollard would alert the driver on each side.

21.3 Methodology

Block Diagram

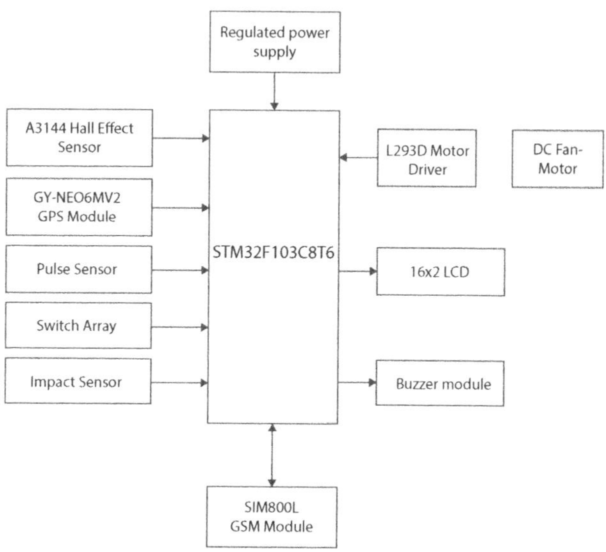

Figure 21.1 Unit 1 of block diagram.

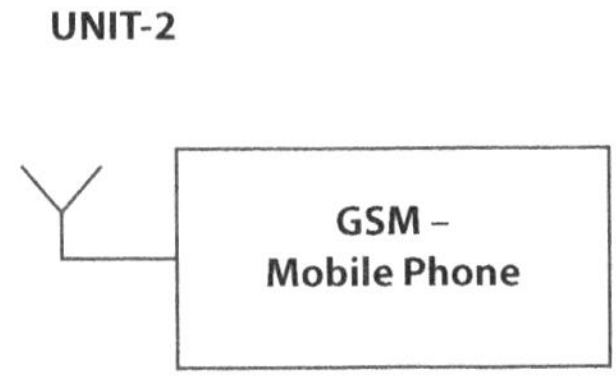

Figure 21.2 Unit 2 of block diagram.

21.3.1 Block Diagram Description

A 12V adapter is used to give power to the power supply. A power supply unit is used to supply power to various components of the project. The power supply unit comes with voltage levels of 12V, 5 V and ground. The ground in the power supply unit is used to give ground connection to all components. Regulated power supply is used to give constant supply even if there is some fluctuation in input. Our project uses the regulated DC power supply. On using the required components, we get the regulated DC power supply of 12v. The IC 7805 is the voltage regulator. The components like LCD, hall effect sensor, and pulse sensor require the voltage of 5V. The IC 7805 voltage regulator provides the input voltage of 5V to these components. Impact sensor is a sensor used to detect collisions in vehicles. If the limit switch is pressed, then it indicates an accident happened. If the limit switch is not pressed, then the vehicle has not met any accident. High or low signal is sent to the microcontroller based on the input to the limit switch. The STM32F103C8T6 ARM STM32 Cortex-M3 minimum system development board is used to build our project. This acts as the heart of the entire system to control all the aspects of the project. The GPS module is used in our system for tracking or finding location. GPS gives coordinate values of the location that helps to trace the location. Pulse sensor gives the heart beat rate value and the signal is sent to the microcontroller. Hall effect sensor measures the RPM of the DC fan and sends the signal to the microcontroller. The GSM module will send SMS to concerned people. Buzzer module is used to give the audible alert. The LCD is used for all display purpose. One switch is used to call the ambulance and another switch is used to confirm no need for the ambulance. The GPS requires the voltage level of 4V so a DC-DC converter is used so that GPS gets the input voltage level of 4V to ensure GPS is not damaged. This forms the working of unit 1 of our project. Another unit involves the mobile phone to receive the message sent through GSM.

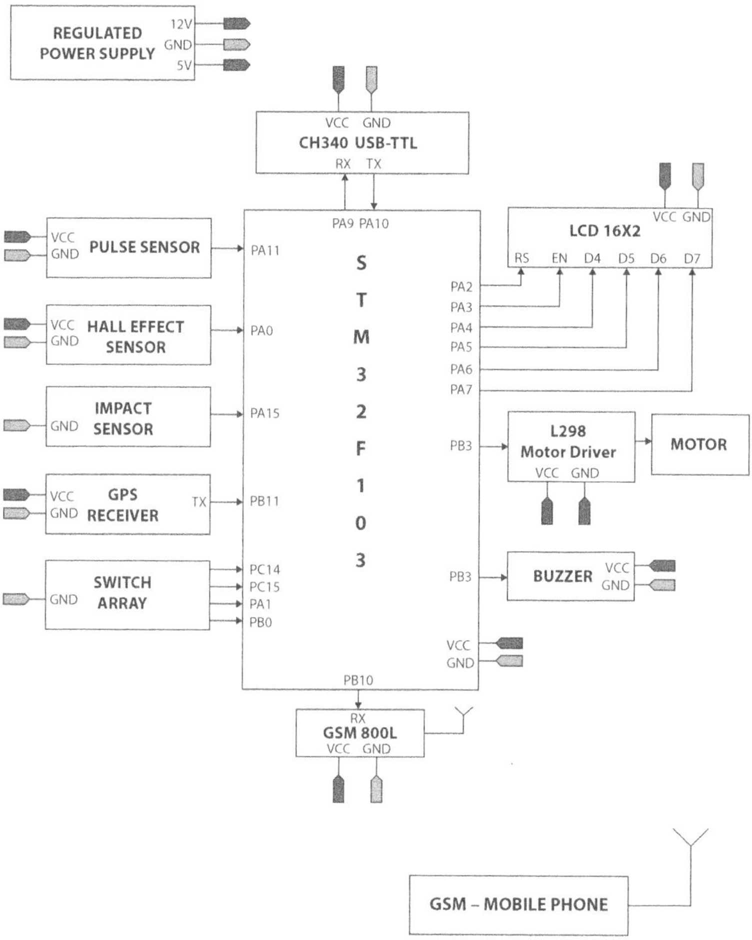

Figure 21.3 Pin connection of the proposed system.

21.3.2 Concept and Overview

In this project, we are using a Bluepill board consisting of STM32103FC8T6 microcontroller. We are building a prototype consisting of sensors like impact sensor, pulse sensor and hall effect sensor. The STM32103FC8T6 microcontroller has inbuilt EEPROM. This EEPROM is used to store values of various parameters like speed of vehicle during accident and coordinates of the accident.

The pulse sensor is used in our project for heart beat monitoring. Thresholds are set for the readings of the pulse sensor. Normal person in a resting position has a heartbeat rate in therange of 60-100 BPM. So, this is the normal reading of the pulse sensor. If the pulse sensorreading does not show the reading in the range of threshold, then the reading is abnormal. If the pulse sensor shows an abnormal reading, then the DC fan remains off, i.e., the vehicle does not start. If the pulse sensor reading is normal, then the vehicle starts is depicted in Figures 21.1 to 21.5.

An alert message is sent to an emergency contact at the same time. Abnormal heart rate might indicate some health issue. So, checking the heart beat rate before startingthe vehicle is done to safeguard the person in case of abnormal heart rates. This is the first working part of our project.

Another part of the project involves the main concept of our project i.e., accident detection and alerting. Limit switch is used as an impact sensor here to detect the accident. If an impact is sensed, then the alert message in the form of SMS is sent through GSM with the live location. GPS helps in getting the live location of the person. Buzzer is provided to indicate the occurrence of the accident to the surrounding vicinity. The person needs to feed emergency contact well in advance for the emergency purposes. SMS is sent both to the ambulance and an emergency contact. This emergency contact could be any relative or friend. If the accident is not severe, the person is conscious and does not require an ambulance, then there is an option for the person to avoid getting ambulance help.

A switch is present for this purpose. This switch is used in our project to send a SMS to theambulance to indicate that the person does not require ambulance help. Hall effect sensor is used to measure the rpm of the DC fan continuously. In real implementation, this conceptis implemented to measure the speed of the vehicle. This system should be integrated with the ECU of the vehicle to implement this project in real life.

21.4 Design and Implementation

First when we power the entire setup using a 12V adapter, the message "ACCIDENT ALERT STM-EEPROM" will be displayed on the LCD. Then the system becomes ready to receive inputs from various components. "System Ready" message gets displayed on the LCD. Then we place the finger on the pulse sensor. When the finger is placed on the pulse sensor, both the pulse rate and saturation level get displayed on the LCD as shown in figure. Now, if the pulse rate is less than the normal value and if we now press the start vehicle switch "Low Pulse Sending SMS" gets displayed on LCD. "SMS Sent" gets displayed on the LCD as soon as the SMS is delivered. Above shown is the SMS sent by GSM to the mobile phone which is an emergency contact number to check on the driver.

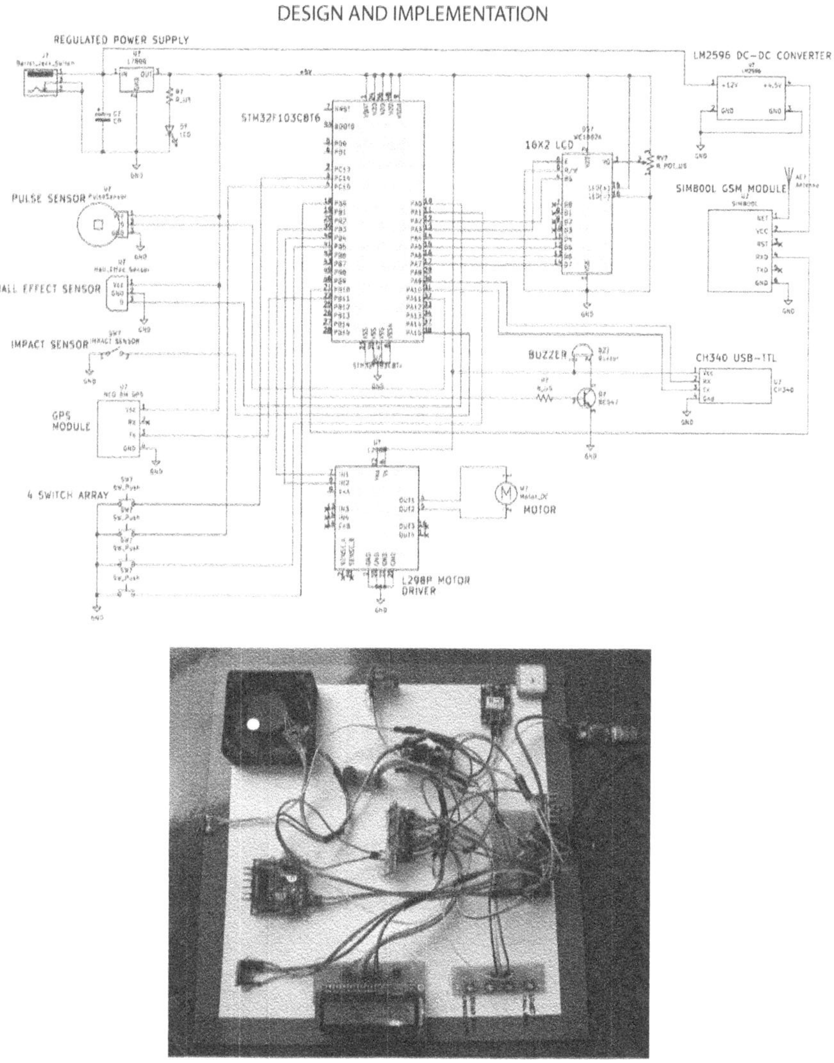

Figure 21.4 (a) Schematic of the proposed system. (b) Hardware setup of the proposed system.

Now, if we place our finger on the pulse sensor and if it is a normal value then if we press the vehicle start switch, the vehicle gets started. The above pictures show the RPM rate of the vehicle. Now if we press the limit switch it means an accident occurred. and the buzzer starts beeping and "Accident detection checking switch" displayed on the LCD. When the call ambulance switch is pressed on the array switch, "Call Ambulance" message gets

Figure 21.5 LCD displays of pulse sensing mechanism.

Figure 21.6 LCD displays of accident detection mechanism.

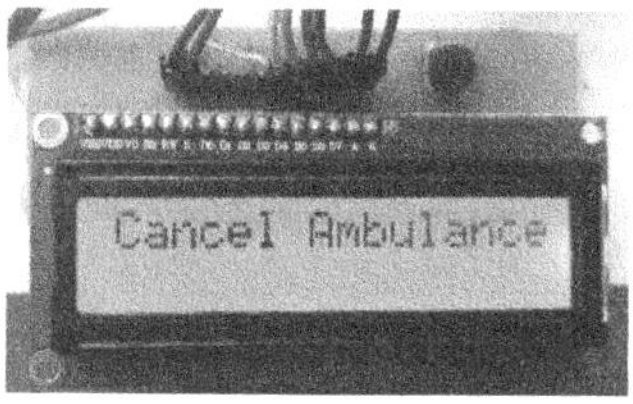

Figure 21.7 LCD display of cancelling the ambulance.

```
Data Written Successfully
Verifying Data>>>
Latitude: 1254.37645
Longitude: 07450.1520
Speed: 37
```

Figure 21.8 Serial monitor display of writing data into EEPROM.

displayed on the LCD. Coordinates of the location i.e., latitude and longitude values get displayed on the LCD and then these coordinates are sent to the ambulance and emergency contact. "Sending SMS" gets displayed on the LCD and later "SMS sent" gets displayed on the LCD to indicate that the message is sent and then the beeping of the buzzer stops. The message sent to the mobile numbers is "Alert accident occurred" with the latitude and longitude values as shown in Figures 21.6, 21.7, 21.8.

Now if we had pressed the limit switch for the accident to occur but it was just a minor accident then we can cancel the arrival of the ambulance by pressing a switch on an array switch. Now the "Cancel Ambulance" message gets displayed on the LCD is depicted in Figure 21.9.

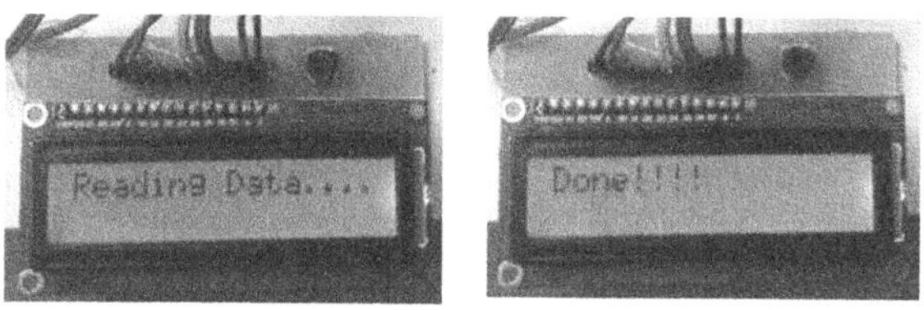


```
Reading Data>>>
Latitude: 1254.37645
Longitude: 07450.1520
Speed: 37
```

Figure 21.9 Serial monitor display of reading data.

```
Heart rate:0.00bpm / SpO2:0%
Heart rate:47.74bpm / SpO2:93%
Heart rate:91.25bpm / SpO2:94%
```

Figure 21.10 Serial monitor display of the pulse oximeter heart rate sensor.

When the accident occurs the coordinates and the speed are saved in the EEPROM of the microcontroller and we can see this on the Serial Monitor when the SMS is sent through GSM to the mobile. On pressing the read data switch on the switch array, "Reading Data" message gets displayed on the LCD. The latitude, longitude and speed get displayed on the Serial Monitor as shown above. Then the "Done" message gets displayed on the LCD to imply that the message is displayed

Shown in Figure 21.10, is the pulse rate display on the serial monitor which shows heart rate is 0.00bpm and Spo2 is 0% when no pulse is sensed. The heart rate is 47.74bpm and Spo2 is 93% which implies abnormal heart rate. The heart rate is 91.25bpm and Spo2 is 94% which implies normal heart rate.

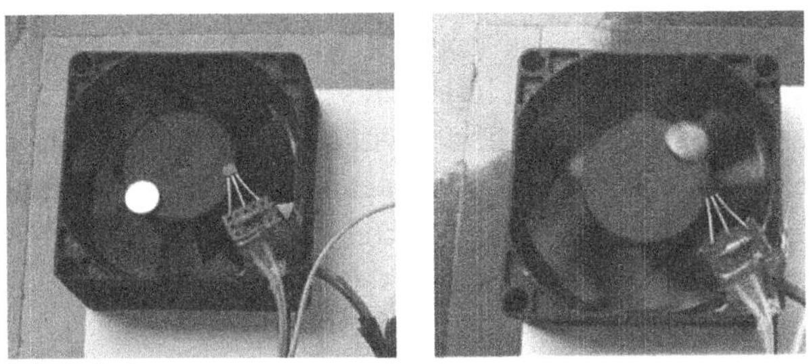

Figure 21.11 Hall effect sensor on DC fan.

Now, to find the speed at which the accident occurred to store in EEPROM of the microcontroller for the future purpose, we should place a Hall Effect sensor on the DC Fan and a magnet is kept on the DC fan to check the RPM (Figure 21.11). And thus, we save the speed value at which the accident occurred.

Now, we open the Coordinates app which is freely available in the Play Store and we need to select the Sexagesimals option on the screen. Now, we should input the coordinates in Degree, Minute and Seconds as shown in figure. Then we will get the accident location on the screen [13].

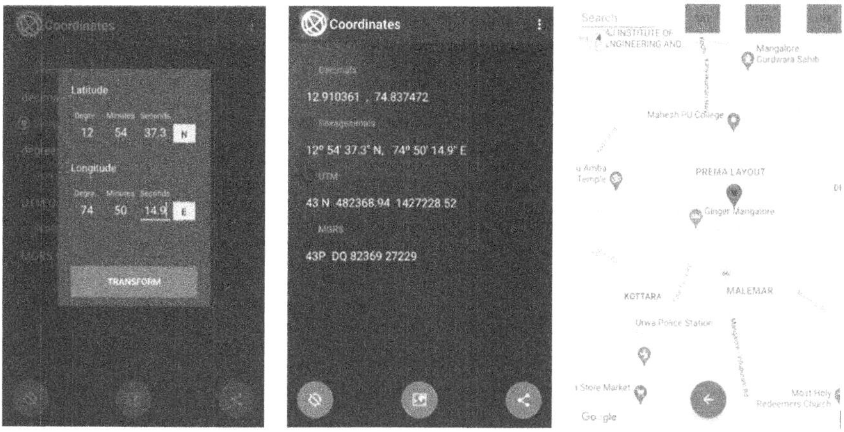

Figure 21.12 Steps involved in getting the exact location.

21.5 Results & Discussion

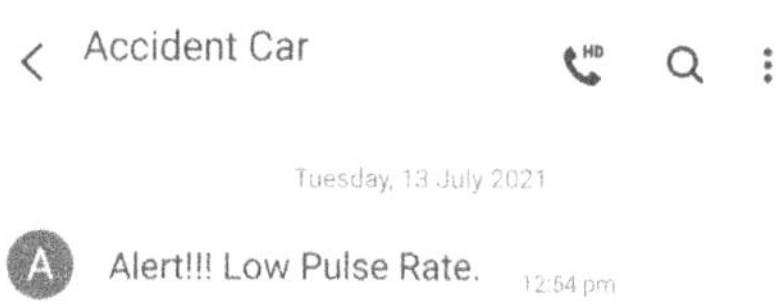

Figure 21.13 Received low pulse alert message.

After placing the finger on the pulse sensor, the pulse sensor will start the pulse sensing mechanism. As a part of pulse sensing mechanism, the values like pulse rate and saturation level are sensed. When a person has a low pulse rate, the above SMS is shared to the emergency contact is depicted in Figures 21.12, 21.13, 21.14.

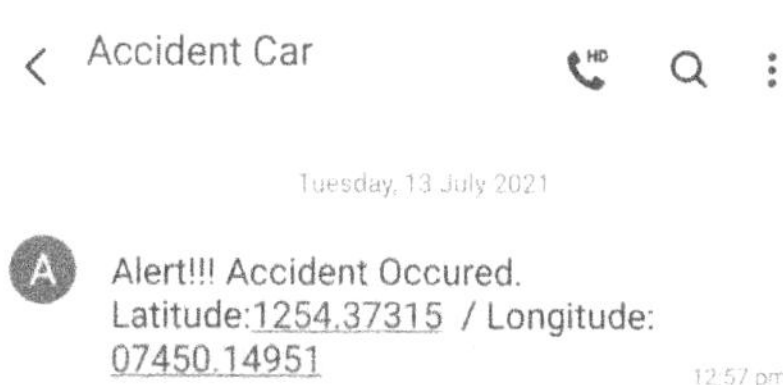

Figure 21.14 Received accident alert message with the location.

When the limit switch senses an impact, then it means an accident has occurred. The above SMS gets sent to both the ambulance and an emergency contact. i.e., both the numbers added in the code. From the coordinate values we obtained in the previous image, we can obtain the actual location of an accident. We entered latitude and longitude values in an app and that app showed the location in the google map and we were able to get the actual location of an accident in the google map.

21.6 Conclusion

When it comes to heart rate monitoring and accident detection and alerting, we noticed that there were fewer technologies and implementations. As a result, we devised a mechanism that accomplishes both of these goals. Because the number of vehicles on the road is growing every day, if the proposed idea is implemented in the vehicle, the driver will have a better chance of surviving an accident and the delay in getting the medical assistance immediately will be resolved. Here the system does not allow the vehicle to start unless the pulse rate is normal which is very important to avoid the risks of accidents due to abnormal pulse rate. Thus, we can conclude that this system will be very beneficial for mankind.

References

1. Chaturvedi, N. and Srivastava, P., Automatic vehicle accident detection and messaging system using GSM and GPS modem. *Int. Res. J. Eng. Technol. (IRJET)*, 05, 03, 252–254, Mar-2018, e-ISSN: 2395- 0056, p-ISSN: 2395-0072.
2. Saeednia, M., and Mendez, M., A consensus-based algorithm for truck platooning. *IEEE Trans. Intell. Transp. Syst.*, 18, 2, 404–415, 2016.
3. Dhake, P., Awnish, A., Deshpande, Pati, N., Smart bus for human safety. *IJARIIE*, 2, 2, 23–27, 2016, ISSN(O)-2395-4396.
4. Hari Babu, K. and Anudeep, P., Wireless reporting system for accident detection at higher speeds. *Peddi Anudeep Int. J. Eng. Res. Appl.*, 4, 9, 17–20, September 2014, ISSN:2248-9622, Version 2.
5. Karthikeyan, S., Srinivasan, S.R., Syed Ali, J., Veeraraghavan, A.K., Smart summoning of ambulance during a vehicle accident. *Second International Conference on Green Computing and Internet of Things (ICGCIoT)*, IEEE, 2018.
6. Prabakar, P., Isaac, S., Sundari, G., An enhanced accident detection and victim status indicating system: Prototype. *2012 Annual IEEE India Conference (INDICON)*, 7-9 Dec. 2012.

7. Gowshika, M. and Jayashree, S.M., Vehicle accident detection system by using GSM and GPS. *Int. Res. J. Eng. Technol. (IRJET)*, 06, 01, 1574–1578, Jan 2019, e-ISSN: 2395-0056, p-ISSN: 2395-0072.
8. Prabha, C., Sunitha, R., Anitha, R., Automatic vehicle accident detection and messaging system using GSM and GPS modem. *Int. J. Adv. Res. Electr. Electron. Instrum. Eng.*, 3, 7, 10273–10277, July 2014, e- ISSN:2278–8875, p-ISSN:2320–3765.
9. Parameswaran, S., Anusuya, P., Dhivya, M., Harshiya Banu, A., Naveen Kumar, D., Automatic vehicle accident detection and messaging system. *Int. J. Eng. Res. Technol. (IJERT)*, 4, 11, 1–3, ISSN: 2278-0181, COCODANTR-2016 Conference Proceedings.
10. Mandal, N.K., Chakraborty, D., Majhi, A.K., Das, T., Banerjee, M., PromothDhar, Rajib, P., Mukherjee, D., Accident detection and alerting system with continuous heart rate monitoring using Arduino for two-wheeler. *Int. J. Latest Trends Eng. Technol.*, 16, 1, 098–105, e-ISSN:2278-621X.
11. Khan, T.R., Ara, R., Islam, Md.S, Mondal, S., Tanseer Ali, M., Improved embedded vehicle safety system with STM32. *2019 22nd International Conference on Computer and Information Technology (ICCIT)*, 18-20 December 2019, IEEE.
12. Haria, S., Gala, V., Anchaliya, S., Maru, T., Car crash prevention and detection system using sensors and smart poles. *Proceedings of the Second International Conference on Intelligent Computing and Control Systems (ICICCS 2018)*, 14-15 June 2018, IEEE Xplore Compliant Part Number: CFP18K74 -ART, ISBN:978-1-5386-2842-3.
13. Karkera, T. and Singh, C., Autonomous bot using machine learning and computer vision. *SN Comput. Sci.*, 2, 251, 2021, https://doi.org/10.1007/s42979-021-00640-6.

About the Editors

Chandra Singh is an assistant professor in the Department of Electronics and Communication Engineering at the Sahyadri College of Engineering and Management. He is pursuing his PhD from VTU Belagavi, India. He has four patents, published over 25 peer-reviewed publications, and is the editor of seven books.

Rathishchandra R. Gatti, PhD, is a professor and Head of the Department of Mechanical Engineering and Robotics and Automation at the Sahyadri College of Engineering and Management, India. He has four patents, published more than 40 papers in peer-reviewed journals, and has edited seven books. He is also the editor of one journal, and he has over 20 years of industry experience.

K.V.S.S.S.S. Sairam, PhD, is a professor and head of the Electronics and Communications Engineering Department at NITTE University, India. He has over 23 years of experience in teaching and research, and he has published over 50 papers in scholarly journals, conferences, and workshops. He is a reviewer for several journals, and he has authored three books.

Ashish Singh, PhD, is an associate professor in the Department of Computer and Communication Engineering at NMAM Institute of Technology, Nitte, India. He has 13 years of teaching experience and has published more than 50 research papers in scholarly journals and conferences.

Index

Also of Interest

RESOURCE MANAGEMENT IN ADVANCED WIRELESS NETWORKS, Edited by A. Suresh, J. Ramkumar, J. M. Baskar and Ali Kashif Bashir, ISBN: 9781119827498. Written and edited by a team of experts in the field, this exciting new volumes provides a comprehensive exploration of cutting-edge technologies and trends in managing resources in advanced wireless networks.

RF CIRCUITS FOR 5G APPLICATIONS: Designing with mm Wave Circuitry, Edited by Sangeeta Singh, Rajeev Kumar Arya, B.C. Sahana and Ajay Kumar Vyas, ISBN: 9781119791928. This book addresses FinFET-based analog IC designing for fifth generation (5G) communication networks and highlights the latest advances, problems, and challenges while presenting the latest research results in the field of mmwave integrated circuits design.

WIRELESS COMMUNICATION SECURITY: Mobile and Network Security Protocols, Edited by Manju Khari, Manisha Bharti, and M. Niranjanamurthy, ISBN: 9781119777144. Presenting the concepts and advances of wireless communication security, this volume, written and edited by a global team of experts, also goes into the practical applications for the engineer, student, and other industry professionals.

ADVANCES IN DATA SCIENCE AND ANALYTICS, Edited by M. Niranjanamurthy, Hemant Kumar Gianey, and Amir H. Gandomi, ISBN: 9781119791881. Presenting the concepts and advances of data science and analytics, this volume, written and edited by a global team of experts, also goes into the practical applications that can be utilized across multiple disciplines and industries, for both the engineer and the student, focusing on machining learning, big data, business intelligence, and analytics.

ARTIFICIAL INTELLIGENCE AND DATA MINING IN SECURITY FRAMEWORKS, Edited by Neeraj Bhargava, Ritu Bhargava, Pramod Singh Rathore, and Rashmi Agrawal, ISBN: 9781119760405. Written and edited by a team of experts in the field, this outstanding new volume offers solutions to the problems of security, outlining the concepts behind allowing computers to learn from experience and understand the world in terms of a hierarchy of concepts.

MACHINE LEARNING AND DATA SCIENCE: Fundamentals and Applications, Edited by Prateek Agrawal, Charu Gupta, Anand Sharma, Vishu Madaan, and Nisheeth Joshi, ISBN: 9781119775614. Written and edited by a team of experts in the field, this collection of papers reflects the most up-to-date and comprehensive current state of machine learning and data science for industry, government, and academia.

MEDICAL IMAGING, Edited by H. S. Sanjay, and M. Niranjanamurthy, ISBN: 9781119785392. Written and edited by a team of experts in the field, this is the most comprehensive and up-to-date study of and reference for the practical applications of medical imaging for engineers, scientists, students, and medical professionals.

SECURITY ISSUES AND PRIVACY CONCERNS IN INDUSTRY 4.0 APPLICATIONS, Edited by Shibin David, R. S. Anand, V. Jeyakrishnan, and M. Niranjanamurthy, ISBN: 9781119775621. Written and edited by a team of international experts, this is the most comprehensive and up-to-date coverage of the security and privacy issues surrounding Industry 4.0 applications, a must-have for any library.

CYBER SECURITY AND DIGITAL FORENSICS: Challenges and Future Trends, Edited by Mangesh M. Ghonge, Sabyasachi Pramanik, Ramchandra Mangrulkar, and Dac-Nhuong Le, ISBN: 9781119795636. Written and edited by a team of world renowned experts in the field, this groundbreaking new volume covers key technical topics and gives readers a comprehensive understanding of the latest research findings in cyber security and digital forensics.

DEEP LEARNING APPROACHES TO CLOUD SECURITY, Edited by Pramod Singh Rathore, Vishal Dutt, Rashmi Agrawal, Satya Murthy Sasubilli, and Srinivasa Rao Swarna, ISBN: 9781119760528. Covering one of the most important subjects to our society today, this editorial team delves into solutions taken from evolving deep learning approaches, solutions allow computers to learn from experience and understand the world in terms of a hierarchy of concepts.

MACHINE LEARNING TECHNIQUES AND ANALYTICS FOR CLOUD SECURITY, Edited by Rajdeep Chakraborty, Anupam Ghosh and Jyotsna Kumar Mandal, ISBN: 9781119762256. This book covers new methods, surveys, case studies, and policy with almost all machine learning techniques and analytics for cloud security solutions.

SECURITY DESIGNS FOR THE CLOUD, IOT AND SOCIAL NETWORKING, Edited by Dac-Nhuong Le, Chintin Bhatt and Mani Madhukar, ISBN: 9781119592266. The book provides cutting-edge research that delivers insights into the tools, opportunities, novel strategies, techniques, and challenges for handling security issues in cloud computing, Internet of Things and social networking.

DESIGN AND ANALYSIS OF SECURITY PROTOCOLS FOR COMMUNICATION, Edited by Dinesh Goyal, S. Balamurugan, Sheng-Lung Peng and O.P. Verma, ISBN: 9781119555643. The book combines analysis and comparison of various security protocols such as HTTP, SMTP, RTP, RTCP, FTP, UDP for mobile or multimedia streaming security protocol.

SMART GRIDS FOR SMART CITIES VOLUME 2: Real-Time Applications in Smart Cities, Edited by O.V. Gnana Swathika, K. Karthikeyan, and Sanjeevikumar Padmanaban, ISBN: 9781394215874. Written and edited by a team of experts in the field, this second volume in a two-volume set focuses on an interdisciplinary perspective on the financial, environmental, and other benefits of smart grid technologies and solutions for smart cities.

SMART GRIDS AND INTERNET OF THINGS, Edited by Sanjeevikumar Padmanaban, Jens Bo Holm-Nielsen, Rajesh Kumar Dhanaraj, Malathy Sathyamoorthy, and Balamurugan Balusamy, ISBN: 9781119812449. Written and edited by a team of international professionals, this ground-breaking new volume covers the latest technologies in automation, tracking, energy distribution and consumption of Internet of Things (IoT) devices with smart grids.

Printed and bound by CPI Group (UK) Ltd, Croydon, CR0 4YY

17/09/2026

14964274-0007